Guiding Waves in Quantum Mechanics

Guiding Waves in Quantum Mechanics

One Hundred Years of de Broglie-Bohm Pilot-Wave Theory

Edited by

Andrea Oldofredi

Centre of Philosophy, University of Lisbon

OXFORD
UNIVERSITY PRESS

OXFORD
UNIVERSITY PRESS

Great Clarendon Street, Oxford, OX2 6DP,
United Kingdom

Oxford University Press is a department of the University of Oxford.
It furthers the University's objective of excellence in research, scholarship,
and education by publishing worldwide. Oxford is a registered trade mark of
Oxford University Press in the UK and in certain other countries.

Published in the United States of America by Oxford University Press
198 Madison Avenue, New York, NY 10016, United States of America

British Library Cataloguing in Publication Data

Data available

Library of Congress Control Number: 2024949716

ISBN 9780198901853

DOI: 10.1093/oso/9780198901853.001.0001

Printed and bound by
CPI Group (UK) Ltd, Croydon, CR0 4YY

The manufacturer's authorised representative in the EU for product safety is
Oxford University Press España S.A. of El Parque Empresarial San Fernando de Henares, Avenida de Castilla, 2 – 28830 Madrid (www.oup.es/en or
product.safety@oup.com). OUP España S.A. also acts as importer into Spain
of products made by the manufacturer.

To Olga, my guiding wave

Editor's Preface

The idea of this volume originated in the summer 2022 while I was organizing the workshop *70 Years of Bohm's Pilot-Wave Theory*, held on 12 October of the same year at the University of Lisbon, in order to celebrate the 70th anniversary of the publication of David Bohm's papers *A suggested interpretation of the quantum theory in terms of "hidden" variables I & II* originally appeared in 1952.

Since these essays provided a fundamental contribution to the debate on the meaning of the quantum formalism – containing its very first alternative interpretation, and thereby leaving a mark in the history of quantum foundations – I invited experts on Bohm's work to speak about the physics, philosophy and history of the pilot-wave theory. On that occasion, we also discussed the recent developments of this approach as well as its future research directions.

Happily every speaker accepted my invitation to contribute to this book. However, although inspired by the mentioned workshop, what you are about to read is by no means a conference proceedings, for only four out of sixteen chapters contain material based on the conference talks. One of the primary goals of this edited collection, in fact, is to provide a discussion on the pilot-wave perspective that is as informative, comprehensive and accessible as possible. In order to achieve such an ideal result, one has certainly to encompass the essential work of David Bohm, but more needs to be explored. The reader will indeed find analyses of de Broglie's first formulation of pilot-wave theory as well as several contributions about the contemporary views on this interpretative framework – which are often conceptually distant from the positions endorsed by the fathers of this perspective.

In addition, it should be underlined that another anniversary must be celebrated in 2024, namely the first 100 years of de Broglie's ideas about the pilot-wave approach, whose gists are contained in his seminal PhD thesis dating back to 1924. There is no better time, then, to ask experts on the de Broglie–Bohm theory to portray its past and present while delineating its future.

The second main aim of the volume, closely related to the previous one, is to represent the *plural* nature of the pilot-wave community, by including the heterogeneous instances and perspectives on this theory endorsed by different authors – although they are sometimes vigorously conflicting. Such divergent viewpoints, however, should represent an opportunity to develop new ideas in order to push forward the research in this interpretation of quantum theory. A healthy exchange of opinions is indeed the essence of science and philosophy; thus, until disagreement is a stimulus for improvement, it will always be very welcome.

The third and final goal of the present book is to show the dynamism of this research programme. As the chapters of this collection will show, the legacy of de Broglie and Bohm is well and alive to this day, for many research paths are being currently explored and investigated. Recently, interesting historical and philosophical results have been emerged, as well as compelling new extensions and applications of the pilot-wave theory to various branches of physics, as for instance quantum electrodynamics, quantum chemistry, and quantum engineering.

Naturally, the editor of this volume does not support every position expressed by the authors, but it could not be otherwise. However, given the pluralist essence of the pilot-wave community mentioned a few lines above, I am glad to offer the readers such a varied landscape of views.

In virtue of its interdisciplinary character, the present volume is naturally intended for a rather heterogeneous readership. The ideal audience is composed of philosophers and historians of physics interested in the metaphysics of quantum mechanics (QM) and its historical developments on the one hand, and physicists working in the field of quantum foundations on the other hand. Moreover, given the presence of introductory, historical and philosophical chapters – cf. for instance Part I, Part II and Part III of the collection – the book can certainly be interesting for metaphysicians, philosophers and historians of science. In addition, let me emphasize that this volume is not only aimed at professional researchers, but also at undergraduate and graduate students in philosophy and physics who are willing either to take the first steps into the de Broglie–Bohm perspective, or to develop their expertise about such a theory. While some chapters may require background knowledge of this interpretational framework, the relevant concepts will be introduced in the first part of the book. Therefore, the majority of the chapters contained here are accessible to any potential reader with an undergraduate level of knowledge of either philosophy or physics.

Hopefully, this collection of chapters will contribute to change the perception of the pilot-wave perspective, which oftentimes is erroneously considered an anachronistic attempt to restore classical ideas. On the contrary, my goal is to show that this is an active and fruitful research programme that embraced the revolution introduced by quantum mechanics while pursuing the higher standards of conceptual clarity demanded by serious physical and philosophical investigations. In this regard, the history of the pilot-wave theory shows that de Broglie and Bohm were innovators, scientists who developed creative ideas in order to achieve a satisfactory formulation of QM. However, the same history also taught us how difficult it has been for them to communicate and disseminate their views and to develop their research. But times have now changed, and this is the right moment to provide a new picture of the pilot-wave theory truly representing the modern and creative spirit of its pioneers. This is the portrait of the de Broglie–Bohm approach that the present editor wants to convey to the readers.

Finally, let me share and renew the same hope that Cushing, Fine and Goldstein expressed more than 25 years ago in the preface of their excellent edited book *Bohmian Mechanics and Quantum Theory: An Appraisal*. Retracing their words, I wish that the present volume may contribute to a serious critical analysis, extension and application of the central ideas of the pilot-wave theory, to test them for consistency, empirical adequacy and fruitfulness in generating new avenues of research.

Editor's Acknowledgements

I owe my interest in the de Broglie–Bohm theory to Nino Zanghì, who had a deep influence on my early career as a young student. Although our trajectories diverged, I am still grateful for his sharp teaching on Bohmian mechanics, and for the time he spent in introducing me to the philosophy of quantum theory.

Moreover, I warmly thank Micheal Esfeld for his unwavering support throughout my doctoral years, when I had the opportunity to explore the fascinating universe of pilot-wave interpretation, and for his trust during my early postdoctoral experience. Michael taught me how to do research properly – improving my views with direct criticisms and intellectual honesty – and how to approach our profession in the best way. For all this, I am very grateful.

I am also indebted to Dirk-André Deckert and Detlef Dürr, with whom I had the pleasure to spend some time during a wonderful research visit in Munich learning about Bohmian extensions to quantum field theory. I am particularly thankful to Dirk for his support during my PhD.

Let's now jump to the present. This volume could not exist without the great work of the authors. I wholeheartedly thank all of them for their efforts, dedication and for having written such interesting chapters. During the development of the book project, I have been very lucky to find collaborative and cooperative colleagues who did their best to help me edit this volume in the smoothest way. A special mention goes to Antonio Vassallo for his constant advice about how to edit a volume, and how to be a good editor.

A word of gratitude must go to the colleagues who kindly agreed to review the chapters, and thereby improving the quality of the whole book: Christian Beck, Craig Callender, Matthias Egg, Federico Laudisa, Dustin Lazarovici, Vera Matarese, Travis Norsen, Sören Petrat, Paula Reichert, Chip Sebens and Ward Struyve. Many thanks to Jean Bricmont, Aurélien Drezet, Olival Freire Junior, Oliver Passon and Antonio Vassallo for your double work as authors *and* referees.

In addition, I thank the reviewers of the book project for their useful suggestions. In particular, huge thanks go to Claudio Calosi for having proposed such a wonderful title.

Furthermore, I want to express my gratitude to Antony Valentini for recommending me to send the proposal to Oxford University Press in the person of Sonke Adlung. This is the best choice I could ever make. Sonke has always provided positive and constructive feedback in order to improve this book project, and it has been a pleasure to work with him. Furthermore, I must thank Giulia Lipparini, who supported me in every single step of the book with constant availability, professionalism, patience and empathy. Giulia, you did an amazing job!

I am thankful to Ricardo Santos, who not only made the conference mentioned in the Preface possible, but also created a friendly and stimulating environment at our Centre of Philosophy, making it a perfect place for conducting research.

Finally, I gratefully acknowledge the Portuguese Foundation for Science and Technology (FCT, Grant No. 2023.07796.CEECIND) for financial support.

After the due professional acknowledgements I would like to express my deepest gratitude to those people who have a special place in my life. I thank my parents, Flora and Rino, who have always constantly supported me despite the difficulties that an academic career entails. Their efforts, dedication and love will always be immensely appreciated.

Lastly, I want to thank Olga, the love of my life, for believing in me unconditionally, for always being there even in the darkest times and for improving my work with clever and sharp criticisms everyday, no matter what. This book would not have existed without her, and my gratitude and love cannot be expressed in mere words.

Andrea Oldofredi
Lisbon, 30 April 2024

List of contributors

Valia Allori, University of Bergamo

Jean Bricmont, Catholic University of Louvain

Siddhant Das, Ludwig Maximilian University Munich

Flavio Del Santo, University of Vienna

Carlos Destefani, Autonomous University of Barcelona

Aurélien Drezet, Institut Neel and University of Grenoble

Michael Esfeld, University of Lausanne

Johannes Fankhauser, University of Innsbruck

Sebastian Fortin, CONICET and University of Buenos Aires

Olival Freire Junior, Federal University of Bahia

Olimpia Lombardi, CONICET and University of Buenos Aires

Cristian Lopez, University of Lausanne

Gerd Christian Krizek, Interdisciplinary Transformation University Austria, Linz

Emilia Margoni, University of Geneva

Pedro Naranjo, Plaza Mayor, 4/1B, 09003, Burgos (Spain)

Xabier Oianguren-Asua, University of Tübingen

Xavier Oriols, Autonomous University of Barcelona

Oliver Passon, University of Wuppertal

Albert Solé, University of Barcelona

Roderich Tumulka, University of Tübingen

Anthony Valentini, Imperial College London

Marij van Strien, Radboud University

Antonio Vassallo, Warsaw University of Technology

Contents

The pilot-wave theory: a plurality of voices

The present volume is a celebration of the de Broglie–Bohm pilot-wave theory, the history of which – as complex as fascinating – reached its first 100 years spanning the *long siècle* of quantum physics.

In particular, the book honours the birth of this theoretical framework whose foundations were laid out in Louis de Broglie's seminal PhD thesis *Recherches sur la théorie des quanta* appeared in 1924, as well as the 72 years of David Bohm's fundamental essays *A suggested interpretation of the quantum theory in terms of "hidden" variables I & II* published in the prestigious journal *Physical Review* in January 1952, which sparked the renaissance of this approach to quantum mechanics (QM).

Looking at the historical developments of quantum theory, one can fairly claim that the pilot-wave programme deserves a notable mention for at least three main reasons. Firstly, well before the advent of a coherent axiomatization of QM, de Broglie proposed a radically new hypothesis concerning the nature of quantum objects and was able to provide them with a consistent, non-classical dynamics governing their motion in space and time. As is well known, these ideas were published in the paper *La mécanique ondulatoire et la structure atomique de la matière et du rayonnement* that appeared in the renowned *Journal de Physique et le Radium* in 1927. Moreover, in the same year, the French scientist presented the content of his essay in the memoir *La nouvelle dynamique des quanta* at the fifth Solvay Congress, discussing it with the most prominent physicists of the time. Interestingly, de Broglie's pilot-wave theory maintained a clear and unambiguous scientific image of the world at the quantum scales – borrowing Wilfrid Sellars' famous expression – while exploring new physical possibilities and implementing innovative concepts. It should be historically appropriate then to consider (and thereby to mention) de Broglie's framework as one of the first attempts to achieve a full theory of QM together with both Born, Heisenberg, and Jordan's matrix mechanics and Schrödinger's wave mechanics.

Secondly, albeit the orthodox formulation of quantum theory imposed itself from the 1930s and the Copenhagen *Weltanschauung*[1] was considered its only feasible in-

[1] Although historians and philosophers of physics argue whether a unitary metaphysical perspective existed among the fathers of quantum mechanics (cf. e.g. Howard 2004), there is however a precise sense in which one may properly speak about a cohesive and unitary 'Copenhagen *spirit*'. In fact, Bohr and many physicists who contributed to the quantum revolution shared a common set of ideas that can be characterized as the inner core of such orthodoxy. Referring to this, Freire Junior underlined that 'in spite of the existence of important differences, both the intellectual backgrounds and the scientific views of people like Bohr, Pauli, Heisenberg, Born, and Jordan, who had been working together on the collective construction of quantum mechanics, had several points in common. All of them endorsed both indeterminism and the assumption of the corpuscular and discrete nature of atomic phenomena. They also firmly believed in the completeness of quantum theory. [...]. [They]

terpretation almost unanimously, David Bohm's 1952 papers provided a clear counterexample to such a widespread belief. For he showed that the a-causal, indeterminate picture of the reality portrayed by QM had no metaphysical necessity, in spite of its extraordinary experimental success. Demonstrating that other interpretations of the quantum formalism were possible, Bohm's works paved the way for the contemporary discussions on the meaning of quantum theory.

Thirdly, it has been largely shown during the second half of the 20th century that the de Broglie–Bohm approach (in its different variants) provides solutions to the technical and conceptual conundra affecting quantum mechanics, clarifying its peculiar features. For instance, this perspective naturally explains both quantum contextuality and the measurement problem in terms of particles' trajectories – a notion notoriously absent in standard quantum theory; moreover, it perfectly agrees with Bell's theorem in virtue of its explicit non-local character.[2]

To this specific regard, in fact, although QM is one of our most empirically sound answers to questions concerning the inherent structure of matter, most physicists and philosophers working on quantum foundations do not consider it the final word about the ontology of the microscopic scales of our universe.

As often argued by John Stuart Bell, the principles of quantum theory contain ill-defined terms – as for instance 'measurement' and 'observer' – leading to a vague and ambiguous description of physical reality. More specifically, Bell claimed that in QM the notion of measurement is promoted to a primitive concept assuming a central role for the theory's empirical consistency. Remarkably, while measurements were *explananda* in classical mechanics, i.e. processes to be explained by the fundamental theoretical entities of the theory (its local beables, cf. Bell 2004, Chapter 7), their status changed into that of *explanans* in QM, becoming an essential piece of its theoretical machinery in order to generate observable experimental outcomes. Consequently, the concept of 'measurement' – as well as the strictly related mechanism of wave function collapse – become essential to secure the empirical adequacy of the quantum algorithm.

On the one hand, however, the physical processes responsible for the actualizations of measurement outcomes are left dynamically unexplained by the theory – that is, the realizations of individual experimental results are not causally determined by the unitary evolution of quantum systems provided by the Schrödinger equation. On the other hand, the very description of measurement interactions according to the quantum framework entails an arbitrary distinction between observed system and measuring device (Bell 2004; Bricmont 2016; Hance and Hossenfelder 2022; Norsen 2017). These are quite unsatisfactory features for a cornerstone theory of contemporary physics.

Retrospectively, eminent scientists were indeed profoundly dissatisfied with the standard formulation of QM as well as with intellectual imperialism of the 'Copenhagen school' and the metaphysical content of this philosophy of QM (cf. Heilbron 1988). For instance, Einstein advanced several arguments against the completeness of quantum

were attached to the revolutionary character of quantum mechanics, and were unsympathetic to any attempt to restore such classical ideals like causality and visualizability in microphysics' (Freire Junior 2015, p. 81). For similar views, see Beller (1999), Heilbron (1988), and Osnaghi *et al.* (2009).

[2]For details on these issues, the reader may refer to Bohm (1952), Bricmont (2016), Dürr and Teufel (2009), and Holland (1993).

theory and proposed a statistical reading of the quantum wave function, whereas Schrödinger attempted to provide $|\psi|^2$ with a realist interpretation as charge density. While Einstein's permanent opposition to the Copenhagen view never translated into a new full-fledged theory and Schrödinger's ideas were shown to be empirically incorrect, the de Broglie–Bohm approach represents the very first alternative formulation of QM, occupying therefore a privileged place in the history of quantum foundations.

Contrary to the orthodox perspective, in fact, the pilot-wave theory undoubtedly showed that an ontologically clear and causal account of quantum phenomena was not only desirable, but also actually realizable, substantiating the interpretational discussions on the meaning of the quantum formalism which is still vigorously debated to the present day.

Before delving into the specific contents of this volume, let us now briefly sketch a few essential steps in the history of the pilot-wave theory in order to illustrate the main ideas that have been proposed by the scientists who shaped this approach to quantum physics. In addition, we also succinctly review the central philosophical discussions that revolved around the de Broglie–Bohm theory. These metaphysical reflections not only contributed to the ongoing debate on the nature of the wave function – the central theoretical entity of QM – but also shed light on the quantum formalism more generally. Lastly, we also mention the primitive ontology programme, a methodological approach to quantum foundations deeply rooted in the pilot-wave interpretation, given its centrality in contemporary literature.

Contentwise, in what follows I kept the overlap with the main chapters to a minimum: whenever a specific topic/issue is addressed in the volume, I directed the readers to the relevant chapter(s). Moreover, since detailed introductions to the de Broglie–Bohm theory are contained in Part I of this book, no pretension of completeness is made here.

Finally, at the end of this chapter, the reader will find an extensive (and hopefully useful) reference section containing the bibliographical sources for every topic mentioned in this historico-philosophical introduction.

A history of ideas: from de Broglie to the present day

As already underlined at the outset of this introduction, the pilot-wave theory originated from the fundamental ideas contained in de Broglie's PhD thesis and his subsequent essays (de Broglie 1924, 1925, 1927). In these works, the French physicist not only hypothesized that electrons and photons were both particles *and* waves – i.e. physical objects showing corpuscular as well as undulatory features, following the idea that wave-like behaviour should be associated with any corpuscular motion – but also proposed a new dynamics according to which particles' velocities were determined by the gradient phase of an associated wave.

Speaking about the wave–particle duality, de Broglie was able to introduce a correspondence between the motion of a body with the propagation of a wave, extending and generalizing Einstein's wave–particle description of light (cf. Boffi 1995; Bacciagaluppi and Valentini 2009). As it is well known, de Broglie's work was heavily influenced by Einstein, who in 1905 proposed to consider the electromagnetic field as constituted by discrete units – the quanta of light (photons in our modern jargon) – employing

Planck's quantum hypothesis to solve the issues related to the photo-electric effect. According to Einstein's hypothesis, then, electromagnetic radiation would display a corpuscular nature. Nonetheless, the vast experimental evidence about diffraction and interference suggested that the electromagnetic field shows wave-like features as well. Such a complementary nature of the latter, however, introduced an 'asymmetry' between the descriptions of matter on the one hand and light on the other. In this regard, the historian of science Niccolò Guicciardini emphasized that

> this asymmetry between the treatment of the electromagnetic field and the treatment of matter was somehow in contrast with Einstein's relativity. According to Einstein, in fact, the famous equation $E = mc^2$ [...] states that the mass m and energy E are two manifestations of the same reality. Given that, once again according to Einstein, a frequency can be associated with the quantum of light via the relationship $E = h\nu$, the idea that it is also possible to associate a frequency with mass arises spontaneously, and therefore it is possible to associate each particle of mass m with a frequency $\nu = mc^2/h$ (Guicciardini and Introzzi 2007, pp. 89–90, my translation from the original Italian).[3]

Between 1923 and 1925, de Broglie made the crucial step of hypothesizing that *matter* can be thought not only in corpuscular terms but also having a wave-like essence.[4] Indeed, to each particle of mass m and momentum p he associated a wave of wavelength $\lambda = h/p$ and a frequency $\nu = E/h$. Consequently, the wave–particle duality became a general feature of nature, dissolving the issue caused by the asymmetry mentioned a few lines above.

Notably, with this new proposal the French physicist was able to explain and predict a wide range of experimental phenomena, as for instance the interference of single photons, the diffraction and interference of electrons – as experimentally verified in the mid-1920s by C.J. Davisson, L. Germer, G. Thomson and A. Reid – and the Bohr–Sommerfeld quantization conditions of atomic energy levels. In particular, according to the wave–particle hypothesis, if one considers an electron orbiting in circular motion with a certain velocity around the nucleus of a given stationary atom, and if one associates a wavelength to such an electron, then only specific orbits will be accessible to it, i.e. those in which the circumference of the orbit is the integral multiple of wavelength associated with the electron. Thus, from the wave–particle duality one naturally arrives at Bohr–Sommerfeld quantization conditions.

In order to emphasize other innovative aspects of de Broglie's approach, three remarks should be made. Firstly, after having generalized the wave particle duality to all material bodies, de Broglie formulated a dynamics for quantum particles that was revolutionary. It in fact presupposed the abandonment of Newton's principle of inertia, in favour of a new one that 'unified Maupertius' variational principle for mechanics with Fermat's variational principle for optics', as underlined by Bacciagaluppi and Valentini, aiming at synthesizing the physics of particles with that of waves (cf. de Broglie 1925, Chapter 2).[5] In particular, as already mentioned at the outset of this section, according to this proposal, particles' velocities are determined by the gradient

[3] Here h represents Planck's constant and c the speed of light.

[4] De Broglie was awarded the Nobel Prize in Physics in 1929 for his hypothesis about the wave nature of electrons.

[5] Cf. Bacciagaluppi and Valentini 2009 p. 37. To this specific regard, the Italian physicist and historian of physics Sigfrido Boffi underlined that de Broglie's research on X-rays, combined with his in-

of the phase of the accompanying wave, contrary to the classical Newtonian framework in which accelerations are determined by forces.

Secondly, another element of originality is found in de Broglie's idea that quantum particles always have definite positions and follow deterministic paths in space, since its rival proposals in the 1920s, matrix and wave mechanics, dropped the notion of particle in their ontologies – and consequently the very notion of trajectory. Referring to this, it is worth mentioning that although we commonly refer to the pilot-wave theory as a 'hidden variable' interpretation of quantum mechanics, such a label is inappropriate for de Broglie's original proposal. In fact, trajectories are described by well-defined dynamical equations of motion in which the variables referring to particles' spatial coordinates appeared explicitly, viz. they are *not* somehow hidden in the formalism of the theory.[6]

Thirdly, it must be underlined that in de Broglie's view the pilot-wave theory was an approximation of a more general project, the 'double solution' programme. As underlined by Colin, Durt, and Willox, as soon as the particle concept was expelled from the theories advanced by Born, Heisenberg, and Jordan on the one hand and Schrödinger on the other, de Broglie tried to 'reinstate the role of the corpuscle' since both particles and waves had an important ontological status in his perspective. To achieve this result, 'he started to develop his double solution program (1925-1927), the (first) core idea of which is that there should be two synchronous, coupled, solutions of the wave equation' (Colin *et al.* 2017, p. 26). In this double solution theory, one finds a ψ-wave, of which only its phase carries physical significance, and a u-wave, that is a solution representing a 'singularity' that corresponds to a particle which follows the flow lines of Schrödinger's wave equation. Remarkably, as an initial step of this double solution programme, de Broglie substituted to each u-wave the position of a corpuscle, and postulated that the particles' velocities are guided by the phase of the wave function ψ. In this manner, he obtained the basis of the pilot-wave theory that he presented at the fifth Solvay conference in 1927.

On that occasion, after having explained the essential elements of his new ideas concerning the corpuscular and undulatory nature of quantum objects, de Broglie takes into consideration the then-cutting edge research in theoretical physics, relating his work with that of Schrödinger and Born. Namely, he provided not only a new dynamics for quantum objects, but also a new interpretation for the wave function. In order to achieve these results, de Broglie described a spinless one-particle system with a wave

terest in Einstein's works on light quanta and relativistic dynamics, generated in him a growing desire to formulate a theory of radiation that could unify wave and corpuscular aspects. In particular, Boffi claims, 'the quantization rules for the motion of electrons in atoms according to Bohr–Sommerfeld's quantum theory introduced the necessity to resort to integer numbers, similarly to what happens in the interpretation of interference phenomena. These facts motivated de Broglie to the study of a formal analogy between analytical mechanics and the theory of wave propagation' (cf. Boffi 1995, pp. 13–14, my translation from the original Italian).

[6] In addition, it would be historically inaccurate to define de Broglie's proposal as a 'hidden variable theory' in the contemporary sense of this expression, since his pilot-wave approach has been developed *before* the establishment of the standard formulation of QM, which conventionally dates back to the early 1930s, when the first axiomatizations of the theory were given by Dirac and von Neumann. Therefore, it would be obviously misleading to claim that such a theory added variables or mathematical structures to the standard quantum formalism.

function $\psi(x,t)$ whose dynamical evolution in the non-relativistic case is governed by the Schrödinger equation:[7]

$$i\hbar\frac{\partial}{\partial t}\psi(x,t) = \Big(-\frac{\hbar^2}{2m}\nabla + V\Big)\psi(x,t). \tag{I.1}$$

In addition, as already said above, he hypothesized that such a particle always has a definite position in space and moves along a deterministic trajectory guided by the ψ wave. The path of the particle is determined by the following law of motion:

$$\frac{dx}{dt} = \frac{1}{m}\nabla S, \tag{I.2}$$

where S is the phase of the guiding wave (cf. Valentini's contribution to this volume). Notably, according to de Broglie's theory, the wave function has a new *double* role: on the one hand it guides the motion of the corpuscles – hence the name 'pilot' wave – on the other it determines the probability, in agreement with Born's rule, that a given particle occupies a certain location in space at an arbitrary time.

In the general discussion that followed de Broglie's report, his proposal received some positive feedback (e.g. from L. Brouillon) as well as compelling criticisms, particularly from Kramers and Pauli who pointed out difficulties for the pilot-wave dynamics when applied to scattering phenomena. Bacciagaluppi and Valentini's detailed analysis of de Broglie's presentation and of the subsequent general discussion underlines that he was not fully able to reply Kramers' objection involving the recoil of a photon hitting a mirror. In order to answer it, the pilot-wave dynamics should have taken into account not only the microscopic system, the photon, but also the mirror. Similarly, Pauli's objection concerning elastic scattering was not completely responded to by de Broglie, who nevertheless gave the gist of the correct solution: the conceptual move necessary to overcome both critical remarks consisted in the extension of his theory to macroscopic objects. Unfortunately, the French scientist did not make this crucial step, and by 1930 he abandoned the pilot-wave approach as well as his double solution theory.[8]

[7]We employ the contemporary notation for the equations of the pilot-wave theory, where i is the imaginary unit, $\hbar$ is the reduced Planck's constant (with $\hbar = h/2\pi$). The terms on the left–hand–side represent respectively the kinetic and potential energy of the particle. For technical details on the original de Broglie's memoir cf. Bacciagaluppi and Valentini 2009, pp. 374–407. Notably, in this report de Broglie also discusses many-body systems, for simplicity we can omit such topic here.

[8]There are various historical reconstructions explaining de Broglie's abandonment of his pilot-wave approach. According to many authors, Pauli's criticisms played a major role (cf. for instance Jammer 1974; Freire Junior 2019; and Talbot 2017), while Bacciagaluppi and Valentini 2009 argue that since de Broglie actively pursued his programme until 1930, Pauli's objections were not so decisive to give up with the pilot-wave theory. In this regard, in the opinion of the present author one may claim that the received criticisms as well as the difficulties in making real progress with the double solution theory convinced de Broglie to quit momentarily his research programme. Nevertheless, he endorsed the pilot-wave approach in later stages of his career, especially after the publication of Bohm's 1952

David Bohm's 1952 pilot-wave theory

After 25 years from the fifth Solvay congress, David Bohm independently rediscovered the pilot-wave theory by re-interpreting the Schrödinger's equation, and extended this framework to its logical conclusions.[9] Indeed, one of his major contributions is having provided a theory of quantum measurements that also included macroscopic devices into the quantum mechanical description. Such a measurement theory is able to explain dynamically and causally how experimental results are actualized and how they emerge from the motion of individual quantum particles guided by a wave field, filling the explanatory gap left by standard quantum theory as well as by de Broglie's original proposal (cf. Bohm 1952, Part II, Section 2).[10] Notably, Bohm answered Pauli objections not only in private correspondence – discussing the problem of inelastic scattering with Pauli himself between the summer and fall of 1951 (cf. von Meyenn 1996, Letter 1264, p. 344, and Letter 1290, pp. 390–392) – but also publicly in Section 7 of Bohm (1952, Part I) and Appendix B of Bohm (1952, Part II).[11]

This remarkable achievement showed the possibility to provide a detailed description of the physical processes causally responsible for measurement outcomes, avoiding conceptual difficulties and technical inconsistencies present in standard QM (Bell 2004; Bricmont 2016; Maudlin 1995a, 1995b). At the basis of such explanation of measurement scenarios we find a theory which is grounded on a clear multi-category ontology provided with a consistent dynamics. According to Bohm's approach, in fact, quantum systems are described in terms of well-localized particles dynamically guided by wave functions, considered in this framework physically real fields, whose time evolution is governed by the Schrödinger unitary law of motion seen above (I.1).

Two key elements should be underlined in order to appreciate the conceptual novelty of Bohm's perspective with respect to both standard QM and de Broglie's ideas. On the one hand, the quantum mechanical postulate for which ψ alone provides the complete description of a certain system is violated. This fact entails that quantum theory can be supplemented with additional ontological commitments towards the existence of corpuscles with definite positions, against the widespread beliefs shared by many eminent physicists in the 1950s – cf. Bricmont's contribution to this volume. Moreover, not only Bohm showed that his re-interpretation of the Schrödinger equation does not entail *any* modification to the quantum mechanical formalism – implying that the very notion of quantum trajectory is already embedded in the mathematical

papers as underlined in Colin *et al.* (2017) and Fargue (2017). Indeed, as soon as a new interest in the pilot-wave theory emerged in France, de Broglie started to work again on the double solution programme, trying to find non-linear modifications of the Schrödinger wave equation. For recent developments of such a theory cf. Drezet (2023).

[9] Cf. Bohm's letter to Pauli sent in summer 1951 contained in von Mayenn 1996, Letter 1264, pp. 345–346. When Bohm wrote his papers on the hidden variable interpretation in 1951, he was not aware of de Broglie's work. Pauli in their epistolary exchange pointed it out to him, who eventually gave the due credit to the French physicist for having introduced the idea of the pilot-wave theory, but also saying that he gave up the theory 'too soon'. On this episode cf. Freire Junior 2019, pp. 71–73.

[10] For an introduction to Bohm's measurement theory, cf. Valentini's contribution to this volume, in particular Section 2.3.2.

[11] Bacciagaluppi and Valentini 2009 also illustrated how to answer Kramers' objection using Bohm's theory of measurement. For technical details, the reader is referred to chapter 10.4, pp. 246–247.

framework of QM[12] – but also that the notion of causality can be re-introduced in the realm of quantum physics.

On the other hand, taking into consideration the original formulation of the pilot-wave theory, we note that the wave–particle duality manifests itself in a new fashion in Bohm's account: whereas according to de Broglie's hypothesis quantum objects have both undulatory and corpuscular features, Bohm introduces a dualistic ontology of particles *and* wave fields, as said above. Those physical situations in which we observe undulatory features of the quantum particles, as for instance in the famous double-slit experiment, are explained via the action of the ψ-field in space. In addition, if one takes into account de Broglie's double solution programme the ontological difference with respect to Bohm's own theory is even more evident: the former is in principle a wave-monistic theory, for reality would be described by a set of two coupled solutions of the wave equation, the ψ-wave and the u-wave (Colin *et al.* 2017, p. 21), whereas in Bohm's theory particles are not defined as 'u-singularities', but as corpuscular entities with definite positions and velocities.

Another crucial difference with respect to the 1927 version of the pilot-wave theory lies in the definition of dynamical laws for quantum objects. In order to define particles' trajectories, Bohm proceeded as follows. Firstly, the wave function is expressed in polar form $\psi = Re^{(iS/\hbar)}$ – where R and S represent two coupled real functions corresponding to the amplitude and the phase of the wave, respectively. Secondly, posing $P(x) = R^2(x)$, where $P(x)$ represents the probability density, one obtains the following equations for R and S:

$$\frac{\partial P}{\partial t} + \nabla \cdot \left(P \frac{\nabla S}{m} \right) = 0, \tag{I.3}$$

$$\frac{\partial S}{\partial t} + \frac{(\nabla S)^2}{2m} + V(x) - \frac{\hbar^2}{4m}\left[\frac{\nabla^2 P}{P} - \frac{1}{2}\frac{(\nabla P)^2}{P^2} \right] = 0. \tag{I.4}$$

The former is the quantum continuity equation for the probability density, whereas the latter is the *quantum Hamilton–Jacobi* equation describing the motion of a particle (or a configuration of particles) with kinetic energy $(\nabla S)^2/2m$ and subjected to the influence of both a classical and a new *quantum* potential:

$$Q = \frac{-\hbar^2}{4m}\left[\frac{\nabla^2 P}{P} - \frac{1}{2}\frac{(\nabla P)^2}{P^2} \right] = \frac{-\hbar^2}{2m}\frac{\nabla^2 R}{R}. \tag{I.5}$$

[12] Hiley, de Gosson and Dennis (2019) interestingly argued that already Dirac tried to define trajectories for quantum particles from the quantum mechanical formalism (see, e.g. Dirac 1945). Moreover, Hiley and Dennis (2018) illustrate how Dirac's analysis of the classical limit of quantum theory contained in his 1930 textbook is similar to Bohm's approach. However, the authors argue that Dirac did not define quantum trajectories only because of the Heisenberg position–momentum uncertainty relation, which forbids the simultaneous assignment of well-defined values for positions and momenta to quantum particles.

This latter element is the distinctive aspect of Bohm's formulation of the pilot-wave theory, absent in de Broglie's original formulation, and it plays a crucial role in order to explain the new features arising in the quantum domain – as for instance the interference pattern of the double slit experiment, the EPRB spin correlations, etc.. The Q potential in fact possesses interesting attributes not present in the classical case: (i) it only depends on the form of the wave function, meaning that it is insensitive to the intensity of ψ, and thereby even a weak quantum field can strongly affect the particles, (ii) its action does not decrease with the distance separating the corpuscles, generating a non-local dynamics, and (iii) it contains information about the whole experimental set-up (Bohm 1952; Bohm and Hiley 1993).[13] In addition, in his 1952 papers, Bohm defined the particles' velocity as $v = \nabla S/m$ to the effect that his proposal is metaphysically unambiguous, for the Bohmian corpuscles always have a precise localization in three-dimensional space and a well-defined velocity independently of any observation or measurement. Referring to this, Bohm interestingly stated that in his theory the uncertainty principle becomes just 'an effective practical limitation on the possible precision of measurements' (Bohm 1952, p. 171). Hence, it should not be interpreted as an irreducible impossibility to conceive the concepts of position and momentum as simultaneously defined quantities, as Heisenberg instead famously claimed.[14] It is relevant to underline that in virtue of the identity $v = \nabla S/m$, one can re-write the quantum continuity equation for the probability density as $\partial P/\partial t + \nabla \cdot (Pv) = 0$. where Pv is interpreted as the mean current of the particles in a given configuration. This equation is particularly important since the Born distribution holds as a consequence of it.

As we already pointed out a few lines above, with this theoretical machinery Bohm was able to provide a causal explanation of quantum measurements, and consequently of the actualization of experiments' outcomes. Alternatively, he solved the quantum measurement problem arising from three assumptions commonly accepted in QM: (i) the completeness of ψ – the idea that the wave function provides the maximal information about quantum systems – (ii) that ψ always evolves according to the linear Schrödinger equation, and (iii) that measurements have a unique determinate outcome (Maudlin 1995a). Interestingly, any pair of these propositions is consistent and entails the falsity of the third one, but their conjunction generates inconsistencies between the quantum postulates and experimental evidence. In essence, in each experimental situation, the pilot-wave theory assigns a position x_{t_0} to the measured system and

[13] For more details cf. Passon and Valentini's chapters in this volume. Let us just note here that Bohm's 1952 theory can be formulated in Newtonian fashion, making explicit the actions of the classical and the quantum potentials on the quantum particles:

$$\frac{md^2x}{dt^2} = -\nabla(V(x) - (h^2/2m)\nabla^2 R/R). \tag{I.6}$$

This dynamical law reproduces the predictions and statistics of QM only if one (i) assumes that the particles are initially distributed according to the Born's distribution, and (ii) takes $v = \nabla S/m$ to be an initial condition (Bohm 1952, pp. 170–171).

[14] The issue whether or not de Broglie–Bohm theory should be considered a hidden variables theory in virtue of its particle ontology is discussed in chapter 3 by Jean Bricmont, where it is also shown how well-known no-go results, e.g. Kochen–Specker theorem, do not prove the inconsistency of this framework.

coordinates y_{t_0} to the particles composing the measuring device – where t_0 refers to the arbitrary initial time before the measurement takes place. A 'measurement' is then defined as an interaction among the particles of the observed system and those forming the experimental apparatus. In this context, such an interaction is described by the dynamical evolution of the quantum particles, which is exclusively governed by the laws (I.1) and (I.4). As a consequence of the Schrödinger dynamics, the system's wave function entangles – i.e. introduces correlations – with the apparatus' wave function, creating disjoint wave packets corresponding to the possible measurement outcomes. The configuration of particles (x_{t_0}, y_{t_0}), which is always well defined and never in superposition, deterministically evolves during the measurement process and enters in only one wave packet, remaining there thenceforth.[15] The chosen branch of the wave function then guides the particle configuration towards the actualization of a definite state of the macroscopic device, pointing to a given specific direction actualizing the measurement result. The latter is described by the particle configuration (x_t, y_t), where t refers to the 'final' time of the experiment, i.e. when the experimental outcome can be unambiguously read off from the macroscopic measurement device. Contrary to the case of standard quantum theory, in Bohm's approach wave functions are not subjected to stochastic jumps: the collapse rule in fact loses its fundamental role in the pilot-wave interpretation, being it only an *effect* of the particles' dynamics, dissolving thereby the inconsistency present between the projection rule and the unitary Schrödinger equation. To this specific regard, it should be here emphasized that in Bohm's theory probabilities do not refer to an inherent indeterminacy of quantum objects, but rather possess an epistemic character: in experimental situations we can neither know, nor manipulate the initial positions of the particles, consequently measurement outcomes will be unpredictable and not controllable. Thus, randomness emerges from our ignorance concerning the initial conditions of a given experiment. Lastly, it should be emphasized that Bohm did not aim at restoring a deterministic picture of physical reality with his interpretation: determinism is indeed inherited from the Schrödinger's equation that is the basis of his model.[16]

Having presented the essentials of Bohm's own formulation of pilot-wave theory, it is worth noting that in his view the causal interpretation, as he later called it, was not a definitive theory of quantum mechanics, but only the first step to reach a satisfactory description of the microscopic regimes of our universe. In fact, looking at Bohm's scientific and philosophical works, he constantly revised, modified and put into questions his own perspectives on the meaning of quantum theory, providing various different conceptions of his own hidden variable model.[17] Therefore, one should not consider his approach as a final word about the ontology of QM, but only as a preliminary step towards a better understanding of the quantum realm. Referring to this, two things should be emphasized: on the one hand, Bohm did not believe that a specific

[15] Referring to this, Bohm underlines that the non-selected wave packets can be ignored for all practical purposes, since they will neither affect the quantum potential nor the particles' momenta (cf. Bohm 1952, p. 182).

[16] Compelling historical discussions about Bohm's view on determinism are given in van Strien as well as Del Santo and Krizek chapters in this volume. The reader may also refer to Oldofredi (2023).

[17] For a comprehensive account of Bohm's approach to quantum foundations, the reader is referred to his scientific biography contained in Freire Junior (2019).

ontology – in particular a corpuscular ontology – should necessarily be maintained at all energy/length scales. Indeed, in the Appendix A to his 1952 papers he provided a guidance equation for *field* coordinates describing the electromagnetic field. On the other hand, already in his correspondence with Pauli in late 1951 and in Section 9 of Bohm (1952, Part I), he made non-trivial statements about the possibility to modify his own interpretation, showing the willingness to develop and generalize it beyond the scales in which non-relativistic QM is applied.[18] Moreover, Bohm was convinced that his approach would have entailed new physics at very short length scales, where he also conceived the possibility to formulate a *sub-quantum* theory implementing a non-linear dynamics. Interestingly, he speculated that such a deeper, more fundamental non-linear law applied at very short distances would allow for a precise description of matter beyond the threshold imposed by current quantum theory, envisioning experiments with unlimited precision able to test the pilot-wave theory, making it *falsifiable* and thereby testable against the predictions of the standard formulation of QM.[19] Although such speculations did not concretize into new results, Bohm's hope to make his theory falsifiable or testable still remains open in today's proponents of the pilot-wave theory (for more details cf. Valentini and Das chapters in this volume and references therein).

With these ideas in mind, in the following years Bohm continued to work on his causal interpretation, and tried to show how Born's rule dynamically emerges from random collisions of (sub)quantum particles. More precisely, in Bohm (1953a) it is argued that if the particles' velocities satisfy the condition $v = \nabla S/m$ and the wave function evolves according to (I.1), then *any arbitrary* probability distribution will eventually approach $|\psi|^2$ due to a large number of random collisions.[20] Therefore, Born's distribution should no longer be postulated as an additional axiom of the theory, but rather it would be a consequence of it. Similarly, Bohm and Vigier (1954) aimed at achieving the same result using a hydrodynamic model – originally proposed by Madelung in 1926 and then extended by Takabayasi and Schönberg. The authors considered a fluid where $|\Psi|^2$ represents its density and $\nabla S/m$ refers to the local stream velocity. Moreover, the equations (I.3) and (I.4) assumed a new meaning, representing respectively the conservation of the fluid and the law determining the velocity potential in terms of both classical and quantum forces. An additional assumption was also

[18] In the 1950s, Bohm claimed in several places that his own approach, exactly as QM, has limited validity. The reader may refer to Bohm (1953b, 1957), as well as to his private correspondence published in Talbot (2017). This is a characteristic trait of Bohm's conceptions of physical theories, which in his opinion always had a specific, restricted domain of application. More details on Bohm's philosophy of science can be found in Oldofredi (2023).

[19] The Bohm–Pauli correspondence reveals interesting ideas that Bohm held about quantum theory. He was in fact convinced that the linearity of the HIlbert space – and of the Schrödinger equation – is a crucial limit to find new physics at more fundamental scales. For historical and technical details cf. respectively Oldofredi (2023) and von Meyenn (1996), Letters 1313, 1314, and 1315.

[20] In Bohm (1953b), p. 274, it is claimed that such a result is 'the direct analogue of Boltzmann's H-theorem in classical statistical mechanics'. The connections between the meaning of $|\psi|^2$ in the de Broglie–Bohm theory and Boltzmann's results in statistical mechanics are indeed very tied: the two contemporary major approaches to de Broglie–Bohm theory – the Dürr, Goldstein, and Zanghì school on the one hand, and Valentini's on the other – implement different 'Boltzmannian' strategies in order to explain the status of the Born's distribution. More information is given in Passon and Valentini's contributions to this volume.

made, i.e. that the fluid fluctuates irregularly as to resemble turbulent motion, where the actual density ρ and the actual velocity v randomly fluctuate respectively around $|\Psi|^2$ and $\nabla S/m$, which are then considered averages quantities. Such irregular motions and fluctuations of the fluid's elements produce a 'mixing' of the fluid itself – in which each 'particle' follows a very irregular path. The authors then showed that a statistical ensemble of such systems with an arbitrary probability density $P(x)$ decays into $P = |\Psi|^2$, concluding that the irregular fluctuations of the fluid provides a physical explanation for the statistical distributions of QM.[21]

Extensions of the causal interpretation: Pauli and Dirac equations

Bohm extended his pilot-wave approach to the treatment of spin employing the Pauli equation as the foundation of the particles' dynamics (Bohm, Schiller, and Tiomno 1955; Bohm and Schiller 1955), as well as to quantum fields by interpreting the Dirac equation in causal terms (see Bohm 1953b). Since these important works are not very much discussed and known in literature, let us take this opportunity to present them in a nutshell.

In the former generalization of the pilot-wave theory, Bohm and his collaborators started their analysis from a hydrodynamic model similar to that studied in Bohm and Vigier (1954), extending it to the description of a spinning electron. They firstly considered the Pauli equation written as follows:

$$i\hbar\frac{\partial\Psi}{\partial t} = \left(-\frac{\hbar^2}{2m}\left(\nabla - ie\frac{\mathbf{A}}{c}\right)^2 + V + \frac{e\hbar}{2mc}(\sigma\cdot\mathbf{B})\right)\Psi, \tag{I.7}$$

where Ψ is a two-component spinor, $\mathbf{A}$ is the electromagnetic vector potential, V is the total potential energy, and $\mathbf{B}$ is the magnetic field. Secondly, they define both the fluid density $\rho = \Psi^*\Psi$ and the fluid velocity.[22] The latter is written as the ratio between the conserved charge ($\rho = \Psi^*\Psi$) and conserved current ($\mathbf{j} = \frac{\hbar^2}{2mi}(\Psi^*\nabla\Psi - \Psi\nabla\Psi^*) - \frac{e}{c}\mathbf{A}$) of the Pauli equation:

$$\mathbf{v} = \frac{\hbar^2}{2mi}\frac{(\Psi^*\nabla\Psi - \Psi\nabla\Psi^*)}{\Psi^*\Psi} - \frac{e}{c}\mathbf{A}. \tag{I.8}$$

Having defined the basic elements of the model, the authors assume a new physical property of the fluid affirming that it possesses an intrinsic angular momentum, i.e. spin, which is obviously connected with the spinorial elements of Ψ. It is worth noting, however, that in this theory spin assumes a very clear and unambiguous meaning: it is in fact claimed that such an intrinsic angular momentum is due to the continuous

[21] It should be underlined that the hydrodynamic model of Bohm and Vigier implements a stochastic dynamics. This fact clearly indicates that Bohm was not interested in preserving determinism at quantum scales.

[22] Another version of this model implementing a direct particle ontology is given in Dewdney *et al.* (1987).

rotational motion of the very small rigid bodies composing the fluid. In other words, to say that a certain particle is in a 'spin up' state with respect to the x axis means that such a particle has a spin vector – specified in terms of Euler angles describing the orientation of a rigid body with respect to a set of coordinates – which points along that axis and making the particle moving in a certain way, whereas the y and z components are not undetermined but zero (since the particle is not rotating with respect to those axes).

Finally, the authors underline that the orientation of such spinning bodies depends on Ψ. Since the causal interpretation of the Pauli equation generalizes Bohm's previous work on the Schrödinger equation – where the wave function ψ was interpreted as a real physical field – they wanted to provide a method to interpret all the four real components of the spinorial wave function 'in a visualizable way by constructing a mathematically equivalent set of fields in euclidean space', as explained in Dewdney *et al.* (1987). Such spinorial wave function is indeed interpreted as a multi-component field that guides the motion of a rigid spinning body, whose centre of mass follows a deterministic trajectory in space generally different from the trajectories implied by the Schrödinger equation, for a spin-dependent force appears in the quantum potential contained in the generalized quantum Hamilton–Jacobi equation obtained from the Pauli dynamics (for more technical details cf. Bohm *et al.* (1955), p. 63, Dewdney *et al.* (1986), Dewdney *et al.* (1987), pp. 537–538, Holland (1993), chapters 9 and 10).

Another interesting metaphysical feature of this model – unfortunately often overlooked in philosophical discussions – is that the authors explicitly claim that particles have a small but non-zero size; therefore, they do not approximate the body of quantum particles to mathematical points, as instead done in the original 1952 papers. As Bohm later remarked, in the causal interpretation of the Pauli equation 'the bodies with which we are concerned are not points' (Bohm 1957, p. 81).[23]

The second generalization of Bohm's causal interpretation is given in Bohm (1953b), where he replied to several objections raised by the Japanese physicist Takehiko Takabayasi. These criticisms gave Bohm the opportunity to clarify several aspects of his work as well as to generalize it. In particular, we find a nice treatment of the single particle Dirac equation interpreted in causal terms (see Passon's chapter, Section 1.4.3 for technical details). As far as the many-body case is concerned, in this article Bohm refers to the Dirac sea picture for the first time, by saying that any quantum field theory (QFT) of electrons is equivalent 'in all of its results to that of a particle theory, in which we restrict ourselves to antisymmetric wave functions, and assume that all negative energy states are occupied in the vacuum' (Bohm 1953b, p. 275). Such a perspective will be further elaborated in more detail in Bohm and Hiley (1993), as well as in recent essays by Colin (2003), Colin and Struyve (2007), and Deckert *et al.* (2019).[24]

[23] More on this specific topic is discussed in the next section.

[24] It must be noted here that Bohm's causal interpretation of the many-body Dirac equation, and more generally his causal interpretation of the Dirac sea picture, is essentially non-relativistic. Bohm was well aware of this feature of his model, but underlined that it is empirically equivalent with its relativistic counterpart. Thus, although Bohm's theory is non-local, faster-than-light signals cannot be sent so that an operational compatibility with relativity is maintained. Moreover, the pilot-wave theory has the merit to have raised new questions in the quantum foundations, namely what a

Notwithstanding the formal consistency of this proposal and its empirical equivalence with respect to the predictions of QM, Bohm's interpretation was unfortunately poorly received, being mostly considered a reactionary attempt to restore a deterministic – i.e. classical – worldview (Freire Junior 2015, 2019; Oldofredi 2023).[25] As a consequence of this fact, Bohm decided to focus on other domains of theoretical physics as well as other philosophical issues (see Freire Junior 2019, Chapters 5 and 6), giving up for some time active research on his causal interpretation.

The second renaissance of the pilot-wave theory and contemporary perspectives

Even though the pilot-wave theory was temporarily abandoned by its pioneers for different reasons, it gained a new life a few years later in virtue of the work of many physicists and philosophers. In particular, this theory received the constant endorsement of John Stuart Bell, who in several essays from the mid-1960s to the late 1980s praised the ontological and physical clarity of the de Broglie–Bohm theory, helping it to obtain more recognition.[26] Furthermore, from the mid-1970 such a framework was revived by a group of Bohm's collaborators and students, in particular C. Dewdney, B. Hiley, P. Holland, P.N. Kaloyerou, A. Kyprianidis, and C. Philippidis, also supported by the long-time champion of the pilot-wave theory, J.P. Vigier – who not only collaborated with Bohm in 1954, but was also one of his most important allies in the 1950s as emphasized by Freire Junior (2019).

Among other accomplishments, these scientists were able to compute and plot actual trajectories for several crucial quantum mechanical experiments, as for instance the double-slit experiment (Philippidis, Dewdney, and Hiley 1979), the Wheeler's delayed choice experiment (Bohm, Dewdney, and Hiley 1985), the EPR scenario (Dewdney, Holland, and Kyprianidis 1987a), the spin superposition in neutron interferometry (Dewdney, Holland, and Kyprianidis 1987b) and the Stern–Gerlach spin-measurements (Dewdney *et al.* 1987) to mention a few.[27] In addition, several ideas were advanced in order to improve Bohm's early attempts to extend the pilot-wave approach to relativistic contexts and quantum field theory, as for instance in Bell (1987, chapter 19); Bohm, Hiley, Kaloyerou (1987); Dewdney *et al.* (1986); Dewdney *et al.* (1992); Holland (1986, 1987, 1988), Bohm and Hiley (1993). The latter book, moreover, provided the

'truly' relativistic theory is, what requirements and conditions must be met by a theory to be said genuinely relativistic, how to properly combine relativistic causality with quantum non-locality in a consistent fashion, etc.. That being said, in the opinion of the present author the issue concerning the combination between relativity and de Broglie–Bohm theory is one of the biggest challenges still open in the pilot-wave community. The reader may find interesting observations and reflections on this topic in Passon, Valentini, Allori and Tumulka's contributions to this book.

[25] Cf. Freire Junior's chapter in the present volume. Moreover, as already pointed out, the contributions of van Strien and Del Santo and Krizek explain why it is historically incorrect to claim that Bohm wanted to restore determinism. A similar point is also contained in Oldofredi (2023), where it is shown that several objections against Bohm's theory were negatively biased and not grounded on a purely scientific basis. In addition, the latter paper argues that Bohm's writings available in the 1950s were sufficient to demystify several beliefs generally held about his pilot-wave approach, in particular those about Bohm's plans to save determinism.

[26] Most of Bell's essay about the de Broglie–Bohm theory are collected in Bell (2004).

[27] Cf. Dewdney (2023) for a historical and personal recollection of these works.

synthesis of Bohm and Hiley's philosophical trajectory combining the ideas of wholeness and implicated order with the causal interpretation. In that volume the authors not only systematized their previous results in a more general conceptual framework, but also discussed new applications of the theory to the Aharonov–Bohm effect, superfluidity and superconductivity as well as the chemical bond.[28] All these mentioned works proved to be fundamental in order for the de Broglie–Bohm theory to acquire a new status in the field of quantum foundations and begin to be considered a serious alternative formulation of (at least non-relativistic) quantum mechanics.

The second renaissance of the pilot-wave theory – occurred between the mid-1970s to the late 1980s – coincided with an epoch theoretical fertility as testified by the above mentioned essays. Moreover, the causal approach continued to develop in the following years, sparking a growing interest in the community working on the foundations of QM. As a consequence, we see the birth of different schools of thought, resulting in a multiplicity of perspectives on the fundamental physical and philosophical assumptions of the pilot-wave theory. Such diverging ideas generated an interesting plurality and diversity of views still present to this day. For instance, while Dewdney *et al.* (1987) and Holland (1993) interpreted spin as a new degree of freedom of the particles in addition to its position – in agreement with the Bohm–Schiller–Tiomno model seen above – Bohm himself rejected such a view, and in Bohm and Hiley (1993) considered it just a component of the wave function, i.e. not a property of the quantum corpuscles. Furthermore, taking into account extensions of the theory to relativistic regimes, Holland (1988) provided an interpretation of the Fermi fields in terms of continuous field variables, in opposition to Bohm, Hiley, and Kaloyerou (1987), who on the contrary claimed that such a field theoretical description could not be given.

The major conceptual differences with respect to Bohm's approach to the pilot-wave theory, however, arose in the early 1990s with the important contributions by Detlef Dürr, Sheldon Goldstein, and Nino Zanghì on the one hand, and Antony Valentini on the other.[29] Under the influence of John Bell's exposition of the pilot-wave theory, the former authors reshaped de Broglie–Bohm theory creating what is now called 'Bohmian Mechanics'. Such a 'Bohmian' framework redefines both the ontology and the dynamics of the pilot-wave theory: on the one hand, ψ is no longer considered a physical field, but rather it is interpreted as a nomological item. On the other hand, the quantum potential was ruled out from the theoretical machinery of the theory, losing any ontological and explanatory weight.[30] Similarly, Valentini's approach detaches

[28] It is worth mentioning that Hiley continuously worked on the pilot-wave theory and still provides interesting contributions to it. To mention a few, Hiley's some notable works concern the algebraic structures and non-commutative geometry in relation with Bohm's theory (Hiley 2002, 2005), the comparison between Dirac's notion of quantum trajectory and that of Bohm, creating the so-called Bohm-Dirac picture (Hiley and Dennis 2018; Hiley, de Gosson, and Dennis 2019), or his recent argument according to which Bohmian trajectories should be interpreted as statistical averages reminiscent of Feynman's paths (Flack and Hiley 2018).

[29] Indeed, almost all the examples mentioned above are variations of Bohm's own formulation of the pilot-wave theory. The approaches that are considered below, on the contrary, show a significant and substantial detachment from the views held by Bohm, his collaborators and those endorsing a quantum potential variant of the causal interpretation.

[30] For conceptual rigour one should call Dürr, Goldstein, and Zanghì's approach *de Brogliean* or *Bellian* mechanics, given that their theory is much closer to de Broglie's proposal or Bell's presentation,

itself from Bohm's, criticizing his new quantum mechanical force introduced in 1952, marking a return to views closer to de Broglie's original proposal.[31]

Notably, in the past 35 years these physicists and their collaborators followed alternative routes to develop the pilot-wave theory, and through the years made – and still make – important contributions to it. For instance, Dürr, Goldstein, and Zanghì (2013, chapter 2) as well as Valentini (1991a, 1991b) provided rival explanations and justifications for the emergence of the Born rule and its status in de Broglie–Bohm theory. Assuming that the Born's distribution holds at the universal level, the former proposal shows that such a distribution is preserved for sub-systems of the universe. On the contrary, Valentini employed a different strategy: rather than taking into account only equilibrium states, as those considered in Dürr, Goldstein, and Zanghì works, he proposed a *dynamical* explanation for the emergence of the Born's rule from *non-equilibrium* states. More details on Dürr, Goldstein, and Zanghì's approach can be found in Passon's essay (Section 1.3.6), whereas Valentini's view is illustrated in the second chapter of this book.[32]

In addition, they proposed further generalizations of such a framework to more fundamental domains of physics, as for instance quantum field theory (Dürr *et al.* 2013, Chapter 10; Dürr *et al.* 2005; Valentini 1996). Referring to this, the Bohmian school of Dürr, Goldstein, and Zanghì improved a stochastic extension of the de Broglie–Bohm theory originally due to John Bell (Bell 2004, Chapter 19). Given the historical relevance of this proposal, let us briefly sketch its basic features.

According to Bell's theory, the physical three-dimensional space is represented by a discrete lattice, and the local beable consists of the fermion number density at each lattice point. The ontology of this model is slightly different with respect to the standard de Broglie–Bohm theory since particles' positions do not possess their usual beable status, and consequently Bell's framework does not provide trajectories for individual fermions moving in space. Another relevant conceptual novelty of Bell's model concerns the dynamics, which is *stochastic*. In order to describe the events of particles' creation and annihilation, he provided an equation of motion specifying the probability for a given configuration of fermions to jump into another configuration with a different number of objects within a certain interval of time. These random jumps correspond to the creation or annihilation of particles. The choice to introduce stochastic elements into the dynamics of the theory is motivated by the random events of particle creation and annihilation empirically observed in QFT. Nonetheless, Bell viewed his theory only as a phenomenological framework without any pretension to be interpreted realistically.[33]

while it differs significantly w.r.t. Bohm's own causal interpretation. However, to avoid confusion, we stick here to the usual label 'Bohmian' mechanics used in literature.

[31]The readers not familiar with Bohmian mechanics may refer to Dürr and Teufel (2009) and Dürr *et al.* (2013) for technical expositions, whereas Goldstein (2017) provides a simpler overview. Valentini's account is nicely summarized in the second chapter of the present volume.

[32]Cf. Callender (2007) for an excellent overview of these approaches.

[33]In this regard, Bell claimed that the stochasticity of his model is unwelcome, since the reversibility of the Schrödinger equation suggests that quantum mechanics is not fundamentally stochastic in nature. Indeed, he suspected that in the continuum limit the dynamics would have been replaced by a deterministic evolution in a continuous space. The reader can find a continuous and deterministic

A generalization of Bell's ideas was advanced by Dürr, Goldstein, Tumulka, and Zanghì, who called their own proposal *Bell-type QFT*, homaging the influence of the Irish physicist. The Bell-type QFT is a theory of particles moving in a continuous three-dimensional physical space, whose trajectories can randomly begin and end at certain space-time points, thus the Bohmian corpuscles can randomly come into existence as well as cease to exist. In this context, creation events naturally correspond to beginnings of particles trajectories, similarly, annihilation events correspond to their end. As a consequence, the dynamics of the model has to introduce variations in the particles' number in order to describe such birth and death processes, which are taken to be inherently stochastic, i.e. happening spontaneously without any external cause. It must be highlighted that between creation and annihilation events the particles follow deterministic trajectories and evolve according to a guiding law specifying the particles' velocities. Destinations and times of the jumps are the stochastic elements of the model, and the probabilities governing them are described by Markov processes: they do not depend on the past histories of the particles' trajectories, but only on the present state of the configuration. Contrary to Bell's proposal, however, in Bell-type QFT the beables are again the particles' positions for both particles and antiparticles. Moreover, it is worth noting that in this latter model bosons assume a particle status exactly as fermions, marking an important difference with Bohm and Bell treatment of photons, as well as with the description provided by the standard model of particle physics.[34]

On the other hand, Valentini once again followed a different strategy with respect to Dürr and collaborators, for he provided a model implementing a field ontology for fermions. One of the main motivations for such an ontology, according to Valentini, comes from the asymmetrical description of fermions and bosons in Bohm's writings on quantum field theory, where fermions are described as particles and bosons as fields. In order to solve this issue the author proposed to represent fermions as fields. As nicely expressed by Struyve, this 'approach starts from a functional Schrödinger representation in which the wave functionals are defined on a space of Grassmann fields (that is, anti-commuting fields). The actual configurations are Grassmann fields, which are guided by the wave functional' (Struyve (2010), p. 5). In this theory the dynamical equations of motion generally describe a non-local evolution of a scalar field in 3-dimensional space, where the velocity at a specific point depends on the instantaneous values of the field at arbitrarily distant points, entailing thereby that space and time are absolute (cf. Valentini 1996, p. 54). Interestingly, in this model quantum particles are described in a low-energy approximation in which the field mimics a standard non-relativistic wave function ψ: if the latter is localized in a small packet of width Δx, 'the field will be a "lump" of size Δx' (*ibid.*). Thus, Valentini redefines in this theory the notion of 'particle' which is not a point-like object lying

Bell model in Colin (2003). Moreover, another deterministic and continuous formulation of a Bohmian quantum field theory is given by the Dirac sea picture already mentioned in this section. Due to editorial constraints I cannot discuss it here, so interested readers may refer to Bohm and Hiley (1993) Chapter 12, Colin and Struyve (2007) and Deckert *et al.* (2019). For a philosophical discussion of the Dirac sea cf. Oldofredi (2018a).

[34] For a new contribution along this tradition, the reader is referred to Tumulka's chapter in this volume.

somewhere in space, and more importantly is not fundamental since it emerges from a field theoretical limit.[35] More details on Valentini's views on quantum field theory are given in the second chapter of this volume, particularly in Section 2.2.5.[36]

Furthermore, in addition to their works on quantum field theory, these physicists provided interesting models to obtain relativistic versions of the causal interpretation (Dürr *et al.* 2013, Chapter 9; Dürr *et al.* 2014; Lienert *et al.* 2020; Valentini 1997). Referring to this, it is interesting to underline that different authors have diverse attitudes concerning Lorentz invariance. On the one hand, there are exponents of the pilot-wave community – as for instance Bohm and Hiley (1993), Holland (1993) and Valentini (1991a) among others – who think that a detailed microscopic description of the quantum mechanical regime must violate Lorentz invariance, whereas Dürr, Goldstein, Zanghì and their collaborators claimed that a genuine Lorentz invariant Bohmian theory is not in principle unattainable (Dürr *et al.* 2013; Dürr *et al.* 2014). In both traditions, however, it seems that the notion of absolute simultaneity must be recovered in order to define a guidance equation for the non-local dynamics of the Bohmian particles. This fact still leaves the question open to define the criteria for a satisfactory combination between special relativity and the pilot-wave theory, and more generally between the axioms of the former with quantum non-locality.[37]

Finally, it should also be underlined that pilot-wave approaches for quantum gravity and cosmology have been recently proposed, as for instance in Goldstein and Teufel (2001); Dürr, Goldstein, and Zanghì (2020); Valentini (2007, 2010, 2023).[38] These theoretical works once again show a variety of research programmes in the pilot-wave community, and the need to extend such interpretation outside the non-relativistic regime. However, although these essays present notable results, they should not be considered as final theories but rather promising steps towards the achievement of a fully satisfactory generalization of the pilot-wave approach to relativistic physics.

On top of this brief overview of the historical development of the pilot-wave theory in its different formulations and variants, it should be emphasized that the de Broglie–Bohm approach found (and still finds) interesting research applications in several branches of physics, from quantum chemistry to quantum engineering. In this

[35] It is very interesting to note that in this pilot-wave model the particles' position become contextual variables. Describing what happens during a position measurement, Valentini affirms that '[t]he initial field may be completely delocalised. During the 'measurement' process the field is constrained to become concentrated in a small region around some point x, the 'measured position'. The value of x is determined by the initial field and apparatus variables (given the initial wave function and Hamiltonian). It has no simple relation to the initial field alone. Thus, in field theory, 'particle position' is a contextual variable' (Valentini 1996, pp. 54–55).

[36] For critical remarks on Valentini's approach see Struyve (2010), whereas discussions of the Bell-type QFT are given in Oldofredi (2018a, 2018b), where some limitations of this theory are also pointed out.

[37] Alternative strategies to obtain a relativistic formulation of de Broglie–Bohm theory can be found in Drezet (2019) and Galvan (2015). It is worth mentioning here that Sutherland (2008) proposed a retrocausal, Lorentz invariant model for Bohmian mechanics in order to explain Bell's theorem without introducing non-locality. Another retrocausal model for Bohmian mechanics can be found in Lewis (2016).

[38] In addition to these mentioned papers, there are other notable essays delineating possible paths to recover a pilot-wave model for quantum gravity, as for instance Pinot-Neto (2005, 2021) and Vassallo and Esfeld (2013). In this volume Vassallo and Naranjo propose extension of pure shape dynamics to de Broglie–Bohm theory, cf. Chapter 14.

regard, the interested reader is referred to Oriols and Mompart (2019), an excellent edited collection of essays where the causal interpretation is applied to a variety of topics such as quantum chemistry, chemical physics and nanoelectronics. Finally, Bohmian approaches based on quantum trajectories found useful applications to model time-of-flight measurements (Leavens 2008; Das and Dürr 2019), to calculate actual trajectories of electron around atoms nuclei (Colijn and Vrscay 2003a, 2003b; Timko and Vrscay 2009), and also in quantum optics (Sanz and Miret-Artés 2011, 2014). Referring to the applications of Bohmian mechanics, the reader may find interesting discussions in the contributions of Fortin and Lombardi, Oriols *et al.*, and Das to the present book.

After having briefly illustrated some of the major steps in the development of the pilot-wave theory, let us now take into account the principal philosophical debates concerning this approach to quantum foundations.

Philosophical reflections on the de Broglie–Bohm theory: ontology and methodology

The philosophical literature on the de Broglie–Bohm theory is so vast that we can only review a small fraction of it here. In what follows, then, we mainly focus on the metaphysical reflections about the ontology of this interpretative framework, with particular emphasis on the dispute concerning the nature of the wave function. In addition, towards the end of this section we briefly analyse the notion of quantum contextuality in the pilot-wave theory, as well as the logical structure of this interpretation. Finally, we mention the methodological and normative ideas of the primitive ontology programme, given its centrality in contemporary discussions on Bohmian mechanics.

Bohm's ontological perspectives: ψ-field and the particle notion

From the early writings on quantum mechanics to his last works, it is clear to Bohm's readers that ontological clarity is an essential feature that physical theories must possess. Indeed, in his opinion the problems of the 'usual interpretation' of QM – identified with Bohr's and Heisenberg's ideas firmly criticized in the 1952 papers – originated from its obscure metaphysical content. Among several issues affecting this theoretical framework, Bohm was deeply dissatisfied by the lack of causal explanations for the actualization of macroscopic measurement outcomes. In particular, as already pointed out, he argued that quantum theory does neither provide a precise characterization of individual systems, nor an accurate description of their dynamical evolution 'without proving that such a renunciation is necessary' (Bohm 1952, p. 168). In fact, he claimed, the empirical robustness and consistency of QM and its contingent mathematical structure are not sufficient reasons to exclude *a priori* other ontologically clearer interpretations of the quantum formalism.

In the previous sections we have seen that in order to overcome these defects of QM Bohm proposed a theory where particles have definite positions at all times and are dynamically guided by a physically real wave field. Despite his interpretation was erroneously deemed by many as 'impossible' – due to von Neumann's no-go theorem against the completion of quantum theory with hidden variables – or viewed as anachronistic – due to its ontological picture and deterministic character – Bohm constantly pondered on the metaphysics of the pilot-wave approach during his whole career, continuously redefining the concepts involved in it.[39]

In order to better understand his metaphysical views on quantum theory, it is useful to make a preliminary remark upon Bohm's thoughts on the aims and scopes of scientific theories more generally.

According to him[40], the historical evolution of physics shows that every theoretical framework proposed so far – from antiquity to contemporary science – provided a reliable description of some *limited portions* of our universe. From this fact, he inductively concluded that physical theories should be considered accurate representations of nature only within specific intervals of energy/length scales, i.e. they have a restricted validity, or alternatively they are not universally valid (cf. footnote 18). In many places, Bohm exemplifies his thesis speaking about the limitations of classical mechanics, special relativity and quantum theory (cf. e.g. Bohm 1953b, 1957).

Similarly, since the history of physics (once again) demonstrates that scientific investigations always disclosed new features and levels of reality, Bohm endorsed a form of what is now called 'metaphysical infinitism', the thesis according to which nature has an infinite (or limitless) structure.[41] Thereby, it is not ontologically reducible to an absolutely fundamental bottom ground: we will be unable to embrace the infinite complexity of nature with a finite set of entities and laws. Thus, no final theory of everything can be ever possibly found that reduces the richness of reality to a limited set of axioms. This conclusion will also be maintained in later stages of Bohm's philosophy, as for instance when he formulated the notion of wholeness and implicate order (cf. Bohm 1980), for he believed that reductionism fails to grasp the holistic and dynamical essence of reality.[42]

In addition to that, Bohm provided compelling reasons to believe that each layer of nature can be studied independently; hence, those theoretical entities and laws describing a certain level should be considered just *relatively fundamental*.[43] Therefore, he claimed, one should inductively infer that future theories will be only valid within certain specific domains of application.

Notwithstanding their local validity, Bohm thought that scientific theories progressively come closer to a true description of reality with increasingly better approximations, expanding our knowledge of the world by discovering new layers of nature – thus,

[39] Bohm (1952), Part II, Section 9 explains why von Neumann's theorem does not apply to the causal interpretation. Moreover, in a private correspondence, Bohm wrote that von Neumann himself recognized the logical consistency of his theory. Cf. Bohm's letter to Pauli in von Meyenn (1996), Letter 1290, p. 392 and Bohm's letter to Melba Phillips in early 1952 in Talbot (2017), p. 147.

[40] In what follows I refer to those general tenets of Bohm's philosophy that remained stable – or almost stable, with minimal modifications – from the 1950s to the 1990s.

[41] Bohm's metaphysical infinitism finds its best expression in the book *Causality and Chance in Modern Physics*, published in 1957, whereas the gist of it can be found in Bohm (1953b). Details about his views can be found in several letters written between 1951 and 1954 and published in Talbot (2017). Bohm's infinitism is moreover studied in Oldofredi (2023). In addition, the reader will find interesting discussions about it in van Strien and Del Santo and Krizek chapters in this volume.

[42] Due to space reasons I cannot introduce here in full detail Bohm's conception of implicate and explicate order, as well as his notions of wholeness. The interested reader may refer to the volumes Bohm (1980) and Bohm and Hiley (1993).

[43] In this specific respect, Bohm believed that different levels of descriptions are closely interrelated, and that relations among theories do clearly exist.

his metaphysical infinitism is not in open conflict with scientific realism.[44] In relation to this, it is worth noting that Bohm was indeed a metaphysical realist, namely he thought that the external world exists independently of our minds, knowledge and possible observations of it. However, given the ideas expressed in many places – as e.g. in his celebrated books *Causality and Chance in Modern Physics*, *Wholeness and the Implicate Order* and *The Undivided Universe* among others – it is not trivial to claim that he endorsed a strong form of scientific realism, for he did not believe that the sentences or statements expressed by scientific theories must be interpreted as literally true. Alternatively, Bohm was convinced that our best theoretical frameworks are *abstractions*, i.e. idealized representations of the real objects and their causal relationships actually constituting our universe. One must therefore not confuse the simplification of a map, although useful and faithful, with the complexity of the territory.

Nonetheless, he claimed that physical theories do in some sense provide an approximately true description of natural phenomena because their laws describe objective, i.e. non-arbitrary, features of the world. While in the 1950s such objective features mainly referred to causal relations, from the 1960s he explicitly claimed that scientific investigations also provide information about structures and processes. The only limitation imposed by his perspective on scientific realism is that such theoretical descriptions are reliable only within definite scales and domains of application, i.e. they are not valid universally, as said a few lines above.[45]

Given this philosophical background, Bohm never thought that the ontology of his pilot-wave theory should have been considered fundamental. To this specific regard, as we already pointed out in the previous sections, he claimed that 'at distances of the order of 10^{-13} cm or smaller and for times of the order of this distance divided by the velocity of light or smaller, present theories become so inadequate that it is generally believed that they are probably not applicable' (Bohm 1952, footnote 6; cf. also Bohm 1957 Chapter 4). In these regimes he expected that new theories will be discovered, thus new ontologies will be explored. Along these lines, he thought that non-relativistic quantum mechanics would become a statistical theory whose physical content emerges from the motion of yet-unknown sub-quantum entities. Hence, he concluded that there are neither theoretical, nor historical motivations to consider QM, or the causal interpretation, as a fundamental theory. Historical evidence for this claim can be found not only in his writings in the 1950s mentioned above, but also in his works on the implicate order approach, where he tried to provide a unified framework for both quantum theory and relativity (Bohm 1980). Nonetheless, the

[44]In his view, a certain theory can be considered 'true' even though (i) it is not concerned with the most fundamental levels of reality, and (ii) is reliable only at specific scales. For instance, Newtonian mechanics can be considered a correct description of the macroscopic realm, albeit its predictions are wrong at microscopic regimes. The conceptual mistake often done by many, Bohm underlines, is to deem this theory as false or wrong because of its failure at microscopic scales. The actual error, in his opinion, is instead to pretend its universal validity. Referring to this, another interesting consequence of Bohm's philosophy is that different ontologies can peacefully coexist at diverse levels, thereby supporting scientific pluralism.

[45]It is interesting to note that in Bohm's views the explanations of physical phenomena given by scientific theories provide knowledge of the external world. Therefore, one can safely conclude that the metaphysical and epistemological aspects of scientific realism are somehow conserved in Bohm's conceptions of physical theories.

pilot-wave theory remains a sound description of reality at quantum non-relativistic scales; thus, it can be considered relatively fundamental in these regimes.

Having framed the limits of this interpretation, we are now ready to focus on the ontological reflections concerning both the nature of the quantum wave function and the notion of particle.

In his works, Bohm provided an innovative understanding of the wave function: rather than being considered a bookkeeping device storing information about probabilities of measurement outcomes, it was interpreted as a real physical field, exactly as the electromagnetic field. Indeed, in several places he drew analogies between these two entities – cf. Bohm (1952, 1957) and Bohm and Hiley (1993) – by saying that the ψ-field exerts a force on the particle similar to the way in which an electromagnetic field exerts a force on a specific charge. In addition, Bohm claimed that

> just as the electromagnetic field obeys Maxwell's equations, the ψ-field obeys Schrödinger's equation. In both cases, a complete specification of the fields at a given instant over every point in space determines the values of the fields for all times. In both cases, once we know the field functions, we can calculate force on a particle, so that, if we also know the initial position and momentum of the particle, we can calculate its entire trajectory (Bohm 1952, p. 170).

Clearly, he was well aware about the important differences between the two – as e.g. the lack of back reaction of the particles on the ψ-field[46] or the absence of sources for the wave function – nonetheless, he always assigned the latter an ontological weight, not considering it just a mathematical symbol devoid of physical significance.

However, over the years Bohm became less satisfied by the interpretation of the wave function given in his 1952 papers, because taking into consideration a many-body system, ψ becomes a field defined in a multi-dimensional configuration space. This fact entails that ψ cannot be considered a field exerting a mechanical force on the particles.

In order to solve this issue while maintaining an ontological reading of ψ, in the 1980s Bohm developed a new interpretation of the wave function in terms of 'information' in collaboration with Hiley. They submitted that while ψ is still interpreted as a real physical field, it does not 'push' the particles in space; rather, it carries information about their motion (cf. Bohm 1990; Bohm and Hiley 1982, 1985, 1993). In this regard, Bohm and Hiley wrote that

> [t]he fact that the wave function is in configuration space clearly prevents us from regarding the quantum field as one that carries energy and momentum that was simply transferred to the particles with which it interacted (thus effectively pushing or pulling mechanically on the latter). This is a further factor in addition to the form dependence of the activity of the field which leads us to consider the interpretation of this field as active information (Bohm and Hiley 1993, p. 61).

It must be clearly underlined that the notion of information employed by the authors is not related with any technical definition of it, as e.g. Shannon's theory. They interpreted the word 'information' literally, in the sense of 'giving form to something',

[46] In Bohm 1957, p. 76, it is claimed that the particles do exert a reciprocal influence on the ψ-field, but given that this influence is extremely small, it can be neglected in the quantum mechanical regime. Bohm hypothesized that the action of the particles on the wave field may be relevant in sub-quantum domains. Such hypothesis however will be changed by Bohm in later years, and eventually dismissed in his last book *The Undivided Universe*.

following its Latin etymology. In this new perspective, the wave function in-forms – i.e. gives form to – the particles' trajectories. Contrary to Bohm's previous works, the wave field does not directly act on the particles in space, but rather it is the form of ψ (in particular the second spatial derivative of the amplitude R appearing in the quantum potential) that prescribes the dynamical evolution of the corpuscles. In order to better understand this point, Bohm offers the following analogy:

> consider a ship on automatic pilot guided by radar waves. The ship is not pushed and pulled mechanically by these waves. Rather, the form of the waves is picked up, and with the aid of the whole system, this gives a corresponding shape and form to the movement of the ship under its own power (Bohm 1990, p. 279).

In this example, the radio waves carry information about the environment and guide the ship through it. Similarly, the ψ-wave – and in particular the quantum potential – contains information about a specific experimental environment and guides the motion of the particles through it.

Notably, this concept of information is strictly related to the notion of *wholeness.* As we already stated in the previous sections, the action of the quantum potential does not in general vanish with distance: namely, it is independent of the intensity of the quantum field but depends on its form. It is its influence on the particles – those composing the observed system as well as the measuring apparatus – that determines and explains observable measurement outcomes. Hence, Bohm claimed that these features of Q imply a holistic relation between the observed system and its environment.

In addition, such a new reading of the ψ-field holds that the information of the wave function is potentially active in every spatial region, but it is actually active only where particles are located. This in turn entails a new meaning for the empty branches of the wave function. These empty wave packets are now defined as *potentially active information* since the particles may interfere again with them in their future paths, and thereby they can become active and effectively guide their motion.[47]

Another compelling philosophical implication of Bohm's mature interpretation of the pilot-wave theory entails a substantial revision of the particle's notion. In Bohm's later views, particles are not conceived as mathematical points. On the contrary, they do possess their own energy and capacity to do work. In addition, they all have complex internal structures, even those objects considered elementary such as electrons. This hypothesis is motivated not only by the inexhaustibility of nature mentioned above, i.e. its irreducibility to a finite set of items and laws, but also by the following suggestion:

> Large aggregates of masses of matter reduce approximately to a simple Newtonian behaviour, whereas the molecules and atoms out of which matter is built have a more complex inner structure. To make this suggestion yet more plausible, we note that between the shortest distances now measurable in physics (of the order of 10^{-16} cm) and the shortest distances in which current notion of space time probably has meaning, of the order of 10^{-33} cm, there is a vast range of scale in which an immense amount of yet undiscovered structure could be contained. Indeed, this range of scale is comparable to that which exists between our own size and that of the elementary particle (Bohm's quote from Talbot (2024), p. 6).

Moreover, according to Bohm's implicate order view, the electron is not even conceived as an individual object but rather as 'a recurrent stable order of unfoldment

[47] For details cf. Bohm and Hiley (1993), Chapters 5 and 6.

in which a certain form undergoing regular changes manifests again and again, but so rapidly that it appears to be in continuous existence' (Bohm 1980, p. 194). In order to better appreciate this point, it is useful to take into account the work contained in Pylkkänen *et al.* (2014). These authors claim that since the 1960s Bohm's interests began to focus on the notion of implicate order as well as on the project of developing a new framework from which QM and relativity would emerge as approximations. Interestingly, they underlined that '[t]his framework suggests a strongly structuralist, process-oriented way of understanding individual quantum systems which is in some ways similar to Ladyman and Ross' structural realism'.[48] To this specific regard, Bohm himself maintained that in his sub-quantum framework the notion of individual particle or more generally of 'separated existing entity' does not arise since each and every object is just 'an abstraction from a totality of process [...]'. For instance, conforming to this perspective in the case of the electron 'what actually exists is a structure of underlying elementary processes or linkages supporting a pattern corresponding to an electron' (Bohm 1965, p. 291). Thus, with Pylkkänen, Hiley and Pättiniemi we should conclude that in Bohm's philosophy the notion of individual particles – and of individual object more generally – became progressively less metaphysically salient from the 1950s to the later stages of his career.

It is also worth noting that even in the 1950s Bohm did not always think of particles as extensionless points. As already mentioned in the previous section, Bohm, Schiller and Tiomno generalized the causal interpretation to the Pauli equation and considered the quantum corpuscles as extended rigid bodies.[49] Moreover, in view of the infinite richness of nature endorsed in those years, combined with the belief that non-relativistic QM was just a statistical theory emerging from a sub-quantum level, it is straightforward to conclude that quantum particles were not considered structureless objects even in Bohm's early views, contrary to what is often said (even today) in relation to the ontology of de Broglie–Bohm theory.

The nomological view of Ψ and recent discussions

Surprisingly – and almost paradoxically – although Bohm's philosophical reflections on the causal interpretation spanned the second half of the 20th century almost in its entirety and influenced generations of scientists and philosophers, and despite the notable contributions of Hiley, Pylkkänen and other scholars working on the quantum potential formulation of the causal interpretation, nowadays Bohm's ideas have been rejected by the majority of contemporary supporters of the pilot-wave theory.[50]

To this specific regard, the work of Dürr, Goldstein, and Zanghì provides an interesting example. As already said, these physicists formulated a version of pilot-wave theory that substantially differs from Bohm's own interpretation. Indeed, they proposed a first-order guiding equation for the Bohmian particles in which the quantum

[48] Cf. Margoni's Chapter in this volume for more details on Bohm's ontological perspectives in the 1960s.

[49] Recalling our previous discussion about spin in the causal interpretation, Bohm stated that 'the bodies with which we are concerned are not points', Bohm (1957), p. 81.

[50] John Bell being an interesting exception. Indeed, his methodological and philosophical reflections on QM have many points of contact with Bohm's views as argued in Oldofredi (2022).

potential does not appear, being considered an idle theoretical entity that should be eliminated from the theory's ontology. In addition, Goldstein and Zanghì proposed a *nomological* interpretation of the universal wave function Ψ, contrary to the field interpretation endorsed by Bohm (cf. Dürr *et al.* 2013, Chapter 12). According to the nomological view, the wave function of the universe is not a physical object existing in space-time that causally guides the particles through it. Rather, it is an abstract entity that governs their motion *generating* the vector velocity field of the corpuscles. To clarify what this means, the authors claim that in Bohmian mechanics Ψ plays a role analogous to the Hamiltonian function in classical mechanics. The latter generates a vector field in classical phase space via Hamilton's equations, similarly the wave function generates a vector field in configuration space via the guiding equation of the Bohmian theory. In the same vein, wave functions describing sub-systems of the universe – for instance those representing experimental situations – are defined as *quasi-nomological* somehow inheriting the nomic character from the universal wave function.

Although Ψ is not a material object, it is still part of the ontology of Bohmian mechanics. Dürr, Goldstein, and Zanghì indeed claim that in each physical theory one can distinguish between theoretical entities representing matter in space, the *primitive ontology* (PO), from those entities without direct physical meaning needed to provide a dynamical structure for the primitive variables.[51] The latter are however part of the ontology of a given theory – which includes both primitive and non-primitive entities (Allori *et al.* 2008, Section 4). Thus, since wave functions belong the second category, they are still provided with some ontological weight. In this specific case, these physicists are realist about laws of nature and seem to endorse a primitivist view, i.e. they consider laws as fundamental items that cannot be reduced to, nor deduced from other elements of the theory's vocabulary. Consequently, even though Ψ is not a material substance, it is an abstract entity belonging to the Bohmian ontology. Since only particles represent real objects in space and time according to these authors, it follows that they do not support a multi-category ontology for material entities, contrary to those endorsing a Bohm-type theory.

Another important difference between Bohm's notion of particle and that employed in Bohmian mechanics should be underlined. During the 1990s, Dürr, Goldstein, and Zanghì considered the Bohmian corpuscles as mathematical points that carry inherent properties such as mass or charge. However, they neither have internal structure/extension as Bohm thought, nor they are derivative entities emerging from a sub-quantum level or from a new type of physical order. On the contrary, the Bohmian corpuscles are considered the primary building blocks of our macroscopic reality – which is ontologically reduced to them and their motion – and thereby are taken to be fundamental entities. In relation to this, it is worth mentioning that in relativistic extensions of Bohmian mechanics or in its generalizations to QFT proposed by these authors, it is always the particles' dynamics that changes, not the corpuscular primitive ontology.

Building on the particle content of Bohmian mechanics, in the early 2000s Goldstein, Tumulka, Taylor, and Zanghì proposed a radical metaphysical hypothesis con-

[51] The notion of primitive ontology is examined in more detail later in this section.

cerning the nature of the Bohmian corpuscles. According to these authors, there are no different particle species but just one family of 'naked particles'. From this viewpoint, thus, there is no intrinsic metaphysical difference between an electron, a positron or a muon; they are instances of the very same family of identical particles. Nonetheless, these objects happen to move differently following diverse trajectories, such dynamical behaviour is what allows us to distinguish them (cf. Goldstein *et al.* (2005a, 2005b) for technical details). In other words, the Bohmian particles do not instantiate inherent properties such as mass, charge or spin. This is reflected in the permutation invariant equations governing their motions, with the effect that one cannot distinguish them a priori, but only by looking at their actual paths in space – the authors in fact say that the particles will move *as if* they were electrons or *as if* they were muons, etc.. In this alternative formulation of the pilot-wave theory called *Identity-based Bohmian mechanics*, such 'intrinsic' attributes are then included in the wave function as dynamical parameters of the theory; therefore, they are not attributes of the material ontology of the world contrary to Bohm's initial proposal.

Along the lines of Dürr, Goldstein, and Zanghì and influenced by the above-mentioned essays of Goldstein *et al.*, the work of Esfeld and Deckert (2017) provides a further radicalization of the primitive ontology concept. The authors provided a new reading of the pilot-wave theory in terms of what they called a 'minimalist ontology of the natural world'. If the notion of primitive ontology is contextualized with respect to a certain physical theory – for instance Bohmian mechanics implements a primitive ontology of particles, while Ghirardi–Rimini–Weber theories may implement matter density fields or flashes – Esfeld and Deckert claimed that 'it is inappropriate to speak of the ontology of this or that physical theory. Ontology is about what there is' (pp. 12–13). They move from the notion of 'primitive ontology of a theory' to ontology *tout court*, developing their metaphysical proposal independently of any particular theoretical framework.

To achieve this result, the authors introduce a background independent atomistic ontology that seeks to recover the predictions of classical mechanics, QM and QFT via the definition of (i) different dynamical laws to be applied at the relevant energy/length scales, and (ii) appropriate typicality measures. More specifically, they formulate a relationalist[52] ontology of propertyless, finite and permanent matter points uniquely individuated by their distance relations – the only property of these points being their position. The change of these relations, which is taken as a primitive fact, constitutes the dynamical aspect of the proposal.

Thereby, Bohmian mechanics becomes a theory about the evolutions of configurations of matter points and their changes in distance relations. Moreover, following Goldstein *et al.*, according to Esfeld and Deckert properties such as mass, charge, etc. are useful mathematical tools used to implement the evolution of material points. Consequently, they are not part of the PO but are included in the dynamical structure of the theory, having an essential explanatory function. Notably, such a minimalist ontology is primitive in a literal sense: it provides the fundamental building blocks of our physical world since the points are (i) ontologically independent from any other en-

[52] According to Esfeld and Deckert space-time and its geometry emerge from the configuration of matter points. Cf. Lazarovici (2018) for critical remarks on this approach.

tity, and (ii) constitute the mereological basis of every physical object in the universe. Hence, contrary to Bohm pluralist approach, Esfeld and Deckert's project excludes the possibility to have different ontologies at different energy/length scales. In particular, they reject the view about the infinite levels of nature endorsed by Bohm: whereas according to him we have historical arguments preventing the possibility to limit the ontological possibilities of nature, Esfeld and Deckert maintain that ontology must be guided by the principles of simplicity and parsimony commonly adopted in science and philosophy. Their proposal, indeed, is one of the simplest ways to explain our macroscopic reality providing the minimal ingredients to recover space-time and matter in motion.[53]

In addition, it must be noted that conforming to this minimalist ontology, the wave function is not a physical object but a part of the dynamical structure of the Bohmian theory. However, Esfeld and Deckert support a different metaphysics about Ψ (and ψ) with respect to (Dürr) Goldstein and Zanghì. They in fact endorse Humeanism about laws of nature.[54] According to this perspective, the world is just a mosaic of particular, local facts – in such minimalist ontology these facts concern matter points and their distance relations. Moreover, in the Humean view there is nothing in the world that constrains the dynamics of these points: a given temporal evolution just contingently happens to occur. The patterns and regularities exhibited by the dynamical change of matter points are conceptualized via the Humean best system account, according to which physical laws are the axioms of 'the system that achieves the best balance between being simple and being informative in describing the evolution of the configuration of matter' (Esfeld and Deckert 2017, p. 45). For instance, in Esfeld and Deckert's opinion, the Humean mosaic is given by the distribution of matter points whose progressive change shows regularities described by the laws of Bohmian mechanics, which are assumed to be the best description of such mosaic at the non-relativistic level. In this scheme, the wave function is an important representational tool appearing in the laws of the theory that efficiently describes the temporal enfolding of the matter points. Thus, in their view ψ is a 'central dynamical parameter figuring in the law of motion for the particles' (*ibid.*, pp. 51–52).

To conclude our brief survey, we need to mention the multi-field account (Hubert and Romano 2018; Romano 2021). This approach offers a realist understanding of the wave function – closer to the spirit of Bohm's original views – given the crucial role played by ψ for the motion of the Bohmian corpuscles. Furthermore, it aims at avoiding the major criticisms raised against the nomological interpretation. To this specific regard, Belot (2012) underlines that the wave function of a given system is generally time-dependent, but the nomological structure of a theory is not supposed to change over time. On the other hand, he claims that Goldstein and Zanghì's view

[53]Such an ontology is defined independently of any physical theory. To describe actual modifications of distance relations at various length scales, one has to provide the correct laws of motion, here is where actual physical theories come into the picture. In their book they provide the relevant dynamical structure for the matter points at various regimes; for instance equation (3.6) regulates the motion of matter points at non-relativistic quantum level, whereas equation (4.4) provides a dynamics for a QFT model based on the Dirac sea picture.

[54]In a previous work, Esfeld *et al.* (2013) proposed a dispositional interpretation of the wave function in Bohmian mechanics. However, guided by a principle of parsimony, Esfeld abandoned dispositionalism in favor of a Humean metaphysics, cf. Esfeld and Deckert (2017), p. 54. For an assessment of dispositionalism in Bohmian mechanics, see Belot (2012).

entails a 'hierarchy of necessity', since some laws would be 'more necessary' than others because

> intuitively at least, the nomological structure encoded in the Schrödinger equation would still have held, even if the part of the nomological structure encoded in the wave-function had been different, but not vice versa – so the wave-function is in a sense more contingent, more like an initial condition, than is the Schrödinger equation (Belot 2012, p. 75).

Along Belot's lines, Romano (2021) argues that the analogy with the Hamiltonian function – one of the main arguments supporting the nomological view – is at best misleading. While the Hamiltonian function is constructed *from* definite properties of classical particles, i.e. their positions and momenta, the wave function in Bohmian mechanics cannot be defined from properties of quantum particles. Being ψ a solution of the Schrödinger equation, it essentially depends on the specific form of the Hamiltonian operator at hand, meaning that the positions and velocities of Bohmian corpuscles are relevant in order to define the Hamiltonian of the particles' configuration under consideration, but they 'will be completely irrelevant for the construction of the wave function' (Romano 2021, p. 6). In addition, if the Hamiltonian is formally and ontologically dispensable – for the dynamics of classical particles can be given, e.g. in Newtonian terms – the wave function is a crucial, indispensable element of the Bohmian theory.[55]

Given the issues affecting the nomological view, Hubert and Romano (2018) provide a new framework in order to maintain a realist interpretation of ψ, overcoming the problems concerning the multi-dimensionality of the wave function. In their view, one can indeed project its values defined in the high-dimensional configuration space into multi-field values defined in three-dimensional space. To achieve this result, the authors associate a definite field value to N-tuples of points of space, i.e. those spatial points corresponding to the exact location of the Bohmian corpuscles. In turn, the particles' coordinates of an actual system 'select' a unique value for the multi-field in three-dimensional space. Since the multi-field value depends on the total particle configuration, quantum non-locality receives a natural explanation in this approach. However, given that ψ is a complex function, the multi-field values must be complex as well. This fact may undermine such a position for complex numbers are not usually associated with actual physical values. To solve this issue, Romano (2021) argues that one should not consider ψ itself as a multi-field, but rather one should take into account its real components. Consequently, the amplitude R and the phase S of the wave function should be interpreted as multi-fields assigning definite real values in physical space to N-tuple of points. This strategy closely follows Bohm's original formulation of the pilot-wave theory, where the wave function is decomposed into its polar coordinates. From this viewpoint, ψ becomes a mathematical tool that expresses the salient information contained in the R- and S-fields and their dynamical coupling.[56]

[55] Cf. Allori (2021) for other critical remarks on the nomological view. This paper provides a functionalist account of the wave function, where ψ is a non-representational entity, i.e. it does neither represent a physical item in space, nor agent's epistemic knowledge, but it is defined via the roles – viz. functions – it plays in the context of QM.

[56] Objections to the multi-field approach are contained in Belot (2012). Hubert and Romano (2018) and Romano (2021) provide answers to common criticisms to their views, in particular the lack of back reaction of the particles on the multi-field (cf. Romano 2021, Section 6.3).

The fate of contextuality and the logic of de Broglie–Bohm theory

After having briefly summarized the debate on the nature of the wave function, another relevant topic addressed in philosophical discussions about de Broglie–Bohm theory concerns the status of quantum operators and the notion of contextuality.

As largely discussed, this theory rejects the postulate according to which ψ provides a complete description of quantum systems, and tries to provide a justification for the Born rule, which according to several scholars in the pilot-wave community should not be viewed as an additional principle. Furthermore, we illustrated how the pilot-wave theory actually *reduces* the number of the axioms of QM, for the collapse postulate becomes a consequence of the particle dynamics.[57] In the light of these facts, let us now discuss the nature of quantum observables in this framework.

According to the principles of standard QM, self-adjoint operators correspond to inherent attributes possessed by quantum systems, and the relation between them is formalized via the eigenstate–eigenvalue link. The existence of definite values for the properties of quantum objects, however, is constrained by the non-commutative algebraic structure of QM, which does not allow for the simultaneous actualization of conjugate properties, such as position and momentum or spin along different axes. Moreover, it is also notorious that quantum measurements do not generally reveal pre-existing values of such observables, which are usually indeterminate before the performance of experiments. It is then claimed that measurement outcomes will strongly depend on the specific experimental arrangement or context that is actually chosen – a phenomenon called *contextuality*. On the philosophical side, in turn, QM entails that the properties of quantum systems do not have definite values prior measurements, contrary to the case of their classical counterparts. These conclusions are supported by important no-go results – as for instance the Bell–Kochen–Specker theorem or even von Neumann theorem (cf. Bub 2010) – showing that QM cannot be completed with local hidden variables always assigning definite (i.e. non-contextual) values to quantum observables.

Natural questions are therefore how to incorporate the notion of contextuality in de Broglie–Bohm theory and what status can be given to quantum operators. In order to answer them, one should simply recall Bohm's theory of measurement. According to it, all measurements of the usual quantum observables are explained in terms of the dynamical evolution in space and time of quantum particles guided by ψ. Thereby, *all* quantum measurements are 'reduced to' position measurements. More precisely, as we have already seen, experimental outcomes ontologically depend on the initial conditions of the experimental set-up – i.e. on the initial particle configuration of the measured system and the measuring apparatus. The equations of motion of the theory will then evolve such initial particle configurations into a final state of the coupled system formed by a measured object and experimental device, providing us with a definite outcome. It is not surprising, then, that the pilot-wave theory features

[57] This virtue of de Broglie–Bohm–type theories should find more resonance in literature, since it is often argued that they *add* structures to standard QM, and for this reason one should prefer simpler alternatives. In this introduction I hope to have offered convincing arguments against this deceptive objection. If not, the reader may refer to Bohm (1952), Maudlin (1995b), Dürr and Teufel (2009), and Bricmont (2016).

contextuality, for the initial particle configuration of a certain experiment (measured system + environment) will determine its final result together with the dynamical laws of the theory.

Two additional remarks should be made here. On the one hand, it has been largely argued that in de Broglie–Bohm theory observables are not considered metaphysically genuine properties of the particles, but rather useful mathematical tools expressing statistics of certain quantum measurements in a compact way (Dürr *et al.* 2013, Chapter 3; Bricmont 2016). Thus, in this framework no pre-determined values are attached to them prior the performance of any measurement, in agreement with QM.

On the other hand, trivially, measurements do not reveal pre-existing values of self-adjoint operators as requested by quantum theory. Therefore, the pilot-wave theory conforms to those no-go theorems showing the contextuality of quantum observables, and the impossibility to complete QM with non-contextual hidden variables other then particles' positions (cf. Bricmont's chapter in this volume for more details about this very issue).

Notably, these points have been exemplified by the reduction of spin measurement to position measurements, for details the reader may refer to Bell (2004, Chapter 17), Dewdney *et al.* (1987); Dürr *et al.* (2013, Chapter 3); Dürr and Teufel (2009, Chapter 12). In these texts the authors showed that measurement outcomes are functions of the theory's ontology and its dynamics, rejecting the naive realist understanding of operators as actual, genuine attributes of quantum objects typically held in the standard formulation of QM, and thereby clarifying its formalism by providing a statistical reading of quantum observables.

Furthermore, along the lines of the above-mentioned works and in relation to quantum contextuality, the present author showed that the logical structure underlying Bohmian mechanics is *classical* (Oldofredi 2020). Thus, it does not introduce any modification to standard propositional calculus. In brief, the paper argued that this theory implements a classical logical structure at the level of its corpuscular ontology. This fact alone is sufficient to explain why the classical connectives maintain their meanings, without the need to define neither quantum connectives, nor alternative interpretations of the usual logical operators. Moreover, it is concretely shown that the distributive law is not violated in the context of Bohmian mechanics, which provides a powerful counterexample to the belief according to which we need a new logic to describe the quantum domain.

The primitive ontology programme

The last topic mentioned in this section is primitive ontology programme – originated from Dürr, Goldstein, and Zanghì's works on the foundations of quantum theory – given its relevance for the contemporary philosophical discussions on Bohmian mechanics.

Following the methodological reflections contained in John Bell's theory of local beables and Bohm's critique to the standard formulation of QM, these authors developed over the years a set of normative guidelines in order to formulate quantum theories immune from the usual quantum paradoxes, the so-called 'quantum theories without observer'.[58] Interestingly, according to them such theoretical frameworks share a common architecture, i.e. the following set of structural features (Allori *et al.* 2008).

Firstly, serious quantum theories must postulate a clear primitive ontology. In order to do so, the mathematical structure of a theory T is divided into two subcategories, as already said above. On the one hand, there are the 'primitive' variables, viz. those theoretical terms provided with a direct physical meaning. They represent matter and refer to real objects localized and moving in three-dimensional space (or in space-time), which is generally considered a real substance as well.[59] The PO of a theory thereby provides the fundamental building blocks with which it describes and explains macroscopic reality. On the other hand, T contains non-primitive (or non-physical) variables, i.e. the mathematical structures without direct physical meaning responsible for the dynamical evolution of the primitive ontology – a notable example is given by the quantum wave function. It is relevant to repeat once again that such non-primitive entities are part of T's general ontology, although they do not belong to the PO.

Secondly, these authors demand that ψ must govern the motion of the PO via either deterministic or stochastic laws. Related to this, the theory should provides a notion of *typical* history for the primitive ontology, 'for example by a probability distribution on the space of all possible histories; from this notion of typicality the probabilistic predictions emerge' (Allori *et al.* 2008, p. 385).

Thirdly, it is required that the predicted probability distribution of macroscopic configurations of matter determined by the PO agrees at any given time with the predictions of standard QM. In such a way, these authors provide criteria and requirements to construct empirically adequate, internally consistent and metaphysically precise quantum theories.

It is worth noting that in Dürr, Goldstein, and Zanghì's account the *primitiveness* of the PO has two different aspects. On the one hand, the PO of a given theory T cannot be further analyzed and/or defined in terms of more elementary notions contained in the vocabulary of T. Hence, the primitive variables define a set of entities that are considered fundamental *in the context of* T. Referring to this, it is relevant to underline that there is no fixed procedure constraining the selection of a primitive ontology of a given theory. Allori emphasizes that

> there is no rule to determine the primitive ontology of a theory. Rather, it is just a matter of understanding how the theory was introduced, it has developed, and how its explanatory scheme works (Allori 2013, p. 65).

In addition, the choice of a certain PO may rely on a variety of meta-empirical virtues subjectively endorsed by a scientist/research group, e.g. explanatory power, unification, pragmatic utility, internal coherence etc..

On the other hand, the PO is primitive because every physical phenomenon/process included in T's domain is ontologically reduced and explained via the dynamical evo-

[58] The expression 'primitive ontology' was firstly employed in the paper *Quantum Equilibrium and the Origin of Absolute Uncertainty* appeared in 1992. Cf. Dürr *et al.* 2013, Chapter 2. For a detailed discussion about the differences between the primitive ontology approach, Bell's theory of local beables, Bohm's own philosophy of science and Esfeld and Deckert minimalism, see Oldofredi (2022).

[59] Cf. Esfeld and Deckert (2017) for a relationalist account of the PO perspective.

lution of the PO. Given that the primitive ontology programme endorses a strong form of ontological reductionism, it is then possible to distinguish between a primitive and a derivative ontology, where the latter is metaphysically dependent on the former (cf. Allori 2013). In this regard, indeed, measurement outcomes are explained uniquely in terms of the motion of the PO in space and time, becoming *functions* of the primitive ontology, as already underlined several times in this chapter.

Referring to this, those theories implementing a clear PO and a set of consistent laws – such as Bohmian mechanics or Ghirardi–Rimini–Weber theories to mention a few – solve the quantum measurement problem in a very elegant way, i.e. by providing conceptually simple physical explanations of those objects and processes responsible for the actualizations of macroscopic measurement outcomes. Moreover, contrary to standard QM, PO theories avoid by construction the presence of ill-defined terms in their principles, relying on a technically sound and clearly interpreted formalism. The main merit of this methodological approach is in fact to ensure both formal and ontological consistency to those frameworks satisfying its normative requirements.

Finally, as the reader can easily note, such an approach seems to endorse opposite metaphysical principles with respect to Bohm's perspective.[60] Not only his metaphysical infinitism is abolished in favour of a strong form of reductionism – which was considered a weak and misleading philosophy by him (cf. Bohm 1957 for details) – but also the most recent developments of the PO programme, i.e. Esfeld and Deckert's minimalist ontology, rejected any form of scientific pluralism arguing that at each energy/length scale the very same ontology must be maintained, contrary to Bohm's ideas.[61]

In sum, in the past 70 years, we assisted to a constant and stimulating discussion of the philosophical implications and ramifications of the pilot-wave theory. Starting from Bohm's infinitism and pluralism, passing through Bell's theory of local beables, one arrives to the contemporary PO programme and its most recent developments mentioned above. Although the initial steps of the pilot-wave theory are in open contrast, and oftentimes in clear contradiction w.r.t. the latest works proposed within the PO community, all these perspectives share the common desire to solve the puzzles affecting standard QM, elaborating theories and models that are not only technically consistent, but also metaphysically intelligible and unambiguous. Thereby, they all strive to achieve the same aim, i.e. providing a better understanding of the quantum world, and a deeper comprehension of our universe at its most fundamental scales.

The contents of the volume

Having reviewed the essential developments of the de Broglie–Bohm theory and the philosophical debates concerning its ontology, we are now ready to introduce the con-

[60]It is not impossible to combine the methodology of the PO programme with Bohm's philosophical ideas. A possible way to do so would be to consider a given PO as only relatively fundamental in a specific domain of application.

[61]Related to this, it must also be noted that the vast majority of contemporary Bohmians favour a monist ontology of particles at the expenses of multi-category ontologies endorsed by Bohm, which are dismissed due to their less minimal and parsimonious character. A counterexample to this general tendency is provided, e.g. by the supporters of a realist reading of the wave function as well as those proposing a revival of the double solution theory.

tents of the present volume. Given the various topics discussed by the authors, the chapters have been divided into four main parts:

Part I – Introduction to the de Broglie–Bohm Theory and Replies to Critics
Part II – Historical Perspectives
Part III – Philosophical Discussions
Part IV – Non-locality, Extensions of de Broglie–Bohm Theory, and Applications

The first part of the book is devoted to introducing the pilot-wave theory and to answer the most common criticisms raised against this theoretical framework. In the first chapter, *Oliver Passon* provides an overview of the pilot-wave theory illustrating the main ideas of Bohm as well as the formulation of Bohmian mechanics given by Dürr, Goldstein, and Zanghì. This accurate overview uses little mathematical formalism, and thus can be followed also by those readers unfamiliar with the technicalities of quantum theory. Moreover, the author introduces several crucial topics as for instance the various justifications of the Born rule, the generalizations of the pilot-wave approach to relativistic regimes, and the most well-known Bohmian quantum field theories. In addition, Passon replies to common criticisms to the theory – as e.g. the surrealistic trajectories objection, the asymmetry between position and momentum basis or the alleged presence of metaphysical surplus structures – showing that these are not compelling arguments against the consistency and coherence of the de Broglie–Bohm theory.

In the second chapter, *Antony Valentini* offers a more technical presentation of pilot-wave theory covering several topics in great detail, for instance quantum measurement theory, the status of Lorentz invariance, as well as the status of the Born rule. Interestingly, the author also provides a personal account of this framework, reviewing many of his important contributions to the theory. In addition, he convincingly answers three different categories of objections usually raised against the causal interpretation: (i) the thesis according to which de Broglie–Bohm theory entails an unacceptably radical new physics, (ii) the idea that it is not radical enough, and finally (iii) the claim that it does not lead to new physics.

The first part of the volume is concluded by *Jean Bricmont*'s contribution. This chapter illustrates what a 'hidden variable theory' is and explains why the pilot-wave theory is not ruled out by those no-go results affirming the impossibility of hidden variables formulations of QM – in particular, the author focuses on the Kochen–Specker theorem and quantum contextuality. After a brief overview of Bohmian mechanics, Bricmont carefully analyses how the theory describes quantum measurements, also clarifying the reasons why contextuality fits perfectly with this approach. In doing so, he shows that measurements do not reveal pre-existing properties of quantum systems, as well as that quantum operators should not be provided with ontological weight. Finally, he discusses Bell's theorem and non-locality, arguing that the non-local character of the particles' dynamics is not a defect of this framework.

The second part of the book, which is devoted to the history of the pilot-wave theory, begins with *Olival Freire Junior*'s chapter. In Chapter 4, the author carefully introduces the figure of David Bohm, his vicissitudes – as e.g. the suspension from Princeton and the 'exile' in São Paulo, Brazil – and his scientific activity. Freire Junior then accurately describes the poor reception that Bohm's ideas obtained from his contemporaries in the 1950s. In this regard, he mentions the heated discussions with Wolfgang Pauli and Léon Rosenfeld, who was a fervent supporter of the philosophy of complementarity and firmly convinced that Bohr's perspective was the only feasible interpretation of QM. Furthermore, this chapter historically reconstructs the renaissance of Bohm's theory occurred in the 1970s – which is due not only to the increasing interest in the quantum foundations in those years, but also to the fruitful collaboration with Hiley, Dewdney and Philippidis already mentioned in this chapter – delineating the evolution of the theory to the present day.

In the fifth chapter, *Marij van Strien* convincingly argues that the restoration of determinism was not in Bohm's agenda, although his 1952 formulation of pilot-wave theory implements a deterministic dynamics. Van Strien's thesis is well documented and carefully argued, casting doubts on what has been believed by eminent physicists in the 1950s as well as by many scholars in the present day. By looking into Bohm's writings and private correspondence, she clarifies that a fundamentally deterministic description of reality is incompatible with his philosophical reflections concerning the ontology of our physical universe, according to which the latter is hierarchically structured in an infinite number of levels. Referring to this, van Strien claims that Bohm found resources in Marxist philosophy in order to elaborate a non-deterministic notion of causality, which is grounded on the idea of infinite complexity of nature.

A complementary claim is made in Chapter 6 by *Flavio Del Santo* and *Gerd Christian Krizek*. The authors begin their chapter with a careful historical analysis of what they call 'the standard narrative', according to which Bohm had a radical shift in his views about QM after the publication of his textbook *Quantum Theory*, abandoning Bohr's perspective and rejecting indeterminism in favour of a deterministic account. In addition, they claim that many historians of physics tried to explain such a sudden and unexpected turn. Notably, the authors point out with sound and documented arguments that this narrative is simply incorrect: Bohm was never interested in providing a deterministic picture of physical reality. Therefore, they conclude that he never underwent a transition from indeterminism to determinism. In this regard, Del Santo and Krizek not only show that Bohm remained faithful to his constant quest for a causal description of physical phenomena, but also that he strongly rejected mechanical determinism.

The third part of the volume contains compelling philosophical discussions about de Broglie–Bohm theory and begins with *Valia Allori*'s contribution. In the seventh Chapter, the author proposes an interesting thesis, namely that both quantum mechanics and special relativity are incomplete frameworks. On the one hand, Allori underlines that QM is not complete since it lacks a clear ontology able to explain measurement results – and more generally our macroscopic manifest image of the world – in terms of well-defined quantum objects and their dynamics. On the other hand, the author claims that Einstein's relativity theory is not complete because its space-time structure cannot account for quantum non-locality. In order to complete both theories, it is argued that one should opt for a relativistic pilot-wave approach. Explaining that the requirement of locality must be abandoned as a consequence of the empirical confirmation of Bell's theorem, and given that Lorentz invariant non-local pilot-wave theories do exist, Allori claims that relativistic Bohmian approaches maintain the spirit of both quantum mechanics and relativity. These models, although not definitive and final, show how to properly combine QM with Einstein's relativity.

In Chapter 8, *Emilia Margoni* offers an original analysis of Bohm's understanding of quantum and relativity theories in terms of process philosophy, i.e. the idea according to which physical reality is not composed of static, immutable entities, but rather dynamical processes describing the variable, diachronic and constantly changing nature of things. After an introduction of the main principles of process philosophy, Margoni argues in detail why Bohm's causal interpretation should not be considered a mechanistic theory. His understanding of the pilot-wave theory, indeed, rejects the reductionist thesis according to which reality is completely explained by a restricted set of unvarying and static objects taken as fundamentally primitive. In addition, the author shows how Bohm's reading of special relativity can accommodate a process type of ontology.

In the ninth Chapter *Johannes Fankhauser* presents a detailed discussion concerning the role and status of trajectories in Bohmian mechanics. On the one hand, he explains very carefully why weak measurements of particles' velocities – i.e. measurements performed in such a way that the quantum states of the particles are minimally perturbed, contrary to what happens in usual 'strong' measurements – do not provide empirical support for the existence of Bohmian trajectories. On the other hand, Fankhauser shows how the well-known Englert *et al.* argument concerning surrealistic trajectories rests on untenable assumptions, and thereby its conclusions do not reflect any weakness of the de Broglie–Bohm theory. The author then concludes by stating that arguments involving weak measurements of velocity or surrealistic trajectories do not prove, respectively refute, the causal interpretation. Both instances highlight the importance of a sufficient account of accessible measurement records and what they are supposed to signify ontologically.

In Chapter 10 *Cristian Lopez* and *Michael Esfeld* provide a compelling analysis of the direction of time, once again in the context of Bohmian mechanics. The authors argue that this theory can accommodate different and metaphysically diverse perspectives such as primitivism and reductionism. According to the former, the direction of time is a necessary posit in one's ontology, thereby it is an irreducible and fundamental feature of the world. On the contrary, the latter (in its conservative, noneliminativist form) affirms that time's direction can be explained and reduced to some non-temporal physical asymmetry. Lopez and Esfeld show how the Bohmian theory can accommodate both a conservative reductionist and a primitivist conception of time's direction. In the first case, Bohmian mechanics explains temporal asymmetries ultimately reducing them to an epistemic asymmetry about particle positions. In the second case, this theory accommodates primitivism if Leibnizian relationalism is also adopted, for the direction of time is the direction of change in the particle configuration. Since change is regarded as primitively directed to explain the difference between variation and change in the particle configuration, the direction of time is also ipso facto primitively directed.

The third part of the volume is concluded with the contribution of *Sebastian Fortin* and *Olimpia Lombardi*. After having provided an introduction to quantum chemistry within the framework of standard QM – speaking about, e.g. orbitals, electronic and molecular structure, and the Born–Oppenheimer approximation – the authors of Chapter 11 explain in detail why the de Broglie–Bohm theory offers a better metaphysical picture of atoms and their structure with respect to standard quantum mechanics. In virtue of its corpuscular content, in fact, this framework is closer to the chemists' conception and modelization of atoms. Moreover, Fortin and Lombardi devote the second part of their chapter to the description of the de Broglie–Bohm account of orbital shapes – taking into consideration the example of the hydrogen and helium atoms as case studies – and chemical bonds, emphasizing the prominent role played by spin-dependent particles' trajectories.

The fourth and last part of the book concerns more technical topics such as quantum non-locality, extensions of the de Broglie–Bohm theory and its applications. In Chapter 12 *Aurélien Drezet* discusses the Einstein–Podolsky–Rosen scenario and Bell's theorem in the context of the de Broglie–Bohm theory. One of the merits of this framework, the author says, is to show that a causal as well as non-local account of the EPR paradox can be given in terms of a clear corpuscular ontology. In relation to this, Drezet interestingly stresses that both the EPR argument and Bell's theorem posit as beables not only hidden variables but also the wave function. This fact allows the author to reformulate both the EPR and Bell results as *theorems* proving the non-local character of QM itself, and not only of hidden-variable approaches as it is often mistakenly believed.

In Chapter 13 *Roderich Tumulka* presents an extension of Bohmian mechanics to quantum electrodynamics (QED). The author firstly introduces the problems affecting the standard formulation of such a theory, clarifying that some of them are inherited from non-relativistic QM – e.g. the lack of a clear ontology – while others pertain to QED specifically, as for example the absence of a Born's rule for electrons, positrons and photons' positions as well as ultraviolet divergences. Secondly, after laying out interesting arguments against a field ontology for a Bohm-type QED, the author outlines the bare bones of a new Bohmian QED also specifying what desirable features such a framework should possess. In a nutshell, Tumulka envisions a Bohmian QED in a multi-time wave function formalism, with particle trajectories for both fermions and photons – although he claims that satisfactory equations for photon trajectories have not been found yet. Moreover, these trajectories depend on a preferred foliation of space-time into spacelike hyper–surfaces. The equations of motion for such Bohmian QED present terms governing pair creation and annihilation, which are interpreted as inherently stochastic events.

In Chapter 14 *Antonio Vassallo* and *Pedro Naranjo* propose an extension of pure shape dynamics (PSD) to de Broglie-Bohm theory. The authors immediately clarify that their model is not a mere translation in relational terms of this interpretation, but a completely novel formulation of it. Indeed, they claim that PSD is apt in order to clarify the meaning and the nature of the wave function in the Bohmian context. This new relationalist perspective will prove useful, the authors say, to eliminate those surplus structures responsible for the metaphysical conundra when it comes to the interpretation of ψ. The authors implement the main tenets of the Leibnizian/Machian approach to physics into a newly formulated pilot-wave theory, redefining not only its state space in terms of a shape space, but also the particles' trajectories and their meaning. Finally, Vassallo and Naranjo provide a compelling set of arguments concerning how the very notion of shape space is essential in providing the wave function a clear meaning on the one hand, and the advantages of their model in dealing with other issues – as e.g. the classical limit – on the other hand.

Chapter 15 is devoted to a detailed analysis of weak values and their interpretation in a Bohmian context. In this essay *Xabier Oianguren-Asua*, *Albert Solé*, *Carlos F. Destefani* and *Xavier Oriols* speak about them from a threefold perspective. Firstly, specifying the protocols that enable experimental determination of weak values; secondly, deriving their correlates in the quantum formalism and finally providing a metaphysical discussion of the meaning of such weak values. Notably, the authors claim that the latter have predictive power and provide new ways to characterize quantum systems. In addition, it is claimed that de Broglie–Bohm theory constitutes a valuable heuristic tool for identifying informative weak values or functions thereof. The claims made in this chapter are exemplified with a case study involving quantum thermalization. The authors show that certain weak values, singled out by Bohmian mechanics as physically relevant, play a crucial role in elucidating the thermalization time of certain systems, whereas standard expectation values are 'blind' to the onset of thermalization.

The final chapter of the volume is devoted to the illustration of spin-dependent formulation of the pilot-wave theory and its application to time-of-flight momentum spectroscopy. In this essay *Siddhant Das* carefully reviews the de Broglie–Bohm theory for single spin-1/2 particles which are based on spinor-valued wave functions. Interestingly, the author does not only discuss the motion of a single non-relativistic spin-1/2 particle trapped in the ground state of a spherical box, but also analyses the relativistic case. Moreover, Das examines a Bohmian description of time-of-flight measurements showing that the pilot-wave theory is capable of recovering the usual quantum mechanical predictions without the reference to any observable. Furthermore, it is interestingly argued that the Bohmian theory avoids the 'arrival-time problem' affecting standard QM. In addition, the author claims that the application of Bohmian mechanics to such experimental situations offers the possibility to illustrate how 'random' outcomes arise in a deterministic framework.

Since we live in a time where interpretational debates on the meaning of the quantum formalism are thriving and vibrant, I sincerely wish, concluding this long introduction, that the present volume will not only provide the reader with the most updated snapshot of the pilot-wave approach – presenting it as a *plurality of voices* and a fruitful research avenue – but also clarify those confusions and misunderstandings that are still unfortunately widespread despite the long history of de Broglie–Bohm theory.

References

Allori, V. (2013). Primitive ontology and the structure of fundamental physical theories. In Albert, D.Z. and Ney, A. (eds.) *The Wave Function: Essays on the Metaphysics of Quantum Mechanics*, pages 58–75. New York: Oxford University Press.

Allori, V. (2021). Wave-functionalism. *Synthese*, **199**: 12271–12293.

Allori, V., Goldstein, S., Tumulka, T. and Zanghì, N. (2008). On the common structure of Bohmian mechanics and the Ghirardi-Rimini-Weber theory. *British Journal for the Philosophy of Science*, **59**(3): 353–389.

Bacciagaluppi, G. and Valentini, A. (2009). *Quantum Theory at the Crossroads. Reconsidering the 1927 Solvay Conference.* Cambridge: Cambridge University Press.

Bell, J.S. (2004). *Speakable and Unspeakable in Quantum Mechanics*, 2nd edn. Cambridge: Cambridge University Press.

Beller, M. (1999). *Quantum Dialogue: The Making of a Revolution.* Chicago: University of Chicago Press.

Belot, G. (2012). Quantum states for primitive ontologists. a case study *European Journal for Philosophy of Science*, **2**(1): 67–83.

Boffi, S. (1995). *Le onde di de Broglie. Quaderni di Fisica Teorica.* Pavia Editore.

Bohm, D. (1952). A suggested interpretation of quantum mechanics in terms of "hidden" variables. I & II. *Physical Review*, **85**(2): 166–193.

Bohm, D. (1953a). Proof that Probability Density Approaches $|\psi|^2$ in Causal Interpretation of the Quantum Theory. *Physical Review*, **89**(2): 458–466.

Bohm, D. (1953b). Comments on an Article of Takabayasi concerning the Formulation of Quantum Mechanics with Classical Pictures. *Progress in Theoretical Physics*, **9**(3): 273–287.

Bohm, D. (1957). *Causality and Chance in Modern Physics.* London: Routledge.

Bohm, D. (1965). Problems in the Basic Concepts of Physics. In *Satyendranath Bose 70th Birthday Commemoration Volume, Part II.* Prof. S. N. Bose 70th Birthday Celebration Committee.

Bohm, D. (1980). *Wholeness and the Implicate Order.* London: Routledge.

Bohm, D. (1990). A new theory of the relationship of mind and matter. *Philosophical Psychology*, **3**(2–3): 271–286.

Bohm, D. and Hiley, B. (1993). *The Undivided Universe. An Ontological Interpretation of Quantum Theory.* London: Routledge.

Bohm, D. and Schiller, R. (1955). A causal interpretation of the Pauli equation (B). *Il Nuovo Cimento*, **1**: 67–91.

Bohm, D. and Vigier, J.P. (1954). Model of the Causal Interpretation of Quantum Theory in Terms of a Fluid with Irregular Fluctuations. *Physical Review*, **96**(1): 208–216.

Bohm, D. and Hiley, B. (1982). The de Broglie Pilot Wave Theory and the Further Development of New Insights Arising Out of It. *Foundations of Physics*, **12**(10): 1001–1016.

Bohm, D. and Hiley, B. (1985). Unbroken Quantum Realism, from Microscopic to Macroscopic Levels. *Physical Review Letters*, **55**(23): 2511–2514.

Bohm, D., Dewdney, C. and Hiley, B. (1985). A quantum potential approach to the Wheeler delayed-choice experiment. *Nature*, **315**: 294–297.

Bohm, D., Hiley, B. and Kaloyerou, P.N. (1987). A causal interpretation of quantum fields. *Physics Reports*, **144**(6): 349–375.

Bohm, D, Schiller, R. and Tiomno, J. (1955). A causal interpretation of the Pauli equation (A). *Il Nuovo Cimento*, **1**: 48–66.

Bricmont, J. (2016). *Making Sense of Quantum Mechanics*. Berlin: Springer.

Bub, J. (2020). Von Neumann's 'No Hidden Variables' Proof: A Re-Appraisal. *Foundations of Physics*, **40**: 1333–1340.

Callender, C. (2007). The emergence and interpretation of probability in Bohmian mechanics. *Studies in History and Philosophy of Science part B: Studies in History and Philosophy of Modern Physics*, **38**(2): 351–370

Colijn, C. and Vrscay, E.R. (2003a). Spin-dependent Bohm trajectories for hydrogen eigenstates. *Physics Letters A*, **300**(4): 334–340.

Colijn, C. and Vrscay, E.R. (2003b). Spin-dependent Bohm trajectories for Pauli and Dirac eigenstates of hydrogen. *Foundations of Physics Letters*, **16**: 303–323.

Colin, S. (2003). A deterministic Bell model. *Physics Letters A*, **317**(5-6): 349–358.

Colin, S. and Struyve, W. (2007). A Dirac sea pilot-wave model for quantum field theory. *Journal of Physics A*, **40**(26): 7309–7341.

Colin, S., Durt, T. and Willox, R. (2017). L. de Broglie's double solution program: 90 years later. *Annales de la Fondation Louis de Broglie*, **42**: 19–71.

Cushing, J., Fine, A. and Goldstein, S. (eds.) (1996). *Bohmian Mechanics and Quantum Theory: An Appraisal.* Dordrecht: Springer.

Das, S. and Dürr, D. (2019). Arrival time distributions of spin-1/2 particles. *Scientific Reports*, **9**: 2242.

de Broglie, L. (1924). *Recherces sur la théorie des quanta*, PhD Thesis, University of Paris.

de Broglie, L. (1925). Recherces sur la théorie des quanta. *Annales de Physique*, **10**(3): 22–128.

de Broglie, L. (1927). La mécanique ondulatoire et la structure atomique de la matière et du rayonnement. *Journal de Physique et le Radium*, **8**(5): 225–241.

de Broglie, L. (1930). *An Introduction to the Study of Wave Mechanics*. E.P. Dutton and Company.

Deckert, D.A., Esfeld, M., and Oldofredi, A. (2019). A Persistent Particle Ontology for Quantum Field Theory in Terms of the Dirac Sea. *British Journal for the Philosophy of Science*, **70**(3): 747–770.

Dewdney, C. (2023). Rekindling of de Broglie–Bohm Pilot Wave Theory in the Late Twentieth Century: A Personal Account. *Foundations of Physics*, **53**, 24.

Dewdney, C., Holland, P. and Kyprianidis, A. (1986). What happens in a spin measurement? *Physics Letters A*, **119**(6): 259–267.

Dewdney, C., Holland, P. and Kyprianidis, A. (1987a). A causal account of non-local Einstein–Podolsky–Rosen spin correlations. *Journal of Physics A: Mathematical and General*, **20**: 4717–4732.

Dewdney, C., Holland, P. and Kyprianidis, A. (1987b). A quantum potential approach to spin superposition in neutron interferometry. *Physics Letters A*, **121**(3): 105–110.

Dewdney, C., Holland, P., Kyprianidis, A. and Vigier, J.P. (1986). A review of extended probabilities. *Physics Reports*, **133**(6): 386–388.

Dewdney, C., Holland, P., Kyprianidis, A. and Vigier, J.P. (1987). Spin and nonlocality in quantum mechanics. *Nature*, **336**: 536–544.

Dewdney, C., Horton, G., Lam, M.M., Malik, Z. and Schmidt, M. (1992). wave–particle Dualism and the Interpretation of Quantum Mechanics. *Foundations of Physics*, **22**(10): 1217–1264.

Dirac, P.A.M. (1945). On the Analogy Between Classical and Quantum Mechanics. *Review of Modern Physics*, **17**(2–3): 195–199.

Drezet, A. (2019). Lorentz-Invariant, Retrocausal, and Deterministic Hidden Variables. *Foundation of Physics*, **49**: 1166–1199.

Drezet, A. (2023). Quantum Solitodynamics: Non-linear Wave Mechanics and Pilot-Wave Theory. *Foundation of Physics*, **53**: 31.

Dürr, D. and Teufel, S. (2009). *Bohmian Mechanics. The Physics and Mathematics of Quantum Theory.* Berlin: Springer.

Dürr, D., Goldstein, S. and Zanghì, N. (2013). *Quantum Physics Without Quantum Philosophy.* Berlin: Springer.

Dürr, D., Goldstein, S. and Zanghì, N. (2020). Quantum motion on shape space and the gauge dependent emergence of dynamics and probability in absolute space and time. *Journal of Statistical Physics*, **180**: 92–134.

Dürr, D., Goldstein, S., Tumulka, R. and Zanghì, N. (2005). Bell-type quantum field theories. *Journal of Physics A: Mathematical and General*, **38**(4): R1–R43.

Dürr, D., Goldstein, S., Norsen, T., Struyve, W. and Zanghì, N. (2014). Can Bohmian Mechanics be made relativistic? *Proceedings of the Royal Society A*, **470**(2161): 20130699.

Esfeld, M., Lazarovici, D. Hubert, M. and Dürr, D. (2014). *British Journal for the Philosophy of Science*, **65**(4): 773–796.

Esfeld, M. and Deckert, D.-A. (2017). *A Minimalist Ontology of the Natural World.* New York: Routledge.

Flack, R. and Hiley, B. (2018). Feynman Paths and Weak Values. *Entropy*, **20**(5): 367–377.

Freire Junior, O. (2015). *The Quantum Dissidents. Rebuilding the Foundations of Quantum Mechanics (1950-1990).* Berlin: Springer.

Freire Junior, O. (2019). *David Bohm. A Life Dedicated to Understanding the Quantum World.* Cham, Switzerland: Springer.

Galvan, B. (2015). Relativistic Bohmian Mechanics Without a Preferred Foliation. *Journal of Statistical Physics*, **161**: 1268–1275.

Goldstein, S., Taylor, J., Tumulka, R. and Zanghì, N. (2005a). Are all particles identical? *Journal of Physics A: Mathematical and General*, **38**(7): 1567–1576.

Goldstein, S., Taylor, J., Tumulka, R. and Zanghì, N. (2005b). Are all particles real? *Studies in History and Philosophy of Modern Physics*, **36**(1): 103–112.

Goldstein, S. (2021). Bohmian Mechanics. In Zalta, E. (ed.) *Stanford Encyclopedia of Philosophy.*

Goldstein, S. and Teufel, S. (2001). Quantum spacetime without observers: ontological clarity and the conceptual foundations of quantum gravity. In Callender, C. and Huggett, N. (eds.) *Physics meets Philosophy at the Planck Scale*, pp. 275–289. Cambridge: Cambridge University Press.

Guicciardini, N. and Introzzi, G. (2007). *Fisica Quantistica. Un'Introduzione.* Roma: Carocci.

Hance, J. and Hossenfelder, S. (2022). What does it takes to solve the measurement problem?. *Journal of Physics Communications*, **6**: 102001.

Heilbron, J. (1988). The Earliest Missionaries of the Copenhagen Spirit. In Ullmann-Margalit, E., (ed.) *Science in Reflection. Boston Studies in the Philosophy of Science*, volume 110, pp. 201–233. Dordrecht: Springer.

Hiley, B. (2002). From the Heisenberg Picture to Bohm: A New Perspective on Active Information and Its relation to Shannon Information. *Proceedings of Conference Quantum Theory Reconsideration of Foundations, Vaxjo, 6–11 June 2005.*

Hiley, B. (2005). Non-Commutative Quantum Geometry: A Reappraisal of the Bohm Approach to Quantum Theory. In Elitzur, A.C., Dolev, S. and Kolenda, N. (eds) *Quo Vadis Quantum Mechanics?.* Berlin: Springer.

Hiley, B. and Dennis, G. (2018). The Dirac-Bohm picture. *arXiv:1809.06072.*

Hiley, B., de Gosson, M. and Dennis, G. (2019). Quantum Trajectories: Dirac, Moyal, Bohm. *Quanta*, **8**(1): 11–23.

Holland, P. (1986). Relativistic algebraic spinors and quantum motions in phase space. *Foundations of Physics*, **16**(8): 701–719.

Holland, P. (1987). Geometry of dislocated de Broglie waves. *Foundations of Physics*, **17**(4): 345–363.

Holland, P. (1988). Causal Interpretation of Fermi Fields. *Physics Letters A*, **128**(1–2): 9–18.

Holland, P. (1993). *The Quantum Theory of Motion. An Account of the de Broglie–Bohm Causal Interpretation of Quantum Mechanics.* Cambridge: Cambridge University Press.

Howard, D. (2004). Who invented the "Copenhagen interpretation"? A study in mythology. *Philosophy of Science*, **71**(5): 669–682.

Hubert, M. and Romano, D. (2018). The wave-function as a multi-field. *European Journal for Philosophy of Science*, **8**(3): 521–537.

Jammer, M. (1974). *The Philosophy of Quantum Mechanics: The Interpretations of QM in Historical Perspective.* John Wiley & Sons.

Lazarovici, D. (2018). Super-Humeanism: A starving ontology. *Studies in History and Philosophy of Modern Physics*, **64**: 79–86.

Leavens, C.R. (2008). Bohm Trajectory Approach to Timing Electrons. In Muga, J.G., Sala Mayato, R. and Egusquiza, Í.L. (eds.) *Time in Quantum Mechanics.* Berlin-Heidelberg: Springer.

Lewis, P.J. (2016). *Quantum Ontology. A Guide to the Metaphysics of Quantum Mechanics.* New York: Oxford University Press.

Lienert, M., Petrat S., and Tumulka, R. (2020). *Multi-time Wave Functions. An Introduction.* Heidelberg: Springer.

Maudlin, T. (1995a). Three Measurement Problems. *Topoi*, **14**: 7–15.

Maudlin, T. (1995b). Why Bohm's Theory Solves the Measurement Problem. *Philosophy of Science*, **62**(3): 479–483.

Norsen, T. (2017). *Foundations of Quantum Mechanics. An Exploration of the Physical Meaning of Quantum Theory.* Cham: Springer.

Oldofredi, A. (2018a). Particle Creation and Annihilation: Two Bohmian Approaches. *Lato Sensu: Revue De La Société De Philosophie Des Sciences*, **5**(1): 77–85.

Oldofredi, A. (2018b). Stochasticity and Bell-type Quantum Field Theory *Synthese*, **197**(2): 731–750.

Oldofredi, A. (2020). Classical Logic in Quantum Context. *Quantum Reports*, **2**(4): 600–616.

Oldofredi, A. (2022). Beables, Primitive Ontology and Beyond: How Theories Meet the World. In Allori, V. (ed.) *Quantum Mechanics and Fundamentality. Naturalizing Quantum Theory between Scientific Realism and Ontological Indeterminacy*, pp. 97–111. Synthese Library, Cham: Springer.

Oldofredi, A. (2023). Orthodox or dissident? The evolution of Bohm's ontological reflections in the 1950s. *European Journal of Physics H*, **48**: 12.

Oriols, X. and Mompart, J. (2019). *Applied Bohmian Mechanics.* Singapore: Pan Stanford Publishing.

Osnaghi, S., Freitas, F. and Freire Junior, O. (2009). The origin of the Everettian heresy. *Studies in History and Philosophy of Science Part B: Studies in History and Philosophy of Modern Physics*, **40**: 97–123.

Philippidis, C., Dewdney, C. and Hiley, B.J. (1979). Quantum interference and the quantum potential. *Il Nuovo Cimento B*, **52**: 15–28.

Pylkkänen, P., Hiley, B. and Pättiniemi, I. (2014). Bohm's approach and individuality. In Guay, A. and Pradeu, T. (eds.) *Individuals Across the Sciences.* Oxford: Oxford University Press.

Pinto-Neto, N. (2005). The Bohm interpretation of quantum cosmology. *Foundations of Physics*, **35**: 577.

Pinto-Neto, N. (2021). The de Broglie–Bohm quantum theory and its application to quantum cosmology. *Universe*, **134**.

Romano, D. (2021), Multi-field and Bohm's theory. *Synthese.*

Sanz, A.S. and Miret-Artés, S. (2011). *A Trajectory Description of Quantum Processes. I. Fundamentals. A Bohmian Perspective.* Berlin-Heidelberg: Springer-Verlag.

Sanz, A.S. and Miret-Artés, S. (2014). *A Trajectory Description of Quantum Processes. II. Applications. A Bohmian Perspective.* Berlin-Heidelberg: Springer-Verlag.

Struyve, W. (2010). Pilot-wave approaches to quantum field theory. *Journal of Physics: Conference Series*, **306**: 012047.

Sutherland, R. (2008). Causally symmetric Bohm model *Studies in History and Philosophy of Science Part B: Studies in History and Philosophy of Modern Physics*, **39**(4): 782–805.

Talbot, C. (2017). *David Bohm: Causality and Chance, Letters to Three Women.* Cham: Springer.

Talbot, C. (2024). *An Ontological Foundation for the Quantum Theory by David Bohm.* Unpublished.

Timko, J.A. and Vrscay, E.R. (2009). Spin-dependent Bohmian electronic trajectories for helium. *Foundations of Physics*, **39**: 1055–1071.

Valentini, A. (1991a). Signal-locality, uncertainty, and the subquantum H-theorem. I. *Physics Letters A*, **156**(1–2): 5–11.

Valentini, A. (1991b). Signal-locality, uncertainty, and the subquantum H-theorem, II. *Physics Letters A*, **158**(1–2): 1–8.

Valentini, A. (1996), Pilot-wave theory of fields, gravitation and cosmology. In Cushing, J.T., Fine, A. and Goldstein, S. (eds.) *Bohmian Mechanics and Quantum Theory: An Appraisal.* Dordrecht: Kluwer.

Valentini, A. (1997). On Galilean and Lorentz invariance in pilot-wave dynamics. *Physics Letters A*, **228**: 215.

Valentini, A. (2007). Astrophysical and cosmological tests of quantum theory. *Journal of Physics A*, **40**: 3285.

Valentini, A. (2010). Inflationary cosmology as a probe of primordial quantum mechanics. *Physical Review D*, **82**: 063513.

Valentini, A. (2023). Beyond the Born rule in quantum gravity. *Foundations of Physics*, **53**: 6.

Vassallo, A. and Esfeld, M. (2013). A Proposal for a Bohmian Ontology of Quantum Gravity. *Foundations of Physics*, **44**(1): 1–18

von Meyenn, K. (ed.) (1996). *Wolfgang Pauli. Scientific Correspondence with Bohr, Einstein, Heisenberg, a.o. Volume IV, Part I: 1950-1952*. Berlin: Springer.

Part I

Introduction to the de Broglie–Bohm Theory and Replies to Critics

1

Why isn't every physicist a Bohmian? Common objections and their response

Oliver Passon

1.1 Introduction

Many years ago, the high-energy physicist Robert D. Cousins asked the question 'Why isn't every physicist a Bayesian?' (Cousins 1995). He was referring to the debate in statistics as to whether the concept of probability should be understood subjectively or frequentistically. In a similar way, one may look at the debate on the interpretation of quantum mechanics (QM) and ask the question raised in the title of this chapter.

And indeed, often the practitioners of either statistics or quantum theory are not aware of these debates, since they follow the teaching tradition in an unreflective way. This holds in particular for quantum physics since here most results and predictions follow from an application of the accepted standard formalism and allow the questions of interpretation to be apparently left aside.[1]

Hence, the foundation of quantum physics seems to be less relevant to the practitioner, and many of them allow themselves the luxury of not taking any specific standpoint on this issue at all. In a similar context, this was clearly expressed by David Mermin:

> Contemporary physicists come in two varieties. Type 1 physicists are bothered by EPR and Bell's Theorem. Type 2 (the majority) are not, but one has to distinguish two subvarieties. Type 2a physicists explain why they are not bothered. Their explanations tend either to miss the point entirely (like Born's to Einstein) or to contain physical assertions that can be shown to be false. Type 2b are not bothered and refuse to explain why (Mermin 1985, p. 41).

Even if one takes this remark with a pinch of salt, Mermin's observation suggest a first answer to our question. Many physicists are neither for nor against a particular interpretation of QM, but simply indifferent or ignorant on the issue.

Another rather contingent or sociological factor adds to the neglect of the de Broglie–Bohm theory in particular. By this we mean that the approach has fallen into oblivion several times. After its discovery by Louis de Broglie in 1927, it was

[1] In statistics likewise, the *numerical* results of Bayesian and frequentistic methods often coincide. But when it comes to inferential statistics and the infamous hypothesis testing one needs to pay close attention to the meaning of probability, cf. Cousins (1995).

quickly dismissed and even de Broglie himself did not pursue this approach any further (cf. Bacciagaluppi and Valentin 2009). Twenty-five years later it was independently re-discovered by Bohm (1952a, 1952b), but again the debate was short-lived and dismissive (see below).[2]

No one worked seriously on the de Broglie–Bohm theory for another 25 years, and it was only mentioned now and then. Notably, Bell (1964) was inspired by its non-locality when he formulated his famous theorem – but this result also received little attention at first. In 1973, the physicist Frederik J. Belinfante published a book called *A Survey of Hidden-Variables Theories* in which Bohm's theory figured prominently. His motivation was to correct a misunderstanding since

> [...] the literature was full of claims that hidden-variables theories would not be possible at all (Belinfante 1973, p. xv).

Curiously, the theory was revived by Bohm's students in the late 1970s because one of them, Chris Dewdney, unexpectedly noticed Belinfante's text in Dillon's bookshop across the street from Birkbeck College (Dewdney 2023).[3] Dewdney learned about his supervisor's theory from this book, that is, Bohm himself did not tell his students about the theory at that time, let alone actively propagated it. No wonder that there were only few adherents of the de Broglie–Bohm theory, if even David Bohm himself did not pursue this approach any further.

Chris Dewdney and Chris Philippidis pioneered the computer-aided calculation of the trajectories. The first (by now iconic) plots of the trajectories and the quantum potential in the double-slit experiment in Philippidis *et al.* (1979) caused much excitement and since then the de Broglie–Bohm theory has not disappeared from the scene.[4]

While the above remarks point to a general neglect of the de Broglie–Bohm theory due to ignorance, or historical and contingent reasons, that cannot possibly be the whole story. Our concern in this chapter is rather with criticism and objections which are explicitly directed against the de Broglie–Bohm theory. In fact, there has been response to, e.g. Bohm's paper from 1952 and Wayne Myrvold notes:

> Bohm's theory did not meet with the acceptance in the physics community that Bohm had hoped for. It was not, however, ignored; several prominent physicists, among which were Einstein, Pauli, and Heisenberg, wrote articles expressing their reasons for not accepting Bohm's theory (Myrvold 2003, p. 7).

In what follows, we also explore these early objections (not all early reactions were hostile though – see e.g. Joseph Keller (Keller 1953).

The objections against the de Broglie–Bohm theory can roughly be divided into two classes. The first applies meta-theoretical or extra-empirical criteria like symmetry or

[2] Because of this independent double discovery, we refer to this approach as the de Broglie–Bohm theory. But given the catchy label 'Bohmian' we sometimes omit the role of de Broglie. The term pilot-wave theory (coined by de Broglie: *theorie de l'onde pilote*) is also used synonymously by us.

[3] This building (82 Gower Street) no longer houses the independent bookshop, but still a branch of a large bookshop chain including a place called Dillon's coffee shop.

[4] Back then, Philippidis *et al.* picked the obscure journal *Il Nuovo Cimento* because they anticipated that any major British or American outlet would not publish it, cf. Dewdney (2023). Meanwhile, Bohm trajectories can be found in prestigious journals like *Nature Communications*, e.g. Foo *et al.* (2022).

simplicity (Section 1.3). The other class of criticism seeks for a more theory-immanent debate, like challenging the consistency or the ability of the de Broglie–Bohm theory to be generalized (Section 1.4). For completeness, we provide a brief summary of the de Broglie–Bohm theory in Section 1.2. A thorough discussion of it can be found, e.g. in Dürr and Teufel (2009), Bohm and Hiley (1993), and Holland (1993).

1.2 The de Broglie–Bohm theory

The de Broglie–Bohm theory describes a non-relativistic N-particle system by its wave function, ψ, and the position, Q_i, of the corresponding quantum objects (e.g. electrons, atoms or the like). The wave function, which is obtained from the ordinary Schrödinger equation, guides the particle motion via the so-called guidance equation:

$$\frac{dQ_i}{dt} = \frac{1}{m_i}\nabla_i S(Q_1, \cdots, Q_N). \tag{1.1}$$

Here, m_i denotes the mass of particle i, ∇_i is the nabla operator applied to its coordinates and S is the phase of the wave function in the polar representation $\psi = Re^{\frac{i}{\hbar}S}$. The global existence and uniqueness of the solutions of the guidance equation has been shown for a wide range of problems (Teufel and Tumulka 2005).

Since the guidance condition is a first-order equation, the initial positions alone fix the motion uniquely.[5] Given a $\rho = |\psi|^2$ distribution as initial positions (the so-called 'quantum equilibrium hypothesis'), Equation (1.1) will reproduce all predictions of ordinary quantum mechanics in terms of position distributions.

Since all measurements can be expressed in terms of position (e.g. pointer positions), this amounts to full accordance with all predictions of ordinary quantum mechanics. Thereby the de Broglie–Bohm theory assigns a distinguished role to position and does not independently assign possessed values to other observables. What might be regarded as the values of quantum observables such as spin, momentum or the like get established only in the context of a corresponding measurement-like experiment.

The motivation of the quantum equilibrium hypothesis has been explored for example in Dürr *et al.* (1992) and Valentini (1991a, 1991b). Most important, the quantum mechanical continuity equation

$$\frac{\partial|\psi|^2}{\partial t} + \nabla\left(|\psi|^2 \cdot \frac{\nabla S}{m}\right) = 0 \tag{1.2}$$

ensures that this condition is consistent, i.e. any system will stay $|\psi|^2$ distributed if the quantum equilibrium hypothesis holds initially.

It follows that in a universe being in quantum equilibrium it is not possible to control the initial positions beyond the $|\psi|^2$ distribution. While ordinary quantum mechanics assumes that probability enters on a fundamental level, the de Broglie–Bohm theory is deterministic and probability enters only as an expression of ignorance, i.e. on an epistemic level. However, given the quantum equilibrium hypothesis this ignorance holds in principle.

[5] Newton's equation of motion is *second* order and therefore requires the specification of positions and momenta for a unique solution.

The important feature of Equation (1.1) is its non-locality. The guidance equation links the motion of every particle to the configuration of the whole system, no matter how distant its different parts are. Technically expressed, this follows from the fact, that the wave function ψ (hence its phase S) at a given time is a function on the configuration space $\mathbb{R}^{3N}$. It is exactly this non-locality which allows the de Broglie–Bohm theory to violate the inequality from Bell (1964) as demanded by experiment.

The guidance condition (Equation (1.1)) can be motivated in different ways and its precise status gives rise to different versions of the de Broglie–Bohm theory. The starting point of Bohm's original presentation of the theory in Bohm (1952a, 1952b) was the decomposition of the Schrödinger equation for the wave function $\psi = Re^{\frac{i}{\hbar}S}$ into a set of two equations for the real functions R and S. The resulting equation for S has a structure similar to the classical Hamilton–Jacobi equation for the action S, which implies $p = \nabla S$ (which is one way to put the guidance equation). The only difference to the classical Hamilton-Jacobi equation is the appearance of an extra term which Bohm called 'quantum potential':

$$U_{\text{quant}} = -\frac{\hbar^2}{2m}\frac{\nabla^2 R}{R}. \tag{1.3}$$

Bohm (and later also, e.g. Bohm *et al.* (1987) and Holland (1993)) regarded the quantum potential as the key ingredient of the theory and derived all its novelty from it.

In contrast to this position, another school of the de Broglie–Bohm theory considers the guidance condition as the fundamental equation and avoids emphasizing the quantum potential. The main proponents of this school are (or have been) Detlef Dürr, Sheldon Goldstein and Nino Zanghì (Dürr *et al.* 1992, 1997; Berndl *et al.* 1995; Daumer *et al.* 1996) who have dubbed their formulation 'Bohmian mechanics'. This view was anticipated by Bell (1987) in his writings on the pilot-wave theory. According to this position, the quantum potential (Equation. (1.3)) deserves no special attention and is rather viewed as an artefact which enters the discussion at best when the classical limit of the theory is treated.

In fact, these different interpretations of Equation (1.1) are related to a substantial difference of view on e.g. the role of 'observables' other than position or the meaning of the wave function. Our discussion of objections against the de Broglie–Bohm theory is in part complicated by this debate on its interpretation. If some criticism applies more strongly or solely to one specific reading of the theory, it does not undermine the concept as a whole. Likewise, the different interpretations sometimes provide different replies to the objections.

Since the rest of our note is concerned with (more or less compelling) objections, we should also make some brief remarks on the merits of the theory. Clearly, in their absence there would be no reason for being a 'Bohmian' in the first place.

The supporters of the pilot-wave theory emphasize its 'clear ontology', i.e. that the vague notion of 'complementarity' and 'wave-particle duality' becomes dispensable. Likewise, the need for a 'classical' description of the experimental arrangement as emphasized by Bohr becomes dispensable. In Bell's terminology, this theory possesses 'beables' (Bell 1976).

Within the de Broglie–Bohm theory, one can consistently entertain the notion of particle trajectories. However, this should not be misunderstood as adherence to classical prejudices but provides a solution of the measurement problem, i.e. the question how to deal with the superposition of macroscopically distinct states of the measurement device. The superposition of the wave function at the end of a measurement causes no difficulty since the *actual* configuration of the system corresponds to the *actual* outcome of the measurement (Lewis 2007).

As mentioned, Bell was one of the few but influential proponents of this theory since the 1960s. He even made a bold suggestion about the reason of its neglect in the teaching:

> Why are they [*i.e. the students*] not told about it? I have to guess here [...] but one of the reasons is surely that this theory takes almost all the *romance* out of quantum mechanics. This scheme is a living counterexample to most of the things that we tell the public on the great lessons of twentieth century science (Bell 1989).

1.3 The meta-theoretical debate

Most authors accept that the de Broglie–Bohm theory and ordinary non-relativistic quantum mechanics make identical predictions, i.e. that no experiment can decide which one to prefer. Even Pauli admitted in a letter to Bohm from December 1951: 'I do not see any longer the possibility of any logical contradiction [...]' (von Meyenn, 1996, Doc. 1313).

But this was only a minimal concession to Bohm. In the absence of any new prediction, the theory was criticized for being not physics but 'metaphysics' (Pauli 1952), and e.g. Heisenberg questioned whether the de Broglie–Bohm theory should be regarded as a new theory at all, or just an 'exact repetition in a different language' (quoted after Myrvold 2003).

However, the above alone does not constitute any reason to reject the pilot-wave theory, and one needs to specify why the ordinary quantum theory should be preferred. We start with the 'meta-theoretical' or 'extra-empirical' arguments.

1.3.1 Ockham's razor

The obvious objection against the de Broglie–Bohm theory is that it postulates the particle position as a new entity while not making any new prediction. However, additional premises which do not enrich the explanatory power should be removed by invoking 'Ockham's razor'.

Given this widely accepted principle, it appears natural to discard the pilot-wave theory since the particle trajectories seem to be exactly such an extra premise. But at the same time this theory eliminates the postulates which are related to the measurement process (not to mention how (un-)compelling these postulates are). Furthermore, the de Broglie–Bohm theory provides a new interpretation of quantum phenomena in which e.g. probability plays no fundamental role. In other words, the descriptive content is identical, but one may question if these theories are equivalent, i.e. whether the precondition for applying Ockham's razor is met.

A variant of this criticism was also voiced by Brown and Wallace (2005). These authors compare the Bohmian solution of the measurement problem with that of the many-worlds interpretation and claim that actually the Bohmian view is a many-world interpretation 'in denial'. By this they mean that the 'empty wave functions' (see Section 1.3.4) have a comparable role to the branching process typical of the many-worlds perspective. In particular, they argue that the Bohmian particles play only a minor role in the solution of the measurement problem; hence, they can be removed according to Ockham's razor. This argument seems to ignore that the Bohmian particles distinguish the branch of the wave function which corresponds to the measurement outcome. Likewise, the problem of empty wave functions can be circumvented (see Section 1.3.4).

1.3.2 Asymmetry in the de Broglie–Bohm theory

Pauli and Heisenberg based their rejection of the de Broglie–Bohm theory mainly on its asymmetry with respect to position and momentum (cf. Myrvold 2003). In the absence of any new prediction, they did not accept this sacrifice.

The reply to this objection can be twofold: (i) the de Broglie–Bohm theory gives position a different status than all other "observables" in order to solve conceptual problems of the ordinary formulation (Daumer *et al.* 1996). After all, symmetry is no end in itself. (ii) Moreover, the Hamiltonian in orthodox quantum theory is not invariant under general unitary transformations, though it is under the usual space time symmetries. Thus, even there it is not the case that all observables are on the same footing.

In fact, the de Broglie–Bohm theory shows another asymmetry, namely with respect to the wave function. While the wave function acts on the particle position, the particles do not react on the ψ-field which is determined independently by the Schrödinger equation.

It is true that this constitutes a peculiar feature of the de Broglie–Bohm theory and in reply to this objection Dürr *et al.* (1997) have suggested that the role of the wave function within the pilot-wave theory should be considered as analogous to the role of the Hamiltonian in classical mechanics. They state

> We propose that the reason, on the universal level, that there is no action of configurations upon wave functions [...] is that the wave function of the universe is not an element of physical reality. We propose that the wave function belongs to an altogether different category of existence than that of substantive physical entities, and that its existence is nomological rather than material. We propose, in other words, that the wave function is a component of physical law rather than of the reality described by the law.

This is a rather spectacular move since exactly the opposite claim, namely that in the de Broglie–Bohm theory ψ is treated like an *ordinary physical field* (and not just encoding our information about future experiments), was made early on by Bohm (1952a, 1952b) and also Bell (1987), p. 162.

1.3.3 Return to classical physics?

A rather unspecific but nevertheless common objection against the de Broglie–Bohm theory is its supposed return to classical notions. For example, Englert (2001) states about the purpose of the theory and its trajectories:

> With reference to these trajectories, atomic processes are then deterministic, and this saves us the mourning that the loss of the deterministic Newton–Maxwellian world view demands of us (translated by the author).

This claim of 'backwardness' is in itself no strong argument against the theory, and one needs to explain why this 'return' is supposed to be artificial or needless. But this argument remains weak, since the de Broglie–Bohm theory possesses so many traits which are common to quantum mechanics and completely 'unclassical' (e.g. wave function on the configuration space, non-locality, etc.) that it does a disservice to anyone seeking for a return to 'deterministic Newton-Maxwellian world view'. Moreover, it is incorrect to emphasise the determinism of the theory as a property that was particularly important to David Bohm (cf. van Strien 2025).

1.3.4 Departure from established principles

While Section 1.3.3 mentioned the objection of 'backwardness', the theory also meets with the contrary criticism. Here, the bizarre features of the de Broglie–Bohm theory are the subject of discomfort.

According to some versions of the pilot-wave theory, the wave function may be compared to other physical fields such as electromagnetic or gravitational fields. As already mentioned in Section 1.3.2, this view was for example held by Bohm (1952a, 1952b) himself and later also by (Bell, 1987, p. 128). This introduces peculiar notions into physics indeed. First of all, the quantum mechanical wave function is defined on the configuration space of the system. This is in sharp contrast to any other physical field. The non-locality is closely related to this feature and is the subject of Section 1.4.2. As mentioned above (see Section 1.3.2),Dürr *et al.* (1997) have proposed that the role of the wave function within the theory should be viewed as nomological which refutes this allegation.

Viewing the wave function as 'nomological' rather than 'physically real' helps also to reply to the objection that the de Broglie–Bohm theory introduces myriads of 'empty waves' into the picture of physical reality. These are the branches of the wave function which do not contain the particle on its trajectory and hence do not correspond to the actual state of the system. Although one can argue that due to decoherence effects these empty branches do not affect the actual system, this feature would be curious.

Another intriguing property of Bohmian trajectories gave rise to a specific objection from Einstein. Since he was one of the famous antagonists of the Copenhagen interpretation, it is interesting to note that he did not endorse the de Broglie–Bohm theory likewise. In a *Festschrift* in honor of Max Born in the year 1953, Einstein discussed a system for which the de Broglie–Bohm theory predicts a vanishing velocity. Einstein discussed a particle in a box as a specific example, but the same behavior appears in any system which is described by a real wave function like, e.g. the energy eigenstates of the harmonic oscillator.[6] According to Einstein, this vanishing velocity

> [...] contradicts the well-founded requirement, that in case of a macrosystem (i.e. for highly excited states) the motion should agree approximately with the motion following from classical mechanics (Myrvold 2003, p. 10).

[6]For a real wave function $S = 0$, hence the velocity (see Equation (1.1)) vanishes.

However, any measurement on the particle would need a change in the arrangement (e.g. one side of the box would have to be removed). The predicted outcome of any such *measurement* of, e.g. the particle momentum, would be the same as in ordinary quantum mechanics. More generally, the Einstein objection illustrates, that within the de Broglie–Bohm theory the representation of any system is provided by the *pair* of wave function and position (ψ, Q_i). To focus on properties of one element can only be misleading.

Einstein's rejection of the de Broglie–Bohm theory is clearly not only related to the problem discussed above. By now famous is his remark in a letter to Born in 1952 about the de Broglie–Bohm theory being 'too cheap' (Einstein *et al.* 1969), letter from 12.5.1952). Squires (1996) writes in the same context that Einstein 'was not interested in attempts to "cure" the theory; rather he wanted to look elsewhere, to start again'.

1.3.5 Underdetermination in the de Broglie–Bohm theory

The de Broglie–Bohm theory reproduces the statistical predictions of ordinary quantum mechanics and underpins them with deterministic and continuous particle trajectories. It is tempting to regard the Bohm trajectories as the actual motion of the quantum objects.

However, Deotto and Ghirardi (1998) have shown that the de Broglie–Bohm theory is underdetermined, i.e. the quantum mechanical current can be 'gauged' by a divergenceless vector field $\mathbf{j}' = \mathbf{j} + \mathbf{a}$ with $\nabla \mathbf{a} = 0$. The corresponding guidance condition yields the same *statistical* predictions, while the *individual* trajectories differ. Hence, it is problematic to consider the Bohm trajectories obtained from Equation (1.1) as representing the 'actual motion' of the quantum particles.

In order to sustain the 'ontological status' of the Bohm trajectories, one has to formulate additional criteria which restrict the possible value of the vector field $\mathbf{a}$. For example, Dürr *et al.* (1992) motivate the guidance equation (1.1) from symmetry and simplicity constraints. The argument in Holland and Philippidis (2003) is based on the assumption that the corresponding problem for relativistic spin $\frac{1}{2}$ particles has been solved uniquely (cf. also Holland 1999). Holland and Philippidis then derive the guidance equation for the non-relativistic limit and reach an equation which contains an additional spin-dependent term. Hence, for spin 0 particles, the original form is recovered.

However, this ambiguity of the de Broglie–Bohm theory does not undermine its conceptual value (e.g. in solving the measurement problem). Since the control of individual trajectories is beyond the principle reach of experiments, one should not put too much emphasize on their particular form anyway.

1.3.6 The status of the quantum equilibrium hypothesis

As mentioned in Section 1.2, the de Broglie–Bohm theory reproduces all predictions of ordinary quantum theory provided that the initial positions of particles described by the wave function ψ are $|\psi|^2$ distributed – the so called 'quantum equilibrium hypothesis'.

One may include this assumption in the very definition of the de Broglie–Bohm theory; however, introducing the quantum equilibrium hypothesis as a postulate provokes the objection that thereby the wave function gets two distinct and logically independent meanings: (i) as the guiding field and (ii) as a probability distribution for the particle position.

This double role for the wave function looks suspicious and unaesthetic. Furthermore, it would remain obscure how random behavior enters into the deterministic de Broglie–Bohm theory. Hence, the very meaning of such a postulate would be not clear at all.

It was therefore among the early efforts of Bohm to clarify the status of the quantum equilibrium hypothesis and to possibly derive rather than postulate it. The paper Bohm (1953) was devoted to this question but could derive the quantum equilibrium hypothesis only for a limited class of systems (Cushing 1994, Valentini and Westman 2005).

A dynamical explanation of the quantum equilibrium hypothesis was also attempted by Valentini (1991, 1991b) but a competing approach is due to Dürr *et al.* (1992). Their starting point is that regarding the de Broglie–Bohm theory as a fundamental theory implies that the behavior of subsystems is determined by the 'wave function of the universe', $\Psi(q)$, and the corresponding configuration.

The term 'wave function of the universe' should not be understood too literally. $\Psi(q)$ denotes rather a hypothetical "Bohmian universe" in order to deal with a closed system in which no property can be attributed to the influence of the environment. Dürr *et al.* show how wave functions can be assigned to subsystems ('effective wave function'). The role of the quantum equilibrium hypothesis on the universal level is *not* probabilistic since we do not have a sample of universes. Instead, it provides a so-called measure of *typicality*. The notion of *typicality* (though not the word) was introduced by Boltzmann in justifying the second law of thermodynamics. This statement holds because an 'overwhelming majority' of initial conditions leads to a behavior in accordance with the second law (cf. Lebowitz 1993). However, the meaning of 'overwhelming majority', i.e. a measure on the corresponding set, needs to be specified. One important requirement for this measure is that it should be 'equivariant', i.e. the notion of typicality should be independent of time. And, in fact, the continuity Equation (1.2) ensures that the measure $|\Psi(q)|^2$ meets this requirement.

Finally, and that is their central result, Dürr *et al.* (1992) can prove that within a 'typical' universe the quantum equilibrium hypothesis holds for all sub-systems. Hence, the typical Bohmian universe – although deterministic – gives the *appearance* of randomness in agreement with quantum mechanics.

This justification of the quantum equilibrium hypothesis has been questioned, e.g. by Dickson (1998). He notes that Dürr *et al.* have not shown that $|\Psi(q)|^2$ provides the *only* equivariant measure. Furthermore, (Dickson, 1998, p. 123) questions that equivariance is a preferred property of measures over the initial distributions at all. This objection challenges the claim that the quantum equilibrium hypothesis can be derived rather than postulated. However, it should be noted that for the justification of classical thermodynamic the question of how to derive apparent randomness from deterministic laws is just as controversial. An up-to-date account of the debate can be found in Norsen (2018).

1.4 The theory immanent debate

Until now we were mainly concerned with meta-theoretical (or extra-empirical) objections which might be viewed as partially subjective. Another strategy has been to seek for a more content-related discussion, e.g. challenging the consistency of the de Broglie–Bohm theory or its ability to be generalized. Most important is the question whether a trajectory interpretation is sustainable in the relativistic domain.

1.4.1 The 'surreal trajectory' objection

In 1992, Englert, Scully, Süssmann and Walther (ESSW) challenged the de Broglie–Bohm theory. They claimed that Bohm-trajectories are not realistic, but 'surrealistic' (Englert *et al.* (1992)).

The corresponding authors analyse the famous delayed-choice double-slit experiment invented by Wheeler (1978) and discussed in the context of the de Broglie–Bohm theory by Bell (1987). Before we turn to the actual ESSW-argument, we first discuss the original set-up.

The delayed-choice double-slit experiment (see Fig. 1.1) consists of a double slit arrangement in which one can freely choose to detect either interference patterns in the region I or particles in the detectors C_1 or C_2. In concrete terms, this means that one can insert a screen in region I in order to detect the interference pattern. This choice can be made after the 'particles' have passed the slits already, hence the name *delayed-choice.*

The whole arrangement is set up in such a way that by symmetry arguments the trajectories of the de Broglie–Bohm theory are not allowed to cross the midplane behind the two-slit screen. They show the 'unclassical' behavior, that the Bohm-trajectories of the particles hitting the upper part behind the screen have traversed the upper slit and vice versa.[7]

One may modify the arrangement by supplying it with additional detectors directly behind the two-slit screen in order to investigate which slit has been traversed. In such a modified version the interference pattern would not occur. Additionally the de Broglie–Bohm-trajectories would be allowed to cross the midplane, since, given the degrees of freedom related to the extra-detector, it is no symmetry plane anymore (Bell (1987), p. 111).

The above-mentioned extra-detectors directly behind the screen were assumed to be 'ordinary detectors', i.e. devices which show a macroscopic change of state (e.g. pointer positions). We now turn to the actual ESSW argument. According to these authors, a problem for the de Broglie–Bohm theory emerges when these extra detectors are chosen to be advanced quantum optical devices, so-called "which-way detectors". These respond on the transition of single particles without affecting the translational part of the wave function[8]. Again, in the presence of these devices, we expect the interference pattern to be destroyed. The special feature of these 'one-bit detectors' is that even their excitation does not alter the symmetry (i.e. $\psi_1(x, y, z) = \psi_2(x, y, -z)$,

[7] The actual experiment has been performed by Hellmuth *et al.* (1987) and shows no difference between normal and delayed-choice mode.

[8] For example, within a micromaser excited Rydberg atoms can radiate off one photon without any other significant change of state.

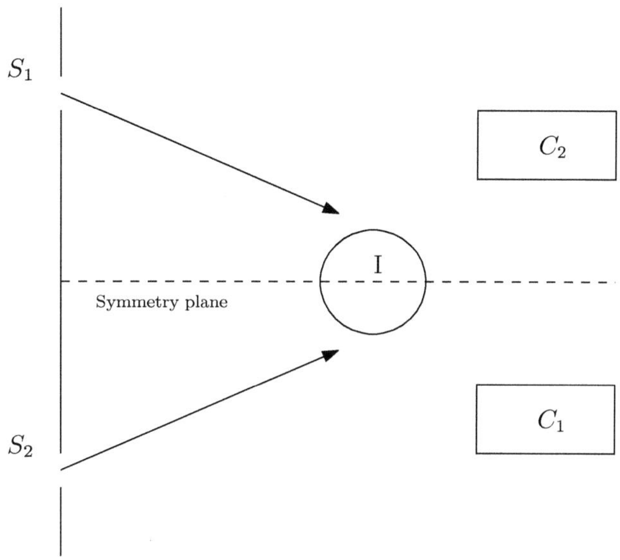

Fig. 1.1: Delayed-choice double-slit experiment. One may either detect particles in the detectors C_i or observe interference in region I by insertion of a screen.

with $z = 0$ being the midplane). Hence the de Broglie–Bohm trajectories are still forced to "bounce off" the midplane. However, the probabilities $|\psi_1|^2$ or $|\psi_2|^2$ are in general not confined to one-half of the screen. According to ESSW, one arrives at the paradoxical situation that the upper which-way detector fires while the screen is hit below the midplane. Englert *et al.* (1992) conclude:

> The Bohm trajectory is here macroscopically at variance with the actual, that is: observed track. Tersely: Bohm trajectories are not realistic, they are surrealistic.

This paper has created a lively debate on the 'surreal trajectory problem' and we do not aim at a complete review (see Solé (2017) and reference therein). One objection against the conclusion of ESSW has been their use of the term 'actual track' in connection with quantum mechanics. ESSW try to defend the orthodox interpretation – but the notion of a 'particle path' is denied within this interpretation. What is meant by "actual track" is not obvious here. However, ESSW claim that even the observed tracks in a bubble chamber are at variance with the Bohm trajectories. This would be a serious objection indeed.

The essential flaw in the reasoning of ESSW is that they consider devices which are not linked to any macroscopic change of state. This feature is crucial because it ensures that their symmetry argument applies. But given that within the de Broglie–Bohm theory only a change in position constitutes a physical fact, such a which-way detector is not regarded as a reliable detector for the actual position of the particle on its Bohmian path.[9] The additional claim of ESSW, that even the tracks in a bubble

[9] Above all, given the assumption of ESSW that the center-of-mass motion is not changed, the state remains in a superposition and is *not* left in a position eigenstate after leaving the which-way

chamber differ from the predicted Bohm trajectories, is therefore unfounded, since a bubble chamber does convert the excitation into a macroscopic displacement.

But the situation which has been considered by ESSW is a bit more subtle: the authors assume that a macroscopic read-out could be connected *after* the particle has been finally detected. However, it remains true that within the de Broglie–Bohm theory the which-way device is not regarded as a detector. A delayed read-out can not turn it into a more trustworthy device. The arrangement which has been considered by Englert *et al.* can be viewed as a special case in which 'empty waves' show an effect if they are still coherent (Hardy 1992). In fact, the non-locality of the de Broglie–Bohm theory makes it possible to explain how the which-way detector can be excited even without any trajectory passing through it.

Recently, Solé (2017) has tried to settle the issue and argues that any mismatch between Bohmian trajectories and the report of which-way detectors is practically impossible (although possible as a matter of principle).[10]

Summing up, we note that these investigations have given fascinating insight into detailed aspects of quantum mechanics in general and the de Broglie–Bohm theory in particular. The results illustrate that the Bohm trajectories exhibit non-classical behavior, as one would expect from a quantum mechanical description.

1.4.2 The non-locality objection

In the current debate, probably the most common reason for rejecting the de Broglie–Bohm theory is its non-locality and its alleged failure to generalize relativistically.

The theory introduced in Section 1.2 is the re-formulation of non-relativistic quantum mechanics; hence, its failure to comply with relativistic requirements cannot be complained about. However, some features of non-relativistic QM (e.g. entanglement) extend naturally into the relativistic domain. And if the non-locality of the de Broglie–Bohm theory would prevent relativistic generalizations, this would indeed call into question the validity of the whole approach. Let's try to disentangle these issues in turn.

The de Broglie–Bohm theory is explicitly non-local, i.e. the motion of each particle is in general a function of the coordinates of the whole system, no matter how distant. This non-locality vanishes only if the wave function factorizes in the contributions of the different quantum objects. Whether this is viewed as an unacceptable feature depends on the attitude towards the problem of non-locality in quantum mechanics in general. In ordinary quantum mechanics, the problem of non-locality appears in at least two places: (i) violation of Bell inequalities and (ii) reduction of the wave function.

Following the work of Bell (1964) and the experimental confirmation of quantum mechanics in EPR-like experiments (Aspect *et al.* (1982)), it became widely (but not universally (Mermin 1998, de Muynck 2001) accepted that quantum mechanics itself

detector. Even within conventional quantum mechanics one may question whether it is possible to infer on anything passing through the detector.

[10] Along similar lines, other arguments have also been advanced in order to show that 'the Bohmian position does not help to understand the result of a measurement', Aharonov and Vaidman (1996). These authors explored Bohm-trajectories in the case of 'weak' and 'protective' measurements but do not claim the *inconsistency* of the de Broglie–Bohm theory (*ibid.* p. 153).

is 'non-local'. If one follows this assessment, the non-locality allegation against the de Broglie–Bohm theory is groundless.

However, the precise meaning of the term 'non-local' is far from being unique, and there exists a vast literature on that topic. A thorough discussion of that issue is far beyond the scope of the present chapter. However, one can reasonably state, that the 'non-locality' of the de Broglie–Bohm theory is more explicit (i.e. dynamical) than the 'non-separability' of ordinary quantum mechanics.

Anyhow, for both, ordinary quantum mechanics and the de Broglie–Bohm theory, it is ensured that the 'non-locality' or 'non-separability' can not be used for superluminal signaling. This requirement (sometimes called 'Einstein-locality') is therefore met by ordinary QM and the de Broglie–Bohm theory. But whether this is enough for full compatibility between quantum mechanics and special relativity has been questioned, e.g. by Leslie Ballentine:

> However it is not clear that the requirements of special relativity are exhausted by excluding superluminal signals. [...] Whether or not there is a deeper incompatibility between quantum mechanics and relativity is not certain (Ballentine 1998, p. 610).[11]

Another indication for 'non-locality' in quantum mechanics is given if one adopts the collapse of the wave function to be a real physical process. After all, the collapse is supposed to reduce the wave function instantaneously and thereby requires a preferred frame of reference.[12] Maudlin (1996), p. 297, argues that the collapse postulate in combination with entangled states necessarily leads to a preferred foliation of space-time – Myrvold (2002) has challenged this claim.

What all this shows is that the satisfactory generalization of the measurement theory into the relativistic domain is still wanting – the measurement problem of quantum theory is even exacerbated in the relativistic theory. Ironically, this specific source of non-locality does not arise in the de Broglie–Bohm theory since here the collapse of the wave function becomes dispensable. But as indicated above, given the non-locality of the de Broglie–Bohm theory, its relativistic generalization is challenging. In what follows, we briefly touch on some of the corresponding work.

1.4.3 Relativistic and quantum field theoretical generalizations

Relativistic generalizations of the de Broglie–Bohm theory do exist.[13] For example, for Dirac particles Bohm and Hiley (1993) have proposed the following guiding equation for a 'Bohm–Dirac theory':

$$\mathbf{v} = \frac{\psi^\dagger \alpha \psi}{|\psi|^2}. \tag{1.4}$$

Here, ψ is a solution of the Dirac equation, $\psi^\dagger$ its conjugate and α is a 3-vector with components that are built from the Pauli matrices:

[11] In the omitted passage, Ballentine refers to the de Broglie–Bohm theory.

[12] An obvious solution to this problem is to suppose that the collapse occurs only along the backward light cone of the measurement interaction (Bloch 1967; Hellwig and Krauss 1987). See, e.g. Barrett (2005) for a discussion of the problems one faces in this approach.

[13] We call them Bohm-like theories in what follows; see Passon (2006) for more details on this notion.

$$\alpha_i = \begin{pmatrix} 0 & \sigma_i \\ \sigma_i & 0 \end{pmatrix}. \tag{1.5}$$

The generalization to the many-particle case is straightforward (Bohm and Hiley, 1993, p. 274ff). However, in the many-particle case this theory is not Lorentz covariant since it has to distinguish the frame of reference in which $\rho = \psi^\dagger\psi$ (the generalized quantum equilibrium hypothesis) holds (Berndl *et al.* (1996)). The (statistical) predictions do nonetheless agree with the standard theory, i.e. the preferred reference frame has no observable effects.

In fact, as shown in Dürr *et al.* (1999), it is even possible to restore Lorentz invariance for the Bohm–Dirac theory by introducing 'additional structure'. Dürr *et al.* introduce an arbitrary space-like preferred slicing of space-time, determined by a Lorentz invariant law.

This work (see also Dewdney and Horton (2002), and Goldstein and Tumulka (2003)) provides some steps towards a Lorentz invariant de Broglie–Bohm theory and a counterexample to the common claim that Bohm's non-locality and Lorentz invariance are in strict opposition. However, many authors (e.g. Dürr *et al.* (2013)) admit that they have not reached yet what Bell (1987, p. 179f), called 'serious Lorentz invariance'. We will return to this question at the end of the subsection.

Finally, there is the widespread suspicion that the concepts of the de Broglie–Bohm theory can not be sustained in the realm of relativistic quantum field theories (see, e.g. the letter to the editors of the *American Journal of Physics* (Gordon 1996) together with the reply). Given that our best current theory of particle physics is a quantum field theory (namely the standard model), this would be a valid argument for the rejection of the de Broglie–Bohm approach.

However, several works on that issue have shown that there seems to be no principle problem to incorporate the concepts and reproduce the predictions of quantum field theories. In what follows, we only sketch some of the corresponding results.

Similar to the situation of relativistic generalizations there are several different approaches to this question. The first, pursued, e.g. by Bohm and Hiley (1993) and Holland (1993), introduces the notion of (bosonic-)field variables as being fundamental together with the particle position for fermions.

For example, the second-quantized real Klein–Gordon field is described by a wave functional $\Psi(\phi(\mathbf{x}), t)$, which satisfies the Schrödinger equation:

$$i\frac{\partial \Psi}{\partial t} = \int d^3x \left(-\frac{\delta^2}{\delta \phi^2} + (\nabla\phi)^2 \right) \Psi. \tag{1.6}$$

The corresponding guidance equation for the field beable $\phi(\mathbf{x}, t)$ reads

$$\frac{\partial \phi}{\partial t} = \frac{\delta S}{\delta \phi}, \tag{1.7}$$

where S is the phase of the wave functional Ψ. However, for fermionic quantum fields, the construction of a Bohm-like theory with field-beables along these lines is not possible since they obey anticommutation relations which have no classical limit (Bohm *et al.* (1987), p. 374). Instead, Bohm, Hiley and Kaloyerou propose to introduce particle beables for fermions according to the Bohm–Dirac theory mentioned above. In fact, other models by Holland and Valentini which try to provide field beables for fermions did not succeed, as shown in Struyve and Westman (2006, p. 1).

Inspired by the difficulty to construct a Bohm-like theory for fermions with field beables, Struyve and Westman (2006) have explored a different direction. They argue that observational equivalence could be achieved by granting a beable status only to the bosonic degrees of freedom. Technically, their work is similar to Bohm's model with field beables for bosons mentioned above. In Struyve and Westman (2006), this approach is carried out for quantum electrodynamics, but it has a natural extension to other gauge theories. Now, what does it mean to achieve 'observational equivalence' in such a model? Roughly speaking, the idea is that also a pointer position (i.e. something which contains fermions) gets correlated with the gauge field. Thus, the field beables will provide an unambiguous and reliable record of the measurement. This is similar to the strategy of the de Broglie–Bohm theory, where the only beable 'particle-position' fixes the outcome of all measurements likewise.

A completely different approach is to stick to particle beables and to model the quantum field theories by jump-processes like the creation of particle-antiparticle pairs. An early (discrete) model of this kind was already formulated by Bell (1987, p. 173ff). In Passon (2006, p. 15ff) this approach is described in some more detail. Dürr *et al.* (2003, 2004) have developed a similar process in the continuum for more or less any regularized quantum field theory and call these models 'Bell-type quantum field theories'.

Using particle beable seems closer in spirit to the original de Broglie–Bohm theory, however, these models include random processes, i.e. are non-deterministic. This fact helps to recognize once again that the determinism of the de Broglie–Bohm theory is not a central feature from the point of view of its adherents.

However, the quantum field theoretical generalizations also suffer from the problem that some generalized quantum equilibrium hypothesis needs to be introduced in a preferred frame of reference, which seems to spoil Lorentz invariance. As argued in a recent summary by Dürr *et al.* (2013), the best solution could be to extract the privileged space-time foliation from the wave function, rather than requiring it as an independent and additional entity. Through a suitable choice of this mechanism, one could reach Lorentz invariance, at lest technically. However, Dürr *et al.* (2013) leave it open whether this would result in 'serious' Lorentz invariance already.

Hence, so far the de Broglie–Bohm theory only allows for a relativistic or quantum field theoretical generalization when the requirement of Lorentz invariance is somewhat relaxed. Does this provide compelling reason for rejecting these Bohm-like theories and in turn also the non-relativistic version? Note that the trouble enters since the quantum equilibrium hypothesis needs a preferred foliation of space-time, i.e. the initial conditions for the 'beables' which are needed to solve the measurement problem. The rejection of the de Broglie–Bohm theory on this ground would only be mandatory if the orthodox formulation had a different and convincing solution to the measurement problem.

One should appreciate that non-locality provides a challenge not only for the de Broglie–Bohm theory but also for ordinary quantum mechanics. Apparently, the relativistic formulations of quantum theory (Dirac or QFT) only obscure this fact by leaving the details of the measurement process unclear. One merit of the de Broglie–Bohm theory, according to its supporters, is to make this feature explicit, but see also Tumulka (2025) and Allori (2025) for the most recent developments in the field.

1.5 Summary

We have collected common criticism against the de Broglie–Bohm theory. Most of them illustrate the peculiar features of this theory, but they do not provide a rigorous disproof. One merit of this discussion is to reveal that even in science a theory can not only be judged by its empirical confirmation.

The most substantial concern is the question of its relativistic and quantum field theoretical generalization. However, several models for such generalizations do exist in which either the preferred foliation of space-time is unobservable or even Lorentz-invariance can be (at least formally) sustained. Nevertheless, important questions remain open. However, whether this problem is outweighed by the conceptual advantages of Bohm-like theories when it comes to the measurement process must again be assessed by everyone.

Finally, let us return to the question we posed in the title of this chapter. It goes without saying that no one has to be a 'Bohmian', but if one chooses not to be, one should not discard the pilot-wave theory for the wrong motivation.

References

Aharonov, Y. and Vaidman, L. (1996). About Position Measurements which do not show the Bohmian Particle Position. In Cushing, J.T., Fine, A. and Goldstein, S. (eds.) *Bohmian Mechanics and Quantum Theory: An appraisal.* Dordrecht: Kluwer Academic Publishers.

Allori, V. (2025). Relativistic Pilot-Wave Theories as the Rational Completion of Quantum Mechanics and Relativity. In Oldofredi, A. (ed.) *Guiding Waves in Quantum Mechanics: One Hundred Years of de Broglie-Bohm Pilot-Wave Theory.* Oxford: Oxford University Press.

Aspect, A., Grangier, P. and Roger, G. (1982). Experimental Realization of Einstein–Podolsky–Rosen–Bohm Gedankenexperiment. A New Violation of Bell's Inequalities. *Physical Review Letters*, **49**: 91.

Bacciagaluppi, G. and Valentini, A. (2009). *Quantum Theory at the Crossroads: Reconsidering the 1927 Solvay Conference.* Cambridge: Cambridge University Press.

Ballentine, L.E. (1998). *Quantum Mechanics.* Singapore: World Scientific.

Barrett, J.A. (2005). Relativistic quantum mechanics through frame-dependent constructions. *Philosophy of Science*, **72**(5): 802–813.

Belinfante, F,J. (1973). A Survey of Hidden-Variables Theories. *International Series of Monographs in Natural Philosophy*, **55**.

Bell, J.S. (1964). On the Einstein Podolsky Rosen Paradox. *Physics*, **1**(3): 195.

Bell, J.S. (1976). The theory of local beables. *Epistemological Letters*, **9**: 11.

Bell, J.S. (1987). *Speakable and Unspeakable in Quantum Mechanics.* Cambridge: Cambridge University Press.

Bell, J.S. (1989). Towards An Exact Quantum Mechanics. In Deser, S. and Finkelstein, R.J. (eds.) *Essays in honor of J. Schwinger 70th birthday.* Singapore: World Scientific.

Berndl, K., Daumer, M., Dürr, D., Goldstein, S. and Zanghì, N. (1995). A Survey of Bohmian Mechanics. *Il Nuovo Cimento B*, **110**: 737–750.

Berndl, K., Dürr, D., Goldstein, S. and Zanghì, N. (1996). EPR-Bell Nonlocality, Lorentz Invariance and Bohmian Quantum Theory. *Physical Review A*, **53**: 2062.

Bloch, I. (1967). Some Relativistic Oddities in the Quantum Theory in Terms of Observation. *Physical Review*, **156**(5): 1377.

Bohm, D. (1952a). A suggested interpretation of the quantum theory in terms of "hidden" variables. I. *Physical Review*, **85**(2): 166–179.

Bohm, D. (1952b). A suggested interpretation of the quantum theory in terms of "hidden" variables. II. *Physical Review*, **85**(2): 180–193.

Bohm, D. (1953). Proof That Probability Density Approaches $|\psi|^2$ in Causal Interpretation of the Quantum Theory. *Physical Review*, **89**: 458.

Bohm, D. and Hiley, B. (1993). *The Undivided Universe: An Ontological Interpretation of Quantum Theory.* London: Routledge.

Bohm, D., Hiley, B.J., and Kaloyerou, P.N. (1987). An ontological basis for the quantum theory. *Physics Reports*, **144**(6): 321–375.

Brown, H.R. and Wallace, David (2005). Solving the Measurement Problem: de Broglie–Bohm Loses Out to Everett. *Foundations of Physics*, **35**: 517–540.

Cousins, R.D. (1995). Why isn't every physicist a Bayesian? *American Journal of Physics*, **63**(5): 398–410.

Cushing, J.T. (1994). *Quantum Mechanics – Historical Contingency and the Copenhagen Hegemony.* Chicago: University of Chicago Press.

Daumer, M., Dürr, D., Goldstein, S. and Zanghì, N. (1996). Naive realism about operators. *Erkenntnis*, **45**: 379–397.

de Muynck, W.M. (2001). Interpretations of quantum mechanics, and interpretations of violation of Bell's inequality. In Khrennikov, A. (ed.) *Foundation of Probability and Physics*, pp. 95–114. Singapore: World Scientific.

Deotto, E. and Ghirardi, G.C. (1998). Bohmian Mechanics Revisited. *Foundations of Physics*, **28**(1): 1–30.

Dewdney, C. (2023). Rekindling of de Broglie–Bohm Pilot Wave Theory in the Late Twentieth Century: A Personal Account. *Foundations of Physics*, **53**(1): 24.

Dewdney, C. and Horton, G. (2002). Relativistically invariant extension of the de Broglie–Bohm theory of quantum mechanics. *Journal of Physics A: Mathematical and General*, **35**(47): 10117.

Dickson, W.M. (1998). *Quantum chance and non-locality in the interpretations of quantum mechanics.* Cambridge University Press.

Dürr, D., Goldstein, S., Munch-Berndl, K. and Zanghì, N. (1999). Hypersurface Bohm–Dirac models. *Physical Review A*, **60**: 2729–2736.

Dürr, D., Goldstein, S., Norsen, T., Struyve, W. and Zanghì, N. (2013). Can Bohmian mechanics be made relativistic? *Proceedings of the Royal Society A*, **470**.

Dürr, D., Goldstein, S., Tumulka, R. and Zanghì, N. (2003). Trajectories and particle creation and annihilation in quantum field theory. *Journal of Physics A: Mathematical and General*, **36**: 4143–4149.

Dürr, D., Goldstein, S., Tumulka, R. and Zanghì, N. (2004). Bohmian mechanics and quantum field theory. *Physical Review Letters*, **93**: 090402.

Dürr, D., Goldstein, S. and Zanghì, N. (1992). Quantum Equilibrium and the Origin of Absolute Uncertainty. *Journal of Statistical Physics*, **67**: 843.

Dürr, D., Goldstein, S. and Zanghì, N. (1997). Bohmian Mechanics and the Meaning of the Wave Function. In Cohen, R.S., Horne, M. and Stachel, J. (eds.) *Experimental Metaphysics - Quantum Mechanical Studies for Abner Shimony*, Volume 1, *Boston Studies in the Philosophy of Science.* Boston: Kluwer Academic Publishers.

Dürr, D. and Teufel, S. (2009). *Bohmian Mechanics. The Physics and Mathematics of Quantum Theory.* Berlin: Springer.

Einstein, A., Born, H. and Born, M. (1969). *Briefwechsel 1916-1955.* Nymphenburger Verlagshandlung: München.

Englert, B.G. (2001). Book review of: Dürr, D. (2001) "Bohmsche Mechanik als Grundlage der Quantenmechanik". *Physikalische Blätter*, **57**(11): 82–86.

Englert, B.G., Scully, M.O., Süssmann, G. and Walther, H. (1992). Surrealistic Bohm trajectories. *Zeitschrift für Naturforschung A*, **47**(12): 1175–1186.

Foo, J., Asmodelle, E., Lund, A.P., and Ralph, T.C. (2022). Relativistic Bohmian trajectories of photons via weak measurements. *Nature Communications*, **13**(1): 4002.

Goldstein, S. and Tumulka, R. (2003). Opposite arrows of time can reconcile relativity and nonlocality. *Classical and Quantum Gravity*, **20**: 557.

Gordon, L. (1996). Does there exist a relativistic Bohm theory (letter to the editor). *American Journal of Physics*, **64**(10): 1227.

Hardy, L. (1992). On the existence of empty waves in quantum theory. *Physics Letters A*, **167**: 11.

Hellmuth, T., Walther, H., Zajonc, A., and Schleich, W. (1987). Delayed-choice experiments in quantum interference. *Physical Review A*, **35**(6): 2532.

Hellwig, K.E. and Kraus, K. (1987). Formal Description of Measurements in Local Quantum Field Theory. *Physical Review D*, **1**(2): 566.

Holland, P. (1993). *The Quantum Theory of Motion. An Account of the de Broglie–Bohm Causal Interpretation of Quantum Mechanics.* Cambridge: Cambridge University Press.

Holland, P. (1999). Uniqueness of paths in quantum mechanics. *Physical Review A*, **60**: 4326.

Holland, P. and Philippidis, C. (2003). Implications of Lorentz covariance for the guidance equation in two-slit quantum interference. *Physical Review A*, **67**: 062105.

Keller, J. (1953). Bohm's Interpretation of the Quantum Theory in Terms of "Hidden" Variables. *Physical Review*, **89**: 1040.

Lebowitz, J.L. (1993). Boltzmann's Entropy and Time's Arrow. *Physics Today*, **46**(9): 32–38.

Lewis, P.J. (2007). How Bohm's Theory Solves the Measurement Problem. *Philosophy of Science*, **74**(5): 749–760.

Maudlin, T. (1996). Space-time in the quantum world. In Cushing, J.T., Fine, A. and Goldstein, S. (eds.) *Bohmian Mechanics and Quantum Theory: An Appraisal.* Dordrecht: Kluwer Academic Publishers.

Mermin, D. (1985). Is the moon there when nobody looks? *Physics Today*, **38**(4): 38–47.

Mermin, D. (1998). What is quantum mechanics trying to tell us? *American Journal of Physics*, **66**: 753.

Myrvold, W. (2002). On peaceful coexistence: Is the collapse postulate incompatible with relativity?. *Studies in History and Philosophy of Modern Physics*, **33**: 435.

Myrvold, W. (2003). On some early objections to Bohm's theory. *Intertional Studies in the Philosophy of Science*, **17**(1): 7–24.

Norsen, T. (2018). On the explanation of Born-rule statistics in the de Broglie–Bohm pilot-wave theory. *Entropy*, **20**(6): 422.

Passon, O. (2006). What you always wanted to know about Bohmian mechanics but were afraid to ask. *Physics and Philosophy*, **3**: 1–25.

Pauli, W. (1952). Remarques sur le problème des paramètres cachés dans la méchanique quantique et sur la théorie de l'onde pilote. In *Louis de Broglie: Physicien et Penseur*, pp. 33-42. Édition Albin Michel: Paris.

Philippidis, C., Dewdney, C. and Hiley, B.J. (1979). Quantum interference and the quantum potential. *Il Nuovo Cimento B*, **52**(1): 15–28.

Solé, A. (2017). Surrealistic Bohmian trajectories appraised. *European Journal for Philosophy of Science*, **7**: 467–492.

Squires, E.J. (1996). Essay review: The unresolved quantum dilemma. *Studies in History and Philosophy of Science Part B: Studies in History and Philosophy of Modern Physics*, **27**(3): 389.

Struyve, W. and Westman, H. (2006). A new pilot-wave model for quantum field theory. *AIP Conference Proceedings*, **844**(1) :321–339.

Teufel, S. and Tumulka, R. (2005). Simple proof for global existence of Bohmian trajectories. *Communications in Mathematical Physics*, **258**: 349–365.

Tumulka, R. (2025). A Vision for a Bohm-Style Theory of Quantum Electrodynamics. In Oldofredi, A. (ed.) *Guiding Waves in Quantum Mechanics: One Hundred Years of de Broglie-Bohm Pilot-Wave Theory.* Oxford: Oxford University Press.

Valentini, A. (1991a). Signal-locality, uncertainty, and the subquantum H-theorem. Part I. *Physics Letters A*, **156**(1–2): 5–11.

Valentini, A. (1991b). Signal-locality, uncertainty, and the subquantum H-theorem. Part II. *Physics Letters A*, **158**(1–2): 1–8.

Valentini, A. and Westman, H. (2005). Dynamical origin of quantum probabilities. *Proceedings of the Royal Society A*, **461**: 253–272.

van Strien, M. (2025). Why Bohm was never a determinist. In Oldofredi, A. (ed.) *Guiding Waves in Quantum Mechanics: One Hundred Years of de Broglie-Bohm Pilot-Wave Theory.* Oxford: Oxford University Press.

von Meyenn, K. (ed.) (1996). *Wolfgang Pauli. Scientific Correspondence with Bohr, Einstein, Heisenberg, a.o. Volume IV, Part I: 1950-1952.* Berlin, Heidelberg: Springer.

Wheeler, J.A. (1978). The 'past' and the 'delayed choice' double slit experiment. In *Mathematical Foundations of Quantum Theory* (ed. A. R. Marlow). Academic: New York.

2

The trouble with pilot-wave theory: a critical evaluation

Antony Valentini

2.1 Introduction

Pilot-wave theory had its beginnings in a series of papers published by Louis de Broglie in 1923, which formed the basis of his celebrated PhD thesis (de Broglie 1924, 1925). Three years later, the complete pilot-wave theory of a many-body system, with a guiding wave in configuration space, was proposed by de Broglie at the 1927 Solvay conference, where it was the subject of lively and wide-ranging discussions (de Broglie 1928; Bacciagaluppi and Valentini 2009). Soon afterwards de Broglie abandoned the theory, largely because he had misgivings about a physics grounded in configuration space (among other reasons). In 1952 the theory was revived, and extended to fields, by David Bohm (1952a, 1952b). Bohm was also able to clarify how the theory accounts for general quantum measurements. The theory is nowadays generally regarded as an alternative, if little used, formulation of quantum physics (Holland 1993; Valentini 2024, 2025a, 2025b, 2025c).

Since its first proposal in the 1920s, and its revival in the 1950s, pilot-wave theory has been subject to numerous objections as well as widespread misunderstandings. In this centenary year of de Broglie's defence of his PhD thesis – concerning which Einstein famously said that de Broglie had 'lifted a corner of the great veil' – it seems appropriate to take stock of the objections and to consider where the theory now stands.

Objections to de Broglie–Bohm theory have been, and continue to be, so many and varied that it would be impractical to consider them all. But most of the objections can be grouped into three main themes. First, that pilot-wave theory is too radical a break from known physics. Second, that pilot-wave theory is not a sufficiently radical break from known physics. And third, that pilot-wave theory is simply the same as known physics albeit expressed in an unusual way. These three kinds of objections are, of course, mutually inconsistent. It is as if the scientific community cannot make up its mind as to the appropriate standard by which to judge or criticise this theory. It may be that, in the current scientific climate, there is no set of shared criteria for how to evaluate competing theories in fundamental physics. As we consider these themes, it will be useful to look at representative examples from specific authors. Our focus, however, will be on general ideas and we make no attempt at an exhaustive survey of the vast literature on objections to this theory.

Objections to pilot-wave theory are often based on simple misunderstandings, and these are hardly worth considering here. Sometimes, however, the objections are interesting and instructive: by considering them, we can improve our understanding of the theory. After a brief review of pilot-wave theory, we consider the three themes we have outlined, and attempt a critical evaluation of what appear to us to be the principal objections to this century-old approach to quantum physics.

2.2 Elements of pilot-wave theory

We begin with a brief review of the elements of pilot-wave theory as we see them.

2.2.1 De Broglie and Bohm

At the 1927 Solvay conference de Broglie considered a system of low-energy spinless particles whose wave function $\psi = \psi(\mathbf{x}_1, \mathbf{x}_2, ..., \mathbf{x}_N, t)$ obeys the Schrödinger equation (units $\hbar = 1$)

$$i\frac{\partial\psi}{\partial t} = -\sum_{n=1}^{N}\frac{1}{2m_n}\nabla_n^2\psi + V\psi\ . \tag{2.1}$$

He postulated that the particles move along trajectories determined by the equation of motion

$$\frac{d\mathbf{x}_n}{dt} = \frac{1}{m_n}\boldsymbol{\nabla}_n S\,, \tag{2.2}$$

where $\psi = |\psi|\,e^{iS}$. These equations define a deterministic dynamics for a many-body system. The particle trajectories are guided by what de Broglie called the 'pilot wave' ψ, which is defined on configuration space and which satisfies the usual Schrödinger equation.

Note that (2.2) is a law for velocity. While de Broglie himself did not put it this way, we can think of ψ as generating an 'Aristotelian force'

$$\mathbf{f}_n = \boldsymbol{\nabla}_n S \tag{2.3}$$

acting on the nth particle – a force that generates velocity (instead of acceleration, as do Newtonian forces) (Valentini 1992, 1997). The configuration-space wave ψ may be regarded as a new kind of causal agent quite different from the forces and (3-space) fields previously encountered in physics.

From (2.1) we find the continuity equation

$$\frac{\partial\,|\psi|^2}{\partial t} + \sum_{n=1}^{N}\boldsymbol{\nabla}_n\cdot\left(|\psi|^2\,\frac{\boldsymbol{\nabla}_n S}{m_n}\right) = 0 \tag{2.4}$$

for the squared amplitude $|\psi|^2$. In principle, however, the wave ψ is unrelated to probability and simply guides the motions of individual systems. If we consider an ensemble of similar systems with the same pilot wave ψ, each system will follow the law of motion (2.2) with the velocity field $\mathbf{v}_n = \boldsymbol{\nabla}_n S/m_n$. Hence an arbitrary ensemble distribution $\rho(\mathbf{x}_1, \mathbf{x}_2, ..., \mathbf{x}_N, t)$ will necessarily obey the continuity equation

$$\frac{\partial \rho}{\partial t} + \sum_{n=1}^{N} \boldsymbol{\nabla}_n \cdot (\rho \mathbf{v}_n) = 0 \,. \tag{2.5}$$

This has the same form as (2.4). Thus, if $\rho = |\psi|^2$ initially, it follows trivially that $\rho = |\psi|^2$ at later times. This is the state of 'quantum equilibrium', in which the Born rule applies. In principle, we may just as well consider initial quantum nonequilibrium ensembles with

$$\rho \neq |\psi|^2 \,, \tag{2.6}$$

but de Broglie confined his considerations to the equilibrium case.

Bohm's formulation in 1952 was rather different. To see the connection with de Broglie's formulation, we may take the time derivative of (2.2) and, making use of (2.1), write

$$\frac{d\mathbf{p}_n}{dt} = m_n \frac{d^2 \mathbf{x}_n}{dt^2} = -\boldsymbol{\nabla}_n (V + Q) \,, \tag{2.7}$$

where $\mathbf{p}_n = m_n \dot{\mathbf{x}}_n$ and with a 'quantum potential'

$$Q = -\sum_{n=1}^{N} \frac{1}{2m_n} \frac{\nabla_n^2 |\psi|}{|\psi|} \,. \tag{2.8}$$

Bohm postulated the equation of motion (2.7), with a Newtonian force $\mathbf{F}_n = -\boldsymbol{\nabla}_n (V + Q)$ acting on the nth particle. In Bohm's dynamics, in principle we have arbitrary initial conditions for positions and momenta. To obtain agreement with quantum theory, Bohm postulated the *initial* condition $\mathbf{p}_n = \boldsymbol{\nabla}_n S$ for the momenta as well as the initial condition $\rho = |\psi|^2$ for the distribution of positions, where both conditions are preserved in time (Bohm 1952a, pp. 170–171). In principle, in Bohm's dynamics we may consider an extended nonequilibrium in phase space with a general distribution

$$\rho_{\text{phase}}(q, p, t) \neq |\psi(q, t)|^2 \, \delta(p - \nabla S) \tag{2.9}$$

(Colin and Valentini 2014). In contrast, in de Broglie's dynamics, $\mathbf{p}_n = \boldsymbol{\nabla}_n S$ is the law of motion and $\rho = |\psi|^2$ is an initial condition, and so the most general initial conditions take the form (2.6) with the momenta fixed by (2.2).

De Broglie's original dynamics is easily generalised to an arbitrary system with a configuration-space wave function $\psi(q, t)$. The general Schrödinger equation $i\partial\psi/\partial t = \hat{H}\psi$ implies a continuity equation

$$\frac{\partial |\psi|^2}{\partial t} + \partial_q \cdot j = 0 \,, \tag{2.10}$$

where ∂_q is a generalised gradient and j depends on $\hat{H}$ (Struyve and Valentini 2009). We may then postulate a de Broglie equation of motion

$$\frac{dq}{dt} = v(q,t) = \frac{j(q,t)}{|\psi(q,t)|^2} \tag{2.11}$$

for trajectories $q(t)$. For an ensemble of systems with the same ψ, a general distribution $\rho(q,t)$ satisfies

$$\frac{\partial \rho}{\partial t} + \partial_q \cdot (\rho v) = 0 \ . \tag{2.12}$$

Again, since ρ and $|\psi|^2$ obey the same continuity equation, we have an equilibrium state $\rho = |\psi|^2$. But in principle we can consider initial nonequilibrium $\rho \neq |\psi|^2$.

The Schrödinger equation $i\partial\psi/\partial t = \hat{H}\psi$ can be derived from a Lagrangian $L = \int \mathcal{L}\, dq$, where

$$\mathcal{L} = \frac{i}{2}\left(\psi^*\dot{\psi} - \psi\dot{\psi}^*\right) - \frac{1}{2}\psi^*\hat{H}\psi - \frac{1}{2}\psi\hat{H}\psi^*$$

is invariant under global phase transformations

$$\psi \to \psi e^{i\theta} \tag{2.13}$$

(where θ is an arbitrary constant). Noether's theorem then implies the local conservation law (2.10) (Struyve and Valentini 2009). Thus the global symmetry (2.13) is the origin of the current j and of the de Broglie velocity (2.11).

This general formalism applies equally to systems of fields. It also applies to systems on a curved spacetime background, as long as the spacetime can be foliated by spacelike hypersurfaces $\Sigma(t)$ labelled by a global time parameter t (widely regarded as a basic requirement for physical spacetimes). The formalism is also readily extended to multi-component wave functions, as required for systems of particles with spin (Bell 1966; 1987, chapters 1, 15 and 17).

2.2.2 Quantum relaxation. Status of the Born rule

It has been argued that, in de Broglie's dynamics, the Born rule $\rho = |\psi|^2$ which we observe today can be understood as having arisen from a past process of 'quantum relaxation', in which initial nonequilibrium $\rho \neq |\psi|^2$ evolves towards the Born rule (on a coarse-grained level), much as in classical physics thermal equilibrium can be understood as arising from a past process of thermal relaxation (Valentini 1991a, 1991b, 1992, 2020; Valentini and Westman 2005).

For a general system with configuration q we may consider coarse-grained densities $\bar{\rho}(q,t)$ and $\overline{|\psi(q,t)|^2}$ (averages over small coarse-graining cells). Relaxation $\bar{\rho} \to \overline{|\psi|^2}$ can then be quantified by an H-function[1]

$$\bar{H}(t) = \int dq\ \bar{\rho}\ln\left(\bar{\rho}/\overline{|\psi|^2}\right), \tag{2.14}$$

with the properties $\bar{H} \geq 0$ and $\bar{H} = 0$ if and only if $\bar{\rho} = \overline{|\psi|^2}$, which obeys an H-theorem (Valentini 1991a)

[1]This is minus the relative entropy of $\bar{\rho}$ with respect to $\overline{|\psi|^2}$.

$$\bar{H}(t) \leq \bar{H}(0) , \tag{2.15}$$

assuming no fine-grained structure ($\bar{\rho}_0 = \rho_0$, $\overline{|\psi_0|^2} = |\psi_0|^2$) at $t = 0$. The quantity $\bar{H}(t)$ strictly decreases when ρ develops fine-grained structure ($\rho \neq \bar{\rho}$), which generally occurs for non-trivial velocity fields (Valentini 1992). Intuitively we may understand quantum relaxation as the mixing of two 'fluids', with densities ρ and $|\psi|^2$ in configuration space, which obey the same continuity equation and are 'stirred' by the same velocity field $\dot{q}$, eventually becoming indistinguishable on a coarse-grained level (Valentini 1991a).

Of course, in a time-reversal invariant dynamics, not all initial nonequilibrium states can relax to equilibrium. Clearly, there are non-relaxing initial conditions, for example with fine-grained structure or with trivial velocity fields. But as far as concerns the history of our actual universe, such initial conditions may be ruled out on empirical grounds (Valentini 2020).

By itself an H-theorem can only provide a general mechanism in terms of which we can understand the approach to equilibrium. Evidence for quantum relaxation comes mainly from numerical simulations, which show efficient relaxation when ψ is a superposition of multiple energy eigenstates (Valentini and Westman 2005; Towler, Russell, and Valentini 2012; Colin 2012; Abraham, Colin, and Valentini 2014; Lustosa, Pinto-Neto, and Valentini 2023). Chaotic features of the trajectories play an important role (Efthymiopoulos, Contopoulos, and Tzemos 2017; Drezet 2021).

For a broad range of examples we find an approximately exponential decay

$$\bar{H}(t) \approx \bar{H}_0 e^{-t/\tau} , \tag{2.16}$$

where the timescale τ depends on the initial quantum state and on the coarse-graining length (Towler, Russell, and Valentini 2012). Relaxation tends to be faster for larger numbers of superposed energy states. To understand when relaxation is likely to have occurred, we must bear in mind that all known laboratory systems have a long and violent astrophysical history stretching back to the big bang. Thus, for example, in the remote past an electron will have been in a superposition of a huge number of energy states, and we may therefore expect that in the early universe relaxation will have taken place rapidly and to high accuracy, thereby accounting for the Born rule today.

What we have said about quantum relaxation applies only to de Broglie's original version of the dynamics. In Bohm's dynamics, by contrast, initial nonequilibrium states (2.9) do *not* relax and are in fact unstable: small deviations $\rho \neq |\psi|^2$ from the Born rule can grow with time (Colin and Valentini 2014). Thus Bohm's dynamics is not only less elegant than de Broglie's, it is actually physically untenable. Even so, the expression (2.7) for the acceleration is sometimes useful. For example, in simple cases, the classical limit can be characterised by the condition $Q \approx 0$ (Holland 1993).

Historically, in the 1920s de Broglie simply assumed the Born rule on the grounds that it is preserved in time by the dynamics. Pauli (1953) and Keller (1953) argued that such a condition could play no fundamental role in a deterministic theory and should have a dynamical origin. In response, Bohm (1953) argued that an ensemble of two-level molecules subject to random collisions would relax to the Born rule. But no general argument for relaxation was provided. Soon afterwards, citing difficulties with understanding relaxation, Bohm and Vigier (1954) moved to a stochastic theory with (subquantum) 'fluid fluctuations' that could drive relaxation for a general system. Bohm's 1953 relaxation argument could work because he assumed the initial momenta $\mathbf{p}_n = \boldsymbol{\nabla}_n S$ and applied de Broglie's law of motion for the trajectories, thereby in effect employing de Broglie's dynamics. Had he instead employed his own Newtonian version, with an initial nonequilibrium (2.9), relaxation would not have been possible, and a move to a stochastic theory would still have been required.

As we have outlined, quantum relaxation does occur in de Broglie's original deterministic theory, and there is no need to propose a stochastic theory to explain the Born rule. In retrospect, it seems unfortunate that Bohm came so close to what we now know about relaxation but then took a different path. Since then, most authors have simply taken the Born rule as an additional postulate (Dürr, Goldstein, and Zanghì 1992; Holland 1993; Dürr and Teufel 2009; Goldstein 2021; Tumulka 2021, 2022). In our view, this is a mistake. In a deterministic dynamics there is a clear conceptual distinction between laws of motion and initial conditions. Laws are fixed features of nature. Initial conditions are contingencies to be determined empirically and are not to be fixed by theoretical or philosophical fiat. Whether our universe began in equilibrium or in nonequilibrium is an empirical question to be determined by cosmological observation (Valentini 2020).

Quantum relaxation in the early universe has been studied for a scalar field on expanding space (Colin and Valentini 2013, 2015, 2016). In Fourier space the equations for a mode $\mathbf{k}$ are just those for a two-dimensional oscillator of mass $m = a^3$ and angular frequency $\omega = k/a$, where $a(t)$ is the scale factor for the expansion (Valentini 2007). For short wavelengths (smaller than the Hubble radius) we find the usual rapid relaxation, while for long wavelengths (larger than the Hubble radius) relaxation is retarded. We then expect quantum noise to be suppressed at large cosmological scales. This has implications for inflationary cosmology, in which the temperature anisotropies in the cosmic microwave background (CMB) derive from perturbations $\phi_{\mathbf{k}}$ of a scalar 'inflaton' field at very early times (Liddle and Lyth 2000, Peter and Uzan 2009). The observed statistical properties of the CMB can constrain the primordial power spectrum $\mathcal{P}(k) = \frac{4\pi k^3}{V} \left\langle |\phi_{\mathbf{k}}|^2 \right\rangle$ for $\phi_{\mathbf{k}}$ (where V is a normalisation volume and $\langle ... \rangle$ is an average over a theoretical ensemble). Thus we can test the Born rule in the very early universe (Valentini 2010a).[2] A large-scale power deficit has been reported (Aghanim *et al.* 2016). But the data neither support nor rule out the deficit (as a function of k) expected from relaxation suppression (Vitenti, Peter, and Valentini 2019). More detailed predictions are needed to confirm or rule out a cosmological quantum relaxation model.

In inflationary cosmology, the matter in our universe was created by a process of inflaton decay at early times (Peter and Uzan 2009). Violations of the Born rule in the early inflaton field can then be passed on to the decay products, opening up the realistic possibility that relic nonequilibrium particles might still exist in our universe

[2] Assuming statistical isotropy and statistical homogeneity, we can constrain the theoretical ensemble by measurements on one universe (Valentini 2020, Section 4).

today (Underwood and Valentini 2015).

Thus, while quantum relaxation can explain why we observe the Born rule in the laboratory, in realistic cosmological scenarios it is also possible for primordial violations of the Born rule to have left their mark on the CMB and perhaps even to have survived to the present day in some systems.

2.2.3 Emergence of locality and uncertainty. Nonequilibrium nonlocal signalling

Pilot-wave theory is nonlocal, as required by Bell's theorem. For a pair of particles A and B with entangled wave function $\psi(\mathbf{x}_A, \mathbf{x}_B, t)$, the velocity $\mathbf{v}_A = \dot{\mathbf{x}}_A(t)$ depends instantaneously on $\mathbf{x}_B(t)$, no matter how far apart the particles may be. Furthermore, if we change the local Hamiltonian $\hat{H}_B$ of particle B, this can induce an instantaneous change in the remote velocity $\mathbf{v}_A$.

Remarkably, such direct action at a distance is erased upon averaging over an equilibrium ensemble. For a joint Born-rule distribution $\rho(\mathbf{x}_A, \mathbf{x}_B, t) = |\psi(\mathbf{x}_A, \mathbf{x}_B, t)|^2$, the marginal distribution $\rho_A(\mathbf{x}_A, t) = \int d^3\mathbf{x}_B\ \rho(\mathbf{x}_A, \mathbf{x}_B, t)$ at A does *not* depend on the local Hamiltonian $\hat{H}_B$ at B. Thus, in quantum equilibrium, locality emerges at the statistical level. It is worth pausing to note how remarkable this is. For individual entangled pairs, changing the Hamiltonian $\hat{H}_B$ at B has an instantaneous effect on the motion $\mathbf{x}_A(t)$ at A. And yet, for an equilibrium ensemble, the marginal $\rho_A(\mathbf{x}_A, t)$ at A is unaffected and in this sense we may say that the underlying nonlocal effects average to zero. However, this masking of nonlocality is a peculiarity of equilibrium. For general nonequilibrium ensembles with $\rho(\mathbf{x}_A, \mathbf{x}_B, t) \neq |\psi(\mathbf{x}_A, \mathbf{x}_B, t)|^2$, the marginal $\rho_A(\mathbf{x}_A, t)$ at A *can* respond instantaneously to a change in the local Hamiltonian $\hat{H}_B$ at B, resulting in statistical nonlocal signalling from B to A (Valentini 1991b, 2025c). On this view, our inability to harness entanglement for superluminal signalling is not a law of physics but a mere peculiarity of quantum equilibrium.[3]

Similar conclusions hold for the uncertainty principle. If we consider, for example, standard measurements of position and momentum, we find that the usual statistical dispersion relation $\Delta x \Delta p_x \geq \hbar/2$ can be broken for nonequilibrium ensembles (Valentini 1991b, 2025c). The uncertainty relations are not laws of physics but mere peculiarities of quantum equilibrium.

2.2.4 Beyond quantum death

A parallel may be drawn between our situation and that of hypothetical beings living in a state of global thermal equilibrium or 'heat death': in the absence of temperature differences, their inability to convert heat into work is not a law of physics but a mere peculiarity of thermal equilibrium. According to pilot-wave theory, we are similarly confined to a highly restrictive state of 'quantum death', whose limitations may appear to us to be fundamental when in fact they are mere contingencies (Valentini 1991a, 1991b, 1992, 1996).

On this view, quantum physics is only a special case of a much wider physics, which may have existed in the very early universe (before relaxation took place), and which may still exist today for some exotic systems (relic particles from very early times,

[3] Similar conclusions hold for general deterministic hidden-variables theories (Valentini 2002a).

or particles radiated by black holes) (Valentini 2007, 2010a, 2023; Underwood and Valentini 2015). In this wider nonequilibrium physics, the Born rule is broken, practical superluminal signalling is possible, and the uncertainty principle can be violated.

Should we one day discover nonequilibrium systems and bring them under our control, there will be radical technological implications. Nonequilibrium ensembles with narrow (sub-Born rule) dispersion could be harnessed to perform 'subquantum measurements', in which we obtain precise information about particle trajectories without collapsing or significantly disturbing the particle wave functions. This would enable us reliably to distinguish single copies of non-orthogonal quantum states, resulting in a breakdown of standard quantum-cryptographic protocols, and enabling new forms of computation (whose power remains to be rigorously determined) (Valentini 2002b, 2025c).

2.2.5 Quantum field theory

High-energy physics is commonly described by a manifestly Lorentz-covariant quantum field theory. The pilot-wave formulation has an underlying preferred rest frame, where nonlocality acts instantaneously with respect to a preferred time parameter t. Lorentz invariance emerges only in equilibrium, along with effective locality and uncertainty.[4]

We can illustrate the theory for a real massive scalar field (Bohm, Hiley, and Kaloyerou 1987; Valentini 1992; Holland 1993). The wave functional $\Psi[\phi, t]$ (for a configuration $q = \phi(\mathbf{x})$) obeys the Schrödinger equation (units $c = 1$)[5]

$$i\frac{\partial \Psi}{\partial t} = \int d^3x \, \frac{1}{2}\left(-\frac{\delta^2}{\delta\phi^2} + (\boldsymbol{\nabla}\phi)^2 + m^2\phi^2\right)\Psi \, . \tag{2.17}$$

This implies

$$\frac{\partial |\Psi|^2}{\partial t} + \int d^3x \, \frac{\delta}{\delta\phi}\left(|\Psi|^2 \frac{\delta S}{\delta\phi}\right) = 0 \tag{2.18}$$

(with $\Psi = |\Psi| \, e^{iS}$) and a de Broglie velocity $v = \delta S/\delta\phi$. We may then write an equation of motion

$$\frac{\partial \phi}{\partial t} = \frac{\delta S}{\delta \phi} \tag{2.19}$$

for trajectories $q(t) = \phi(\mathbf{x}, t)$. For an ensemble of fields with the same pilot wave Ψ, a general distribution $P[\phi, t]$ satisfies

$$\frac{\partial P}{\partial t} + \int d^3x \, \frac{\delta}{\delta\phi}\left(P \frac{\delta S}{\delta\phi}\right) = 0 \, . \tag{2.20}$$

As usual, the Born rule $P[\phi, t] = |\Psi[\phi, t]|^2$ describes an equilibrium ensemble, for which we obtain the usual results of quantum field theory. For more general ensembles, with $P[\phi, t] \neq |\Psi[\phi, t]|^2$, we obtain deviations from quantum field theory.

[4] For a review of pilot-wave high-energy physics see Valentini (2024).

[5] Functional derivatives $\delta\Psi/\delta\phi(\mathbf{x})$ are defined by $\delta\Psi = \int d^3\mathbf{x} \; [\delta\Psi/\delta\phi(\mathbf{x})] \, \delta\phi(\mathbf{x})$ for infinitesimal $\delta\phi(\mathbf{x})$.

If Ψ is entangled with respect to widely-separated field elements $\phi(\mathbf{x})$, $\phi(\mathbf{x}')$, the time evolution of the field at $\mathbf{x}$ can depend instantaneously on the local Hamiltonian applied at $\mathbf{x}'$. For a nonequilibrium ensemble, the marginal distribution at $\mathbf{x}$ can respond instantaneously to a change in the local Hamiltonian at $\mathbf{x}'$, resulting in statistical nonlocal signalling from $\mathbf{x}'$ to $\mathbf{x}$. Such signals define an absolute simultaneity labelled by t. However, as usual, the nonlocal signals are erased upon averaging over an equilibrium ensemble.

For an equilibrium ensemble the statistical predictions follow from (2.17). Transforming to the Heisenberg picture, we obtain a Lorentz-covariant operator field equation

$$\partial_t^2\hat{\phi} - \nabla^2\hat{\phi} + m^2\hat{\phi} = 0 \ , \tag{2.21}$$

with the effective symmetries of Minkowski spacetime.

2.3 An unacceptably radical physics?

We now consider our first broad category of objections to pilot-wave theory: that it constitutes an unacceptably radical physics. This kind of objection tends to centre on three main themes:

1. De Broglie's dynamics is based on a law for velocity, not acceleration, resulting in 'bizarre' trajectories.
2. The trajectories seem to conflict with expectations from the quantum theory of 'measurement'.
3. The underlying dynamics conflicts with standard symmetries and conservation laws.

2.3.1 A nonclassical dynamics

It is often not appreciated that, in the 1920s, de Broglie constructed a new, non-Newtonian dynamics with a law of motion for velocity (Bacciagaluppi and Valentini 2009, Chapter 2). Among other motivations de Broglie had argued that, because a diffracting photon does not move in a straight line, Newton's first law of motion must be abandoned (de Broglie 1923). As a guide to formulating his new dynamics, de Broglie sought to unify the variational principles of Maupertuis ($\delta\int m\mathbf{v}\cdot d\mathbf{x} = 0$, for a particle with velocity $\mathbf{v}$) and of Fermat ($\delta\int dS = 0$, for a wave with phase S), resulting in the guidance equation (2.2).

Unsurprisingly, the trajectories predicted by de Broglie's first-order dynamics do not conform with expectations and intuitions developed from our experience with Newton's second-order dynamics with a law for acceleration. Some authors find this objectionable. It is as if a theory of particle trajectories is acceptable only if the trajectories behave in a manner that is recognisably Newtonian – which of course misses the point that in 1923 de Broglie abandoned Newton's first law and set out to construct a new theory of motion.

Let us consider some illustrative examples of this kind of objection.

Bohm and Hiley argued that pilot-wave theory is unsatisfactory because it predicts a vanishing velocity for an electron in the ground state of hydrogen, a result that is 'counter to our physical intuition' since 'the equilibrium state in an atom should be the result of a dynamic process and should not represent a completely static situation' (Bohm and Hiley 1989, p. 99). It was further argued that this 'unsatisfactory feature' may be avoided by adopting a stochastic modification of the theory (as originally considered in 1954 by Bohm and Vigier in order to explain relaxation to the Born rule). Clearly, the physical intuition being appealed to here is grounded in classical mechanics, where a bound state with a central force may be understood in terms of orbits with non-zero velocity and acceleration. But there is no reason why such intuitions should apply to a dynamics based on velocities. For the ground state of hydrogen, the Aristotelian force ∇S vanishes and de Broglie's law of motion tells us that the electron will be at rest.

On a related note, Holland has suggested that Bohm's second-order dynamics with a quantum potential has an 'explanatory power' that is lost if one makes use of de Broglie's first-order dynamics (Holland 1993, p. 78). The thinking seems to be that Newtonian forces are explanatory while Aristotelian forces are not. But causal explanations need not be given in terms of Newtonian forces that determine acceleration, they can just as well be given in terms of Aristotelian forces that determine velocity. For example, if a particle slows down and comes to rest at a certain point in space, the causal explanation provided by de Broglie's dynamics is simply that the Aristotelian force ∇S vanishes there.

As another example, Cohen *et al.* (2020) express concern that pilot-wave theory violates Newton's first law (the principle of inertia), and propose that this and related disturbing features be avoided by adopting a retrocausal model. There is, however, no logical reason why a non-Newtonian dynamics should respect Newtonian principles – on the contrary, we should expect the opposite.

These examples illustrate a tendency among some physicists to insist on thinking in classical terms about a theory that is fundamentally nonclassical. It cannot be sufficiently stressed that, in the 1920s, de Broglie constructed a radically new and non-Newtonian dynamics that violates many basic tenets of classical theory, including the 'principle of inertia'. To expect the trajectories of pilot-wave dynamics to conform to intuitions derived from classical physics is as unjustified as it would be to expect particle trajectories in general relativity to conform to intuitions derived from Newton's theory of gravity.

As yet another example we may consider the question, frequently encountered in private discussion, of why the often erratic, accelerating de Broglie–Bohm trajectories of charged particles do not give rise to large amounts of radiation. Here there is an implicit assumption that in pilot-wave theory a moving charge will interact with the electromagnetic field as if it were a classical particle. In fact, if one considers for example a low-energy charged particle interacting with the quantised electromagnetic field, we will have a pilot wave $\Psi[\mathbf{x}, \mathbf{A}, t]$ that is a function of the particle position $\mathbf{x}$ and a functional of the vector potential $\mathbf{A}$. The motion of the particle will be given by de Broglie's guidance equation for $\dot{\mathbf{x}}$, not by the Lorentz force law. Similarly, the time evolution of $\mathbf{A}$ will be given by de Broglie's guidance equation for $\dot{\mathbf{A}}$, not by Maxwell's equations. If we consider the motion of a single charged particle, our intuitions from classical electrodynamics must be set aside. If we go on to consider an equilibrium ensemble, the statistical results will be the same as in standard quantum physics – including for any radiation that may or may not be emitted.[6]

Assumptions taken from classical physics sometimes surface in more subtle ways. Some authors find it disturbing that while the pilot wave influences the motion of the system the evolution of the pilot wave itself is not affected by the system. For example, Anandan and Brown (1995) contrast this situation with general relativity, where the geometry of spacetime not only acts on matter but is acted upon by it as well, arguing that pilot-wave theory does not satisfy the 'action-reaction principle' (that a real entity must be capable of being influenced as well as of influencing), thereby casting doubt on the reality of the assumed pilot wave. But we have said that the pilot wave may be regarded as a new kind of causal agent, which exists in configuration space, and which is unlike anything previously encountered in physics. We ought then to beware of comparisons with the classical theory of fields and gravitation, grounded as they are in spacetime. Is the 'action-reaction principle' a general epistemological principle, or is it a mere prejudice inherited from classical physics? To understand the pilot wave, perhaps we must learn to think in ways that are far removed from the physics of Newton, Maxwell, and Einstein.

On a related note, some authors find it disturbing that in pilot-wave dynamics there is no natural definition of a conserved energy. Again, this kind of concern arguably arises from irrelevant intuitions and expectations drawn from classical physics (Section 2.3.3.2).

2.3.2 A nonclassical theory of measurement

As first noted by Bohm (1952b), in pilot-wave theory so-called quantum 'measurements' usually do not indicate the correct value of some pre-existing property of the system (position measurements being a notable exception). This has often proved to be a source of confusion. The logical conclusion to draw is that, in general, quantum 'measurements' simply are not correct measurements. But in that case, why are they still called 'measurements'? And why do we consider them at all? Furthermore, how could we go about performing measurements correctly? The literature tends to be silent on these issues. Bell (1990) highlighted the misuse of the word 'measurement' in quantum physics and suggested it be replaced by the more neutral word 'experiment'. Here we attempt to clarify the nature of measurement in pilot-wave theory.

2.3.2.1 What is a measurement?. Generally speaking, to perform a correct measurement requires some understanding of the relevant physics. For example, to design and construct a moving-coil ammeter to measure electrical current we need some basic knowledge of mechanics and of electromagnetism. We need to know that the magnetic field inside the device will exert a force on the wire carrying the current, thereby displacing the pointer from which we can read off the value of the current. Otherwise we would not be able to design the device correctly, nor would we know how to use it correctly. As philosophers often put it: observation is theory-laden.

[6] In, for example, a two-slit experiment with electrons, the quantum expectation value $\langle \ddot{\mathbf{x}} \rangle$ of the acceleration is non-zero (because of the potential barrier with two slits), and so in conventional quantum theory a small amount of radiation will be emitted.

If our understanding of the physics is mistaken, the 'measurement procedure' is unlikely to be a correct measurement. The interpretation of the experimental results will then be mistaken. For example, in 1887 Michelson and Morley designed and performed an experiment to measure the velocity of the earth relative to the ether. The apparatus – an interferometer – was constructed on the basis of pre-relativistic physics. The null result seemed to imply that the earth was at rest in the ether. Today we would say that their understanding of the physics was mistaken, and for that reason they simply were not measuring what they thought they were measuring.

All this begs the question: today when we perform 'quantum measurements' on microscopic systems, are we really measuring what we think we are measuring? According to pilot-wave theory, the answer is generally 'no'. So-called quantum 'measurements' are usually not correct measurements, and this is because physicists frequently misunderstand the basic physics of quantum systems.

As Einstein put it, in conversation with Heisenberg in 1926: 'It is the theory which decides what we can observe' (Heisenberg 1971, p. 63). How we design and perform measurements depends on which theory we believe to be a correct description of the physical world. If we assume, for the sake of argument, that pilot-wave theory is a correct account of currently-known physics (at least to a good approximation), we should look to pilot-wave theory to design and interpret measurements. What, then, does pilot-wave theory have to say about the experimental operations widely known as 'quantum measurements'?

2.3.2.2 Perspective on quantum 'measurement'. I. A so-called quantum 'measurement' for an observable $\hat{\omega}$ occurs by coupling the system (with configuration q) to an apparatus pointer y via an interaction Hamiltonian

$$\hat{H}_{\mathrm{I}} = a\hat{\omega}\hat{p}_y \ , \tag{2.22}$$

where a is a coupling constant and $\hat{p}_y = -i\partial_y$. We take a so large that it dominates the Hamiltonian. We assume that the initial system wave function is a superposition $\psi_0(q) = \sum_n c_n\phi_n(q)$ of eigenstates ($\hat{\omega}\phi_n = \omega_n\phi_n$) and that the initial pointer packet $g_0(y)$ is narrowly-peaked around $y = 0$. The initial joint wave function $\Psi_0(q, y) = \psi_0(q)g_0(y)$ then evolves to

$$\Psi(q, y, t) = \sum_n c_n\phi_n(q)g_0(y - a\omega_n t) \ . \tag{2.23}$$

At sufficiently large t the different 'branches' separate in y-space.

According to quantum mechanics, a random 'collapse' to the nth branch occurs with probability $p_n = |c_n|^2$. According to pilot-wave theory, there is no collapse. Instead, once the branches have separated, the configuration $Q(t) = (q(t), y(t))$ can occupy only one branch, in which case the velocity $\dot{Q}$ is determined only by that branch. Each term in (2.23) is separable in q and y, so the system velocity $\dot{q}$ is determined only by $\phi_n(q)$. Thus the system has acquired an effective 'collapsed' wave function

$$\psi_{\mathrm{coll}} = \phi_n(q) \ . \tag{2.24}$$

Note that the 'measurement of $\hat{\omega}$' amounts to an approximate position measurement for y, so as to indicate the occupied branch.

For an equilibrium ensemble, with distribution $P(Q,t) = |\Psi(Q,t)|^2$, collapse to ϕ_n occurs with probability $p_n = |c_n|^2$ (this being the probability for $Q(t)$ to occupy the nth branch). For a nonequilibrium ensemble, with $P(Q,t) \neq |\Psi(Q,t)|^2$, in general $p_n \neq |c_n|^2$. The usual Born rule is obtained only in equilibrium.

What has the above experiment actually achieved? According to pilot-wave theory, the system has simply acquired an effective wave function ϕ_n with eigenvalue ω_n. This does not usually imply that the system initially had a physical property with value ω_n, nor does it necessarily imply that the system now has a physical property with value ω_n. This is easily seen from inspection of simple examples.[7]

Consider a particle with position x in one dimension, coupled to a pointer y as in (2.22). For $\hat{\omega} = \hat{x}$, it is found that the particle remains at rest during the interaction, while the displacement of y is proportional to x. For this case, then, the final value of y correctly indicates the (pre-existing) value of x. However, for $\hat{\omega} = \hat{p}$, and taking for example $\psi_0(x) \propto e^{ipx} + e^{-ipx}$, the final pointer reading indicates outcomes $\pm p$, whereas the true initial momentum $p_0 = \partial S_0/\partial x = 0$, so in this case the experiment has not correctly measured the momentum. Instead, the particle has been accelerated to a final momentum $\pm p$ (depending on the final effective wave function $e^{\pm ipx}$). In this case we can regard the experiment as a correct *preparation* of a state in which the particle has momentum $\pm p$.

It might be thought that the last conclusion could hold generally, that a quantum 'measurement' of $\hat{\omega}$ could be reinterpreted as a preparation of a system with an actual property ω. But this is generally not correct either. For example, for kinetic energy $\hat{\omega} = \hat{p}^2/2m$, again with $\psi_0(x) \propto e^{ipx} + e^{-ipx}$, we have an initial eigenstate of $\hat{\omega}$ with eigenvalue $\omega = p^2/2m$. The 'measurement' always gives the outcome $p^2/2m$, whereas the true initial kinetic energy $p_0^2/2m = 0$. Furthermore, at the end of the experiment the effective wave function of the particle is still $\psi_0(x)$, with eigenvalue $\omega = p^2/2m$, and the true kinetic energy is still zero. Thus, in this case, the experiment is neither a correct measurement nor a correct preparation.

Finally, for a particle in three dimensions, we may cite the example of a standard Stern–Gerlach experiment, in which an incoming spin-1/2 particle traverses an inhomogeneous magnetic field. The upwards or downwards deflection of the particle is usually taken to indicate a measurement outcome 'spin up' or 'spin down' (along the axis defined by the magnetic field). However, in Bell's pilot-wave model of spin-1/2, the particle itself does not have any spin-like degree of freedom, it simply has a trajectory guided by a two-component wave function $\psi_s(\mathbf{x},t)$ $(s = +,-)$ (Bell 1966). The components ψ_+ and ψ_- separate as they traverse the magnetic field. Depending on its initial position, the particle ends in the support of one or other component. The upwards or downwards displacement of the particle is regarded, in quantum mechanics, as effecting a 'measurement of spin'. However, in this case, the measurement is not merely incorrect, it is actually meaningless, for in Bell's model the particle does not have a spin, neither before the experiment nor afterwards.

[7]Over an equilibrium ensemble the distribution of final pointer positions of course agrees with standard quantum mechanics. The experimental results are the same, it is only the interpretation and significance of those results that differs.

2.3.2.3 Perspective on quantum 'measurement'. II. According to pilot-wave theory, then, so-called 'quantum measurements' are really experiments that cause the system to acquire a final effective wave function ϕ_n. They do not usually tell us the value of some pre-existing property of the system (nor do they necessarily tell us the final value of some property of the system). Why, then, are they called 'measurements'? From a pilot-wave perspective, that terminology is entirely inappropriate (except for position measurements). In physics generally, the word 'measurement' refers to an experiment whose outcome tells us the value of a pre-existing property – such as the current flowing in a wire, or the mass of an object. If a certain experiment does not tell us the correct value of a pre-existing property, then it is simply nonsensical to refer to it as a 'measurement'.

It may seem unacceptably radical to conclude that what are widely believed to be measurements are not really measurements. But in fact we can turn the point around: what is, or historically was, the justification for believing that 'quantum measurements' are correct measurements in the first place?

The answer has a philosophical element. As is well known, Bohr and Heisenberg believed that the language or framework of classical physics inevitably plays a fundamental role in our description of experiments. In this both physicists were influenced by the philosophy of Immanuel Kant who, in the late eighteenth century, argued that Newtonian physics was in some sense built into the structure of the human mind, and that we are unable to discuss the physical world without reference to Newtonian concepts (Heisenberg 1958; Murdoch 1987). For Bohr and Heisenberg, this meant that any discussion of how to design and construct a measuring apparatus, and of how to use it to perform a measurement, had to be framed in terms of classical physics. As Bohr once put it:

> The unambiguous interpretation of any measurement must be essentially framed in terms of the classical physical theories, and we may say that in this sense the language of Newton and Maxwell will remain the language of physicists for all time (Bohr 1931).

We now know that this is not correct. Quantum measurements can be framed (for example) in terms of pilot-wave theory, which is highly nonclassical. But in the early twentieth century the views of Bohr and Heisenberg dominated the interpretation of quantum physics. As a result, quantum measurements were modelled on a formal analogy with classical measurements. Ultimately, then, the belief that quantum measurements are correct measurements rested on the (perceived) general validity of classical measurement theory. And yet: we know that classical physics does not apply to quantum systems.

In effect, Bohr and Heisenberg insisted on applying a superseded (classical) physics to design appropriate measurement procedures for quantum systems. But as we have noted, without a correct understanding of the relevant physics we are unlikely to be measuring what we think we are measuring. This happened to Michelson and Morley in the late nineteenth century, and (in our view) it happened to Bohr and Heisenberg in the early twentieth century. Because quantum measurements are designed on the basis of an incorrect (classical) physics, it is only to be expected that the resulting experiments are not correct measurements (Valentini 1992, 1996, 2010b).

Let us consider an example. Why is the Stern–Gerlach experiment widely regarded as a (correct) measurement of 'spin'? From the early historical development of quantum physics, the answer is clear enough. The Stern–Gerlach experiment was assumed to measure spin correctly for a quantum particle because the same experiment *would* measure spin correctly if the particle were instead a classical spinning particle with magnetic dipole moment $\boldsymbol{\mu}$. From classical electrodynamics we know that the (classical) Hamiltonian interaction $H_{\rm I} = -\boldsymbol{\mu} \cdot \mathbf{B}$, with a magnetic field $\mathbf{B}$ that varies with z, will generate a net force pushing the particle up or down the z-axis depending on the direction of $\boldsymbol{\mu}$. Classically, the deflection of the particle correctly measures $\boldsymbol{\mu}$. If we believe that quantum measurements should be modelled on classical measurements, we are led to assume that the same experiment will correctly measure $\boldsymbol{\mu}$ even for a quantum particle. Formally, we can simply promote classical variables to operators, so that $H_{\rm I} = -\boldsymbol{\mu}\cdot\mathbf{B}$ in Hamilton's equations is replaced by $\hat{H}_{\rm I} = -\hat{\boldsymbol{\mu}}\cdot\mathbf{B}$ in the Schrödinger equation (where $\hat{\boldsymbol{\mu}}$ is the operator dipole moment).

Generally speaking, the interaction Hamiltonian $\hat{H}_{\rm I} = a\hat{\omega}\hat{p}_y$ for a quantum measurement of an observable $\hat{\omega}$ formally resembles the *classical* interaction that *would* describe a correct measurement of ω if the system were classical. To see this, consider a classical system with configuration x and canonical momentum p_x, coupled to a pointer y by $H_{\rm I} = a\omega p_y$ with $\omega = \omega(x, p_x)$. For large a Hamilton's equations

$$\frac{d\omega}{dt} = \{\omega, H_{\rm I}\} = 0 \quad , \quad \frac{dy}{dt} = \{y, H_{\rm I}\} = a\omega$$

(where $\{,\}$ is the Poisson bracket) imply trajectories

$$\omega(t) = \omega_0 \; , \qquad y(t) = y_0 + a\omega_0 t \; .$$

From the pointer displacement $a\omega_0 t$ we can deduce ω_0. Thus the assumed interaction describes a correct measurement of ω – for a classical system. But there is no reason why an analogous experiment should be a correct measurement of ω for a nonclassical system. In fact, we should expect the contrary. And yet, for quantum systems a correct measurement of ω is widely believed to be implemented by the replacement

$$H_{\rm I} = a\omega p_y \longrightarrow \hat{H}_{\rm I} = a\hat{\omega}\hat{p}_y \; . \tag{2.25}$$

This is *called* a 'measurement of ω' because it formally resembles a classical measurement of ω. But for a nonclassical system, such an experiment is unlikely to be a correct measurement.

2.3.2.4 Further remarks. A similar theme underlies debates about measurements of the path taken by a quantum particle. In quantum mechanics, if the response of a 'which-way' detector is appropriately delayed, the apparent measured path can disagree sharply with the path expected from pilot-wave theory (Englert *et al.* 1992; Valentini 1992, pp. 22–24; Dewdney, Hardy, and Squires 1993; Aharonov and Vaidman 1996). It has been claimed on this basis that de Broglie–Bohm trajectories are 'surrealistic' and illusory (Englert *et al.* 1992). But this claim assumes the validity of conventional quantum measurement theory. One can just as well turn the argument around, and use such examples to illustrate further how quantum 'measurements' tend to be mistaken (Valentini 1992, pp. 22–24). In quantum equilibrium, pilot-wave theory predicts the same probabilities for detector firings as standard quantum mechanics. But the interpretation of those firings can be quite different, and in particular their interpretation as quantum 'which-path measurements' can be incorrect.

Citing similar delayed which-way experiments, Aharonov and Vaidman (1996) criticised pilot-wave theory for not satisfying the following 'principle': an operation that constitutes a correct measurement of position in quantum theory must always constitute a correct measurement of position in pilot-wave theory. But, bearing in mind that observation is theory-laden, there is no motivation or foundation for such a principle. One could just as well claim that a correct measurement of position in Newtonian gravity must always constitute a correct measurement of position in general relativity – a claim that is falsified on a daily basis by the need to take general-relativistic corrections into account for the accurate functioning of the satellite-based Global Positioning System. There is no reason to expect measurement operations from older theories to apply to a new domain. It is the theory that tells us how measurements are to be performed, and different theories will usually entail different measurement procedures for many observables.

We have already seen how pilot-wave theory itself can tell us which operations are correct measurements and which are not. There is a further class of operations which can in principle be considered, in which nonequilibrium systems are employed to perform measurements. It can be shown that, if we are given an ensemble of apparatus pointers y with initial wave function $g_0(y)$ and with an arbitrarily narrow initial nonequilibrium distribution $\pi_0(y) \neq |g_0(y)|^2$ (much narrower than the standard quantum width), then by coupling to a particle position x for a very short time it is possible to measure x without disturbing its initial wave function $\psi_0(x)$ (to arbitrary accuracy) (Valentini 2002b). By performing a sequence of such 'subquantum measurements' it is possible to track the particle trajectory $x(t)$ without significantly affecting the wave function, in gross violation of quantum measurement theory (which would entail an effective collapse of $\psi_0(x)$ already after the first measurement).

As so often in pilot-wave theory, a proper perspective on quantum 'measurement' requires that we consider the more general physics of quantum nonequilibrium. It is apparent that the usual restrictions – the uncertainty principle, effective wave function collapse, and so on – are not fundamental but mere features of the equilibrium state. In principle, they can be avoided, if only we are able to discover and control nonequilibrium ensembles. Clearly, in pilot-wave theory, the so-called quantum theory of 'measurement' has no fundamental status. It is merely a convenient equilibrium phenomenology, or book-keeping device, employed to describe certain kinds of experiments.

2.3.3 Emergent Lorentz symmetry and conservation laws

Other perennial objections concern the status of Lorentz symmetry and of energy-momentum conservation in pilot-wave theory.

2.3.3.1 The question of Lorentz invariance. In pilot-wave field theory there is a preferred state of rest with an absolute time t. Lorentz invariance is an emergent symmetry of quantum equilibrium ensembles. We saw an example in Section 2.2.5. In the Heisenberg picture the operator field equation (2.21) has the effective symmetries of Minkowski spacetime, even though the underlying field dynamics (for individual systems) is not Lorentz covariant.

Many commentators find this too radical. It is often asserted that, if pilot-wave theory is to be taken seriously, we must construct a fundamentally Lorentz-invariant formulation. There have been numerous attempts, though without success. A preferred rest frame is required in order to define the dynamics and the quantum equilibrium distribution consistently (Hardy 1992; Berndl *et al.* 1996). It is possible to write the equations so that the definition of the preferred frame is formally Lorentz covariant (Dürr et al. 1999; Tumulka 2007; Dürr *et al.* 2014). This can easily be achieved, for example, if the preferred frame coincides with the mean rest frame (Vigier 1985; Dürr *et al.* 2014). However, even if the equations are formally Lorentz covariant, there is still a physical preferred frame.

There seem to be several misunderstandings behind the insistence on fundamental Lorentz invariance. Firstly, it should be recognised that, as we enter deeper domains of physics, the kinematics often changes hand in hand with the dynamics – as witnessed in particular by the historical transitions from Aristotelian to Newtonian physics, from Newtonian physics to special relativity, and from special relativity to general relativity. It is hardly coherent to be open to fundamental changes in dynamics while being closed to fundamental changes in kinematics, especially since kinematics is really a description of universal features of dynamics (features that apply to all bodies irrespective of their mass or composition). Secondly, when judging a theory that attempts to describe a new physical domain, it seems unsound to assert that the theory is unacceptable unless it extends all the symmetries of presently-known physics into the deeper domain. Scientifically speaking, we can only require that known symmetries emerge phenomenologically at the level at which they are in fact known to apply. Thirdly, because (from a pilot-wave perspective) quantum theory is only an equilibrium theory, it is hardly surprising that the state of statistical equilibrium described by the Born rule should possess more physical symmetries than does the underlying physics (Valentini 1992).

The widespread nervousness at abandoning fundamental Lorentz invariance may be due in part to a tendency to think of spacetime as a given background on which we write the laws of physics. On this view, it might be said that physical systems respect Lorentz invariance 'because' spacetime has a (local) Minkowskian structure. But an examination of the historical and experimental basis for relativity physics suggests that this view be turned around: in fact, we believe that spacetime is Minkowskian *because* in our experience physical systems are seen to respect Lorentz invariance. Should we eventually discover conditions under which Lorentz invariance is broken, we will simply have to revise the structure we assign to spacetime.

In fact, the basic physics of pilot-wave theory strongly suggests that Lorentz invariance cannot be valid at the fundamental level. There are two main reasons. First, the theory predicts nonequilibrium nonlocal signals, which in principle provide an operational definition of absolute simultaneity with a preferred global time t (Valentini 1991b, 2008). Second, the first-order (velocity-based) dynamics of pilot-wave theory naturally defines an Aristotelian kinematics – or Aristotelian spacetime – with a preferred state of rest (Valentini 1997).

The last point is clear even at low energies. Galilean invariance is a fictitious symmetry – analogous to the fictitious symmetry of Newtonian mechanics under transformation to a uniformly-accelerated frame. Under a Galilean transformation to a frame with velocity $\mathbf{v}$ the Aristotelian force $\mathbf{f}_n = \boldsymbol{\nabla}_n S$ transforms to $\mathbf{f}'_n = \mathbf{f}_n - m_n\mathbf{v}$ – just as, in Newtonian mechanics, under a transformation to a frame with acceleration $\mathbf{a}$ the Newtonian force $\mathbf{F}_n$ transforms to $\mathbf{F}'_n = \mathbf{F}_n - m_n\mathbf{a}$. In both cases the extra forces are mass proportional and affect all particles equally, hence they are 'fictitious forces'. Such formal symmetries have only mathematical significance, and are not part of the physical spacetime symmetry group (Valentini 1997).[8]

In our view, the widespread (artificial) restriction of pilot-wave theory to equilibrium makes fundamental Lorentz invariance seem much more plausible than it really is. For as long as the Born rule remains valid, nonlocal effects between entangled systems average to zero and it is not possible to send practical nonlocal signals, averting a head-on clash with the kinematics of special relativity. In equilibrium we have what Shimony (1984) called a 'peaceful coexistence' with relativity. However, outside that restricted domain, practical superluminal signalling becomes a reality, with the obvious clash with Lorentz symmetry. Minkowski spacetime can only be an effective spacetime valid in quantum equilibrium. For more general ensembles, the true (Aristotelian) structure of spacetime will become apparent.

It is instructive to draw a comparison with attitudes to Lorentz invariance in other areas of physics. There are plausible models of particle physics (Kostelecký and Mewes 2002) and of quantum gravity (Hořava 2009) in which Lorentz invariance breaks down at high energies. As yet there is no evidence for such breakdown, but it seems reasonable to explore the possibility both theoretically and experimentally. In high-energy physics, Lorentz invariance is merely one important symmetry of the Lagrangian, among others, which may or may not be fundamental. In contrast, attitudes in quantum foundations tend to be more dogmatic. This is ironic because it is in quantum foundations that we have the strongest evidence against fundamental Lorentz invariance (the evidence for nonlocality, deduced from violations of Bell's inequality).

In pilot-wave field theory, the emergent spacetime symmetries depend on the form of the Hamiltonian. For example, if in the Schrödinger equation (2.17) we replace $(\boldsymbol{\nabla}\hat{\phi})^2$ by $-(\boldsymbol{\nabla}\hat{\phi})^2$, we obtain the operator wave equation (2.21) with $-\nabla^2\hat{\phi}$ replaced by $\nabla^2\hat{\phi}$. The effective symmetries then correspond to Euclidean spacetime. Experimentally, for Hamiltonians found in nature, the emergent spacetime symmetries correspond to Minkowksi spacetime. Perhaps there are deeper reasons for this. If instead we consider models in which Lorentz invariance breaks down at high energies, then

[8]It is sometimes claimed that Galilean invariance plays a role in selecting the form of the guidance equation (2.2) (Dürr, Goldstein, and Zanghì 1992). In fact, the de Broglie velocity originates quite generally from the global phase symmetry (2.13) on configuration space and not from any symmetry of spacetime.

in pilot-wave theory we will find that Lorentz invariance emerges as an equilibrium symmetry only at low energies.

Finally, on a point of historical perspective, it is worth noting that renormalisability was once widely regarded as a fundamental constraint on viable theories of high-energy physics. Today, from the point of view of effective field theory, renormalisability is generally regarded as a mere low-energy screening phenomenon (Weinberg 1995). For all we know, Lorentz invariance might eventually suffer a similar fate.

2.3.3.2 Conservation laws. Similar reasoning applies to the standard conservation laws which, like boost invariance, are intimately related to the second-order nature of classical physics. Because pilot-wave dynamics is first-order in time, there is no natural definition of a conserved energy (arising as a first integral of second-order equations of motion). We recover the usual conservation laws only for equilibrium ensembles.

This is easily illustrated for a single particle. From Bohm's acceleration equation (2.7), and defining a total energy $E = \mathbf{p}^2/2m + V + Q$, classically speaking Q acts like an external source of energy and momentum which are transmitted to the particle at respective rates $\partial Q/\partial t$ and $-\boldsymbol{\nabla} Q$. These rates vanish, however, when averaged over an equilibrium ensemble:

$$\int d^3\mathbf{x}\ |\psi|^2 \frac{\partial Q}{\partial t} = \int d^3\mathbf{x}\ |\psi|^2 \boldsymbol{\nabla} Q = 0\ . \tag{2.26}$$

If we consider instead a nonequilibrium ensemble, the means $\langle E \rangle$ and $\langle \mathbf{p} \rangle$ satisfy (Valentini 1992)

$$\frac{d\langle E \rangle}{dt} = \frac{d\left\langle \hat{H} \right\rangle}{dt} + \int d^3\mathbf{x}\ (\rho - |\psi|^2) \frac{\partial Q}{\partial t}\ ,$$
$$\frac{d\langle \mathbf{p} \rangle}{dt} = \frac{d\langle \hat{\mathbf{p}} \rangle}{dt} + \int d^3\mathbf{x}\ (\rho - |\psi|^2)(-\boldsymbol{\nabla} V - \boldsymbol{\nabla} Q)\ ,$$

where $\left\langle \hat{H} \right\rangle$ and $\langle \hat{\mathbf{p}} \rangle$ are the usual quantum averages. We recover the usual statistical conservation laws only in the equilibrium limit $\rho = |\psi|^2$.

It is noteworthy that, even in standard quantum mechanics, energy-momentum conservation generally holds only at the statistical level and not for individual systems (with the exception of eigenstates).

In Bohm's dynamics, Q acts like an external energy source. Holland (1993, p. 120) suggests that this is unsatisfactory and that the theory should be repaired by incorporating a back reaction from the configuration q to the wave ψ. In our view this concern arises from rewriting de Broglie's dynamics in an artificial pseudo-Newtonian form, which as we have noted is in any case unstable and therefore physically untenable.

There is no general reason of principle why de Broglie's non-Newtonian dynamics, based on velocities, should have a fundamental conserved energy, just as there is no reason why it needs to respect fundamental boost invariance. After all, even in classical general relativity there is no general law of (global) energy conservation (Misner, Thorne, and Wheeler 1973, Chapters 19 and 20). This is not a difficulty for general relativity; nor is it a difficulty for pilot-wave theory. As long as energy conservation, and the other standard conservation laws, are recovered at the quantum and classical levels, pilot-wave theory agrees with experiment and remains viable.

2.4 An insufficiently radical physics?

It is sometimes said that pilot-wave theory is insufficiently 'radical'. In most scientific contexts – from cosmology and particle physics to genetics and molecular biology – such an objection would seem peculiar. After all, the aim of science is to explain the phenomena in as simple and comprehensible a fashion as possible. When commentators state that pilot-wave theory is not sufficiently radical, they are often misunderstanding the theory, or else emphasising certain features (such as discreteness, relational structures, or subjectivism) which they favour and which this theory happens not to share.

Historically, however, there was one noteworthy rejection of pilot-wave theory which in its time did amount to a compelling case that pilot-wave theory was insufficiently radical. After its revival by Bohm in 1952, pilot-wave theory was famously dismissed by Einstein – in a private letter to Born – as 'too cheap' (Einstein 1952). In a similar vein, at a seminar on pilot-wave theory delivered by Max Dresden at Princeton in the 1950s, Einstein commented that he found the theory 'too trivial to be true'.[9] Einstein was not one to have groundless scientific opinions. What might have been the reasoning behind these remarks?

2.4.1 Einstein and pilot-wave theory

It is not widely known that, from as early as 1909 and up until 1927, Einstein himself developed ideas that were similar to pilot-wave theory. However, he encountered difficulties with separability and locality – the requirements that reality should separate into spatially-disjoint parts that interact only locally. Despite frequently returning to these ideas, he never published his results.

Around 1909 Einstein was working on a theory in which each light quantum is a singularity surrounded by a spatially extended vector field that affects the motion of the quantum (Bacciagaluppi and Valentini 2009, pp. 197–200). The electromagnetic field was supposed to emerge from the collective behaviour of large numbers of such systems. In 1925, at a colloquium in Berlin, Einstein discussed the idea that every particle (including electrons, following de Broglie) is accompanied by a 'guiding field' in 3-space (Pais 1982, p. 441; Howard 1990, p. 72). According to Wigner (1980, p. 463), who was present, Einstein 'was fond' of the idea but never published it because he found that energy and momentum were conserved only statistically and not for individual collisions. Each particle had its own guiding field, so there could be no entanglement. The theory was separable. The resulting correlations were not strong enough to guarantee energy-momentum conservation for single events (Howard 1990, p. 73; Bacciagaluppi and Valentini 2009, pp. 200–204).

In May 1927 Einstein wrote a paper proposing a many-body theory with a guiding field in configuration space.[10] In this model the 'tensor of ψ-curvature' $\psi_{\mu\nu} \equiv \nabla_\mu \nabla_\nu \psi$

[9] Max Dresden, private communication, Notre Dame, September 1997.

[10] De Broglie first arrived at his own many-body pilot-wave theory in a paper published in the same month (May 1927) (de Broglie 1927; Bacciagaluppi and Valentini 2009, pp. 55–67).

determines a velocity field $\dot{q}^{\mu}$ for the trajectories. The paper was submitted for publication but withdrawn before it appeared in print (Kirsten and Treder 1979, p. 135; Pais 1979, p. 901). The manuscript survives, however, in the Einstein archive (Howard 1990, pp. 89–90; Belousek 1996). Why the paper was withdrawn is not known. The manuscript has a note 'added in proof' pointing out a difficulty with separability – velocities were not independent for unentangled subsystems – and suggesting this could be solved by the replacement $\psi \to \ln \psi$ in the velocity field. We now know that the model is flawed: while the replacement $\psi \to \ln \psi$ indeed removes the difficulty with separability, the continuity equation is generally not satisfied (so $|\psi|^2$ is not an equilibrium state), and $\dot{q}^{\mu}$ can be imaginary (Holland 2005). In retrospect, Einstein's construction seems strangely over-complicated compared with de Broglie's (Bacciagaluppi and Valentini 2009, pp. 234–239). But his concern with separability is significant.

Einstein also became concerned that locality would not be a natural feature of a theory grounded in configuration space. As he put it a few months later, in the general discussion at the fifth Solvay conference:

> ... the feature of forces of acting only at small *spatial* distances finds a less natural expression in configuration space than in the space of three or four dimensions. (Bacciagaluppi and Valentini 2009, p. 442)

These concerns and difficulties mirror those Einstein had with quantum theory itself. From his calculations of statistical fluctuations in blackbody radiation, as early as 1909 Einstein understood that spatially-separated light quanta could not behave independently, and he encountered similar difficulties in the early 1920s for photons in a cavity with perfectly reflecting walls (Howard 1990). From his work on Bose statistics, in 1925 Einstein again concluded that one would have to think of photons as subject to mysterious mutual influences (to explain their tendency to occupy the same state). According to Howard (p. 80), this is why Einstein's contributions to quantum theory came to an end in 1925: he saw that it was leading to nonseparability, which he found unacceptable.[11] As for locality, as is well known, this requirement led Einstein to conclude that quantum theory must be incomplete.

There is then a strong case for concluding that Einstein abandoned pilot-wave theory for the same reasons he rejected quantum theory: he could not see how these ideas could ever yield an account of a physics that was fundamentally separable and local. When Bohm revived such ideas in the 1950s, unsurprisingly, Einstein dismissed them as 'too cheap' and 'too trivial'. Einstein had long since concluded that a completely different starting point was required.[12] His favoured idea was that quantum theory would emerge from a unified classical field theory of gravitation and electromagnetism. By

[11] It seems these concerns remained with Einstein until the end of his life. Just one week before his death in April 1955, in a wide-ranging conversation with Dennis Sciama, Einstein raised an objection to quantum theory involving a mirror and blackbody radiation (Dennis Sciama, private communication, Trieste, 1991).

[12] For example, in 1954, in a letter to Aron Kupperman, Einstein wrote: 'I think it is not possible to get rid of the statistical character of the present quantum theory by merely adding something to the latter, without changing the fundamental concepts about the whole structure' (Fine 1986, p. 57). Similarly, in Einstein's Autobiographical Notes, we read that quantum theory 'offers no useful point of departure for future development' (Einstein 1949, p. 87).

imposing the condition that the solutions to the (nonlinear) field equations be regular everywhere, Einstein hoped that appropriate quantisation conditions would emerge (Einstein 1949). Einstein's approach was reasonable in its time. But today a large body of evidence indicates that Einstein was simply wrong about both separability and locality. Up to some caveats, Bell's theorem rules out local theories (Bell 1964), while the PBR theorem shows that the wave function is a physical object associated with individual systems (Pusey, Barrett, and Rudolph 2012) – an object that is generally nonseparable.

Einstein's objection, that a theory grounded in configuration space does not offer a natural description of a separable and local physics, can in fact be turned around. On current evidence, we may argue *in favour* of a fundamental dynamics in configuration space, since this can provide a natural understanding of the nonseparable and nonlocal universe in which we seem to live.

2.4.2 The radical nature of pilot-wave dynamics

At this point it is worth emphasising the essential radical features of pilot-wave dynamics, which may be summarised as follows.

First, indeed, the dynamics is grounded in configuration space. It seems that this is not merely a mathematical convenience, as in for example Lagrangian dynamics, but a fundamental physical fact. The motion of the system configuration $q(t)$ is determined by a pilot wave $\psi(q,t)$, where the wave is regarded as a physical object even though it exists in configuration space and not in 3-space.[13] This is a radical claim by any standards. Heisenberg (1958, p. 130) thought it unacceptable, asserting that a wave in configuration space cannot be a real object because (so Heisenberg claimed) 'things' exist in ordinary 3-space and not in configuration space. Pilot-wave theory suggests otherwise: the true arena of physics lies in configuration space, a view that is reinforced by the fundamental role played by the global phase symmetry (2.13).

Second, the pilot wave generates an Aristotelian dynamics for $q(t)$ with a preferred state of rest. The theory undoes some of the key insights of Galileo and Newton, insights which served as the foundation of physics for more than three centuries. In particular Newton's first law, and the closely-related relativity of uniform motion, must be abandoned.

Third, the dynamics is highly nonlocal when viewed in terms of ordinary 3-space. Widely-separated entangled systems interact instantaneously across space. Unlike in Newtonian gravitation, the effects of such interactions are undiminished by distance. It seems that entangled systems are connected directly, by means of the pilot wave in configuration space, quite irrespective of spatial separation.

Fourth, as emphasised by the later Bohm, the theory is 'non-mechanical', in the sense that interactions between systems are not given by fixed functions (for example, functions of distance) but depend on what the pilot wave happens to be (Bohm, Hiley,

[13] Suggestions that ψ is a purely mathematical (nomological, or law-like) feature of the de Broglie equation of motion (Dürr, Goldstein, and Zanghì 1997; Goldstein and Zanghì 2013) are unconvincing, since ψ contains a lot of independent and contingent structure (Valentini 1992, 2010b, 2020; Brown and Wallace 2005), noting that even in quantum gravity and quantum cosmology there is an infinity of solutions for ψ.

and Kaloyerou 1987). In Newtonian gravity, for example, the acceleration of a particle is a fixed (inverse-square) function of the distance between it and other particles. In pilot-wave theory, in contrast, for a pair of entangled particles the velocity of one particle depends on the position of the other, but the functional dependence is set by whatever ψ happens to be. This is radically different from previous dynamical theories.

All of these features make pilot-wave dynamics radically different from generally accepted theories of motion. And there is more. If we had access to nonequilibrium systems, nonlocality would be directly visible, and we could harness entanglement for practical nonlocal signalling (Valentini 1991b). We could also perform subquantum measurements and circumvent the uncertainty principle. Other standard quantum constraints, such as the inability to distinguish single copies of non-orthogonal quantum states reliably, would break down (Valentini 2002b). Quantum physics would prove to be only a special case of a wider physics. Technologically speaking, the discovery of quantum nonequilibrium would probably be as significant as the discovery of electricity and magnetism.

If all this were shown to be true, it would by any measure amount to one of the most radical conceptual shifts in the history of physics. And yet, it is often claimed that pilot-wave theory is 'not radical enough'. Why?

For a variety of reasons, the true nature of pilot-wave theory has been overlooked by many authors. A notable exception was de Broglie himself, who in 1923 abandoned Newton's first law and set out to develop a non-Newtonian dynamics with a law for velocity. Subsequently, in the period 1928–30, de Broglie was so perturbed by the need for a pilot wave in configuration space that he abandoned the theory. But other authors have tended to present the theory in ways that disguise its radical features. In 1952 Bohm presented the theory in terms of Newtonian physics, with a law for acceleration including a 'quantum potential', which made the theory appear to be based on simplistic ideas taken from a superseded physics. The later (post-1952) de Broglie school insisted on a physics grounded in 3-space, based largely on ideas taken from classical field theory. Bohm's pseudo-Newtonian formulation dominated the field until Bell (1987) promoted the theory in terms of de Broglie's original first-order dynamics. Finally, almost all authors have taken the Born rule as a postulate (Dürr, Goldstein, and Zanghì 1992; Holland 1993; Dürr and Teufel 2009; Goldstein 2021; Tumulka 2021, 2022). As a result, much of the interesting physics of the theory has been screened off. In short, pilot-wave theory has often been viewed through the distorting lens of classical physics, and with most of its features hidden behind a seemingly impenetrable veil of quantum noise. Perhaps for these reasons, there has been a widespread failure to see the theory for what it really is: a radically new physics unlike any that has gone before.

2.5 A complete absence of new physics?

Pilot-wave theory is frequently seen as a mere reformulation of quantum theory, with no genuinely new physics to offer. Some authors make the related claim that, conceptually speaking, pilot-wave theory is in effect a theory of Everettian many worlds. These claims are misleading, firstly because the theory is widely and erroneously restricted to equilibrium ensembles, and secondly because of a failure to understand pilot-wave theory on its own terms.

2.5.1 The misleading restriction to equilibrium

There is a long history of pilot-wave theory being dismissed as a rather odd formulation of quantum theory with superfluous trajectories whose details can never be observed and which play no useful role in applications. For example, Heisenberg (1958, pp. 132–133) claimed that pilot-wave theory was just Copenhagen quantum theory recast in an artificial language. He argued that the hidden variables are an 'ideological superstructure', essentially on the grounds that the trajectories play no role in the description of what can actually be seen. Similar criticisms have been made by numerous authors since. For example, Brown (1996, pp. 194–195) raised the question of what the trajectory actually does in pilot-wave theory: its role in singling out the realised branch of the wave function seems so minor (compared to the role played by the wave function itself) as to cast doubt on its existence.

However, as we have discussed, the details of the trajectories are invisible to us only because we are confined to equilibrium. Pilot-wave theory itself can hardly be faulted for that. In principle there is a wider nonequilibrium physics in which subquantum measurements can be performed and the details of the trajectories *can* be observed. The recurring objection, that the theory contains unobservable structure, carries weight only when the theory is mistakenly identified with the equilibrium theory only.

Unfortunately, Bohm himself seems not to have recognised that the usual limitations on measurement are merely a peculiarity of the equilibrium state. Instead, he hoped to avoid these limitations – thereby inducing departures from quantum theory – by altering the dynamics of the theory (Bohm 1952a, p. 179). Among other ideas Bohm suggested that his single-particle equation of motion be modified to read

$$m\frac{d^2\mathbf{x}}{dt^2} = -\boldsymbol{\nabla}V - \boldsymbol{\nabla}Q + \mathbf{F}(\mathbf{p} - \boldsymbol{\nabla}S), \tag{2.27}$$

where $\mathbf{F}(\mathbf{p} - \boldsymbol{\nabla}S)$ vanishes when $\mathbf{p} = \boldsymbol{\nabla}S$ and is supposed to take a form such that over time $\mathbf{p}$ rapidly approaches $\boldsymbol{\nabla}S$. Over very short times, Bohm expected to find $\mathbf{p} \neq \boldsymbol{\nabla}S$ and departures from quantum theory. Heisenberg (1958, p. 132) poured scorn on this proposal, sarcastically comparing it to the hope that $2 \times 2 = 5$ so that we could improve our financial situation. But in fact, there is no need for arbitrary changes to the dynamics: as it stands the theory already contains an extended nonequilibrium physics, in which the uncertainty principle is broken and the underlying trajectories can be measured.

The widespread and misleading restriction to equilibrium has been exacerbated in recent decades by the 'Bohmian mechanics school' of pilot-wave theory (Dürr, Goldstein, and Zanghì 1992; Dürr and Teufel 2009; Goldstein 2021; Tumulka 2021, 2022). This school, which despite its terminology adopts de Broglie's version of the dynamics, has been particularly influential among philosophers.[14] The school has made a number

[14] For example, Albert (2015).

of distinctive claims – for example concerning the nature of the wave function, and the theory of fermions – which in our view are incorrect and which we have discussed in detail elsewhere (Valentini 2020). Here our particular concern is with the claim that the Born rule has a fundamental status as a preferred measure of 'typicality' (or probability) for the whole universe. The school has argued that the observed Born rule for subsystems can be explained by applying the Born-rule measure to the initial universal configuration. In our view, the argument is essentially circular.[15] A non-Born rule measure for the initial universal configuration will imply violations of the Born rule for subsystems at early times (Valentini 1996, 2001, 2020). As we have noted, in a deterministic physics there is a clear conceptual distinction between the dynamical laws, which are fixed, and the initial conditions, which are contingencies to be determined empirically. Whether the universe began in equilibrium or in nonequilibrium is a question for experiment. Unfortunately, the insistence on a Born-rule measure for the initial configuration has added to the widespread misconception that pilot-wave theory has no observable new physics to offer.

In fact, pilot-wave theory is a much richer theory than the restriction to equilibrium will allow. The discovery of relic nonequilibrium particles (Valentini 2007; Underwood and Valentini 2015, 2020) would have radical technological implications, including the breaking of quantum cryptography and new forms of computation (Valentini 2002b), as well as superluminal signalling (Valentini 1991b). The possibility of primordial nonequilibrium offers a possible explanation for reported large-scale anomalies in the cosmic microwave background (Valentini 2010a; Colin and Valentini 2015; Vitenti, Peter, and Valentini 2019), while the theoretical creation of quantum nonequilibrium by evaporating black holes offers a new approach to the long-standing problem of information loss (Valentini 2004, 2007; Kandhadai and Valentini 2020). Furthermore, recent work suggests that, in a universe described by quantum gravity, with a non-normalisable Wheeler–DeWitt wave functional, there simply is no well-defined Born-rule measure at the beginning of the universe, so that the universe necessarily begins in nonequilibrium and the argument originally given by Dürr *et al.* cannot be applied (Valentini 2021, 2023).

2.5.2 Many worlds in denial?

It has been claimed that 'pilot-wave theories are parallel-universes theories in a state of chronic denial' (Deutsch 1996, p. 225).[16] On this view, the pilot wave $\psi(q,t)$ itself already contains many (parallel) worlds, and since the trajectory $q(t)$ merely singles out one of those worlds it may as well be dropped. A detailed rebuttal has been given (Valentini 2010b). Here we summarise the main points.

First, because pilot-wave theory allows violations of the Born rule, while Everettian quantum theory does not, they are necessarily distinct physical theories with (potentially) different empirical consequences. Therefore it cannot be correct to claim that the former is just the latter in disguise (or in denial). Second, quantum nonequilibrium can enable subquantum measurements (Valentini 2002b), which can track $q(t)$ with negligible disturbance of $\psi(q,t)$, thereby distinguishing the occupied branch from the

[15] Similar circular arguments have been applied to classical coin tossing (Dürr and Struyve 2021).
[16] Similar claims were made by Zeh (1999, p. 200) and by Brown and Wallace (2005, p. 527).

empty ones. Interpreting pilot-wave theory on its own terms, the empty branches are merely concentrations of a complex-valued field on configuration space, not parallel worlds.

But the claim falls short even if pilot-wave theory is (artificially) restricted to equilibrium. This is because, in realistic models, the wave functions are delocalised and the claim cannot be properly formulated – at least not in a strong form, as we shall clarify.[17]

To convey the point, we first consider an unrealistic example where two localised and non-overlapping packets $\psi_1(q,t)$, $\psi_2(q,t)$ each trace out an approximately classical trajectory (as might occur in appropriate conditions via Ehrenfest's theorem). If the wave function is a superposition

$$\psi(q,t) = \frac{1}{\sqrt{2}}\left(\psi_1(q,t) + \psi_2(q,t)\right) , \tag{2.28}$$

then the de Broglie–Bohm trajectory $q(t)$ will occupy only one packet, say ψ_1, and will trace out an approximately classical path $q_1(t)$. But the empty (localised) packet ψ_2 also traces out an approximately classical path $q_2(t)$. Since we regard ψ as a physical object, we have *two* (approximate) classical paths $q_1(t)$, $q_2(t)$ existing simultaneously. This is what we call a 'strong form' of the claim. As noted, if we had access to nonequilibrium systems and subquantum measurements, we could distinguish the empty from the non-empty packet by monitoring $q(t)$, so even in this strong form the claim ultimately fails. But what if we restrict our attention to the equilibrium theory? Even then, such an argument cannot apply to a realistic model of the macroscopic domain, because initial narrow packets rapidly spread and delocalise (in particular for chaotic systems, which are ubiquitous).

To illustrate this we might consider a second unrealistic example, with two packets $\psi_1(q,y,t)$, $\psi_2(q,y,t)$ that are highly localised and non-overlapping with respect to a pointer degree of freedom y, and which are highly *de*localised with respect to the system degrees of freedom q. With respect to q the packets might be modelled as WKB wave functions, giving rise to approximately classical de Broglie–Bohm trajectories $q(t)$. If the wave function is now a superposition

$$\psi(q,y,t) = \frac{1}{\sqrt{2}}\left(\psi_1(q,y,t) + \psi_2(q,y,t)\right) , \tag{2.29}$$

the total trajectory $Q(t) = (q(t), y(t))$ will occupy only one packet, say ψ_1, and the system will again trace out an approximately classical path $q_1(t)$. But now, because the empty packet ψ_2 is delocalised with respect to q, it does *not* trace out an alternative (approximate) classical path. We cannot point to localised 'ψ-stuff' executing an alternative trajectory, so there is no parallel classical world to be in denial about.

In our second example, at best we might point to alternative trajectories $q(t)$ associated with different initial conditions $q(0)$ (within both ψ_1 and ψ_2). In fact, for

[17] There should be no need to explain why a microscopic system, whose quantum state is a superposition of eigenstates with eigenvalues ω_n, does not occupy parallel realities simultaneously. As we have seen, from a pilot-wave perspective the ω_n generally have no ontological status. A misguided 'eigenvalue realism' may give the mistaken impression that a mathematical superposition of eigenstates somehow corresponds to a physical superposition of distinct realities (Valentini 1996, 2010b).

any state $\psi(q,t)$ of any system, there is a whole set of associated trajectories $q(t)$ obtained by integrating de Broglie's equation of motion. But to view this is 'many worlds in denial' would be a very weak claim indeed. One might just as well argue that, in the classical theory of a charged test particle in an external electromagnetic field, the alternative trajectories $\mathbf{x}(t)$ (arising from different initial conditions $\mathbf{x}(0)$) constitute parallel realities – and that the conventional single-particle interpretation amounts to 'many worlds in denial'. Most physicists would regard the alternative trajectories $\mathbf{x}(t)$ as having a purely mathematical existence, there being only one actual trajectory $\mathbf{x}(t)$. Similarly, for a delocalised ψ-field, the alternative trajectories have a purely mathematical existence and there is only one actual trajectory $q(t)$.

In our view, in realistic models with environmental decoherence (Zurek 2003), the situation is similar to that of our second example with delocalised states ψ. In such models, tracing over environmental degrees of freedom yields a reduced density operator for the system with matrix elements $\rho_{\text{red}}(q,q',t)$. One may then define a quasi-probability distribution in phase space:

$$W_{\text{red}}(q,p,t) \equiv \frac{1}{2\pi}\int dz\; e^{ipz}\rho_{\text{red}}(q-z/2,q+z/2,t) \tag{2.30}$$

(the reduced Wigner function). In certain conditions, $W_{\text{red}}(q,p,t)$ is approximately non-negative and its time evolution approximates that of a classical phase-space distribution. In a pilot-wave treatment, under certain conditions the trajectories of the system become approximately classical (Appleby 1999). Now, for a simple system such as an oscillator, a localised packet $W_{\text{red}}(q,p,t)$ can trace out a thin 'tube' approximating a classical trajectory in phase space (Zurek, Habib, and Paz 1993). A strong form of the claim might arise from a superposition of two or more such packets with macroscopic separations (leaving aside doubts about the ontological status of W_{red}). But a realistic world model will contain chaotic systems. An initial minimum-uncertainty packet will rapidly spread over macroscopic regions (Zurek 1998), yielding a highly delocalised distribution $W_{\text{red}}(q,p,t)$ (Habib, Shizume, and Zurek 1998; Zurek 2003, pp. 745–47).

Thus, in a realistic model, there will be no well-defined localised 'ψ-stuff' tracing out classical (chaotic) trajectories. As usual, in a pilot-wave treatment different initial conditions will yield different trajectories. But there is no reason to ascribe anything other than mathematical status to the alternative trajectories – just as in the analogous classical case of a charged particle in an external field. In our view, in realistic models, an *apparent* multiplicity of worlds arises from a reification of purely mathematical trajectories.

2.6 Outlook

As we take stock of a century of objections and misunderstandings, it is difficult to avoid the impression that pilot-wave theory may well be one of the most misunderstood theories in the history of physics. As we have argued, the theory is often deeply misunderstood even by some of its keenest supporters. To explain why this has happened would require a much broader historical and philosophical outlook than we could hope to provide here. A full answer would surely include the following two points. First, most physicists did not understand that in the 1920s de Broglie formulated a new theory of motion, beginning in 1923 with his radical abandonment of Newton's first law. Second, de Broglie's 1927 pilot-wave theory was in several respects far ahead of its time, and it has taken us almost a century to catch up.

Pilot-wave theory is nonlocal, and it was not until 1964 that we came to understand that this was not a peculiarity of the theory but a much more general feature (Bell 1964). Pilot-wave theory is also 'contextual' with respect to quantum measurements, which we also now know to be a general feature (Bell 1966, Kochen and Specker 1967). Nonequilibrium pilot-wave theory allows superluminal signalling (Valentini 1991b), which again has been shown to be more general (Valentini 2002a). Finally, pilot-wave theory has an ontological quantum state ψ associated with individual systems, and we now know that this too is a general feature (Pusey, Barrett, and Rudolph 2012). There can be little doubt that pilot-wave theory teaches us important lessons which may well remain valid even if the theory itself should turn out to be incorrect.

Despite this deepening of our understanding, the objections and misunderstandings continue. Even so, over the last thirty years pilot-wave theory has been able to provide new approaches to several important problems in fundamental physics and cosmology. First, the theory has been extensively applied to quantum cosmology, with the advantage of having no need to invoke an outside observer (Pinto-Neto 2005, 2021; Pinto-Neto and Fabris 2013). In a bouncing cosmology, for example, there is a contracting era followed by an expanding phase ('the big bang'), which can be described objectively in terms of trajectories that existed at early times when no observers were present. It is difficult to see how textbook quantum mechanics could even begin to describe such processes. Second, as we have noted, pilot-wave theory provides a potential explanation for puzzling, though still controversial, anomalies in the cosmic microwave background (Valentini 2010a; Colin and Valentini 2015, 2016). There seem to be several anomalies in the data, associated with large spatial scales, which could find a common origin in primordial quantum nonequilibrium, even if as yet no definite conclusions can be drawn (Vitenti, Peter, and Valentini 2019). Third, pilot-wave theory offers a new approach (still under development) to one of the deepest problems of theoretical high-energy physics: the puzzle of information loss in evaporating black holes, which could potentially be resolved by a breakdown of the Born rule in Hawking radiation (Valentini 2004, 2007; Kandhadai and Valentini 2020). Finally, pilot-wave theory has much to say about long-standing conceptual problems in quantum gravity (Pinto-Neto and Santini 2002), including the status of the Born rule and the treatment of probability generally (Valentini 2021, 2023).

At the end of the day, theories must be judged by their explanatory power and by confrontation with experiment. There is no doubt that pilot-wave theory has much to offer in terms of novel approaches to certain deep puzzles in fundamental physics and cosmology (and not just the notorious measurement problem). The theory is also able to make predictions which can be tested experimentally. There are plausible prospects for observing deviations from standard quantum physics, whether in relic particles from the early universe or in radiation from exploding primordial black holes. Despite its detractors, the theory still has much to contribute to the ongoing development of theoretical physics. The next generation of theorists might do well to see past the continuing objections and take advantage of the novel perspective, and of the intriguing opportunities, provided by this distinctive approach to quantum physics.

References

Abraham, E., Colin, S. and Valentini, A. (2014). Long-time relaxation in pilot-wave theory. *Journal of Physics A*, **47**: 395306.

Aghanim, N. *et al.* (Planck Collaboration) (2016). *Planck* 2015 results. XI. CMB power spectra, likelihoods, and robustness of parameters. *Astronomy and Astrophysics*, **594**: A11.

Aharonov, Y. and Vaidman, L. (1996). About position measurements which do not show the Bohmian particle position. In Cushing, J.T., Fine, A. and Goldstein, S. (eds.) *Bohmian Mechanics and Quantum Theory: An Appraisal.* Dordrecht: Kluwer.

Albert, D. (2015). *After Physics.* Cambridge, Massachusetts: Harvard University Press.

Anandan, J. and Brown, H.R. (1995). On the reality of spacetime geometry and the wavefunction. *Foundations of Physics*, **25**: 349.

Appleby, D.M. (1999). Bohmian trajectories post-decoherence. *Foundations of Physics*, **29**: 1885.

Bacciagaluppi, G. and Valentini, A. (2009). *Quantum Theory at the Crossroads: Reconsidering the 1927 Solvay Conference.* Cambridge: Cambridge University Press.

Bell, J.S. (1964). On the Einstein-Podolsky-Rosen paradox. *Physics*, **1**: 195.

Bell, J.S. (1966). On the problem of hidden variables in quantum mechanics. *Reviews of Modern Physics*, **38**: 447.

Bell, J.S. (1987). *Speakable and Unspeakable in Quantum Mechanics.* Cambridge: Cambridge University Press.

Bell, J.S. (1990). Against 'measurement'. In Miller, A.I. (ed.) *Sixty-Two Years of Uncertainty: Historical, Philosophical, and Physical Inquiries into the Foundations of Quantum Mechanics.* New York: Plenum Press. [Reprinted: Bell, J.S. (1990). Against 'measurement'. *Physics World*, **3**(8): 33.]

Belousek, D.W. (1996). Einstein's 1927 unpublished hidden-variable theory: its background, context and significance. *Studies in History and Philosophy of Modern Physics*, **27**: 437.

Berndl, K., Dürr, D,. Goldstein, S. and N. Zanghì (1996). nonlocality, Lorentz invariance, and Bohmian quantum theory. *Physical Review A*, **53**: 2062.

Bohm, D. (1952a). A suggested interpretation of the quantum theory in terms of "hidden" variables. I. *Physical Review*, **85**(2): 166.

Bohm, D. (1952b). A suggested interpretation of the quantum theory in terms of "hidden" variables. II. *Physical Review*, **85**(2): 180.

Bohm, D. (1953). Proof that probability density approaches $|\psi|^2$ in causal interpretation of the quantum theory. *Physical Review*, **89**: 458.

Bohm, D. and Hiley, B.J. (1989). nonlocality and locality in the stochastic interpretation of quantum mechanics. *Physics Reports*, **172**: 93.

Bohm, D., Hiley, B.J. and Kaloyerou, P.N. (1987). An ontological basis for the quantum theory. *Physics Reports*, **144**: 321.

Bohm, D. and Vigier, J.P. (1954). Model of the causal interpretation of quantum theory in terms of a fluid with irregular fluctuations. *Physical Review*, **96**: 208.

Bohr, N. (1931). Maxwell and modern theoretical physics. *Nature*, **128**: 691.

Brown, H. R. (1996). Mindful of quantum possibilities. *British Journal for the Philosophy of Science*, **47**: 189.

Brown, H.R. and Wallace, D. (2005). Solving the measurement problem: de Broglie–Bohm loses out to Everett. *Foundations of Physics*, **35**: 517.

Cohen, E., Cortês, M., Elitzur, A. and Smolin, L. (2020). Realism and causality. I. Pilot wave and retrocausal models as possible facilitators. *Physical Review D*, **102**: 124027.

Colin, S. (2012). Relaxation to quantum equilibrium for Dirac fermions in the de Broglie–Bohm pilot-wave theory. *Proceedings of the Royal Society A*, **468**: 1116.

Colin, S. and Valentini, A. (2013). Mechanism for the suppression of quantum noise at large scales on expanding space. *Physical Review D*, **88**: 103515.

Colin, S. and Valentini, A. (2014). Instability of quantum equilibrium in Bohm's dynamics. *Proceedings of the Royal Society A*, **470**: 20140288.

Colin, S. and Valentini, A. (2015). Primordial quantum nonequilibrium and large-scale cosmic anomalies. *Physical Review D*, **92**: 043520.

Colin, S. and Valentini, A. (2016). Robust predictions for the large-scale cosmological power deficit from primordial quantum nonequilibrium. *International Journal of Modern Physics D*, **25**: 1650068.

de Broglie, L. (1923). Quanta de lumière, diffraction et interférences. *Comptes Rendus (Paris)*, **177**: 548.

de Broglie, L. (1924). Recherches sur la théorie des quanta. PhD thesis, University of Paris.

de Broglie, L. (1925). Recherches sur la théorie des quanta. *Annales de Physique*, **3**(10): 22.

de Broglie, L. (1927). La mécanique ondulatoire et la structure atomique de la matière et du rayonnement. *Le Journal de Physique et le Radium*, **8**(6): 225.

de Broglie, L. (1928). La nouvelle dynamique des quanta, in *Électrons et Photons: Rapports et Discussions du Cinquième Conseil de Physique*, Paris: Gauthier-Villars. [English translation: Bacciagaluppi and Valentini (2009).]

Deutsch, D. (1996). Comment on Lockwood. *British Journal for the Philosophy of Science*, **47**: 222.

Dewdney, C., Hardy, L. and Squires, E.J. (1993). How late measurements of quantum trajectories can fool a detector. *Physics Letters A*, **184**: 6.

Drezet, A. (2021). Justifying Born's rule $P_\alpha = |\Psi_\alpha|^2$ using deterministic chaos, decoherence, and the de Broglie-Bohm quantum theory. *Entropy*, **23**: 1371.

Dürr, D., Goldstein, S., Münch-Berndl, K. and Zanghi, N. (1999). Hypersurface Bohm-Dirac models. *Physical Review A*, **60**: 2729.

Dürr, D., Goldstein, S., Norsen, T., Struyve, W. and Zanghì, N. (2014). Can Bohmian mechanics be made relativistic?. *Proceedings of the Royal Society A*, **470**: 20130699.

Dürr, D., Goldstein, S., Tumulka, R. and Zanghì, N. (2004). Bohmian mechanics and quantum field theory. *Physical Review Letters*, **93**: 090402.

Dürr, D., Goldstein, S., Tumulka, R. and Zanghì, N. (2005). Bell-type quantum field theories. *Journal of Physics A*, **38**: R1.

Dürr, D., Goldstein, S. and Zanghì, N. (1992). Quantum equilibrium and the origin of absolute uncertainty. *Journal of Statistical Physics*, **67**: 843.

Dürr, D., Goldstein, S. and Zanghì, N. (1997). Bohmian mechanics and the meaning of the wave function. In Cohen, R.S., Horne, M. and Stachel, J. (eds.) *Experimental Metaphysics: Quantum Mechanical Studies for Abner Shimony*. Dordrecht: Kluwer.

Dürr, D. and Struyve, W. (2021). Typicality in the foundations of statistical physics and Born's rule. In Allori, V., Bassi, A., Dürr, D. and Zanghì, N. (eds.) *Do Wave Functions Jump? Perspectives of the Work of GianCarlo Ghirardi*. Cham: Springer.

Dürr, D. and Teufel, S. (2009). *Bohmian Mechanics: The Physics and Mathematics of Quantum Theory*. Berlin: Springer.

Efthymiopoulos, C., Contopoulos, G. and Tzemos, A.C. (2017). Chaos in de Broglie-Bohm quantum mechanics and the dynamics of quantum relaxation. *Annales de la Fondation Louis de Broglie*, **42**: 133.

Einstein, A. (1949). Autobiographical Notes. In Schilpp, P.A. (ed.) *Albert Einstein: Philosopher-Scientist*. Illinois: Open Court.

Einstein, A. (1952). Letter to Max Born, 12 May 1952. In *The Born-Einstein Letters*, trans. I. Born, 1971, p. 192. London: Macmillan.

Englert, B.-G., Scully, M.O., Süssmann, G. and Walther, H. (1992). Surrealistic Bohm trajectories. *Zeitschrift für Naturforschung* **47a**: 1175.

Fine, A. (1986). *The Shaky Game: Einstein, Realism and the Quantum Theory*. Chicago: University of Chicago Press.

Goldstein, S. (2021). Bohmian mechanics. In Zalta, E.N. (ed.) *The Stanford Encyclopedia of Philosophy* (Fall 2021 Edition).

Goldstein, S. and Zanghì, N. (2013). Reality and the role of the wavefunction in quantum theory. In Albert, D. and Ney, A. (eds.) *The Wave Function: Essays in the Metaphysics of Quantum Mechanics.* New York: Oxford University Press.

Habib, S., Shizume, K. and Zurek, W.H. (1998). Decoherence, chaos, and the correspondence principle. *Physical Review Letters*, **80**: 4361.

Hardy, L. (1992). Quantum mechanics, local realistic theories, and Lorentz-invariant realistic theories. *Physical Review Letters*, **68**: 2981.

Heisenberg, W. (1958). *Physics and Philosophy: the Revolution in Modern Science.* New York: Harper & Brothers.

Heisenberg, W. (1971). *Physics and Beyond.* New York: Harper & Row.

Holland, P.R. (1993). *The Quantum Theory of Motion: an Account of the de Broglie–Bohm Causal Interpretation of Quantum Mechanics.* Cambridge: Cambridge University Press.

Holland, P.R. (2005). What's wrong with Einstein's 1927 hidden-variable interpretation of quantum mechanics?. *Foundations of Physics*, **35**: 177.

Hořava, P. (2009). Quantum gravity at a Lifshitz point. *Physical Review D*, **79**: 084008.

Howard, D. (1990). 'Nicht sein kann was nicht sein darf', or the prehistory of EPR, 1909–1935: Einstein's early worries about the quantum mechanics of composite systems. In Miller, A.I. (ed.) *Sixty-Two Years of Uncertainty: Historical, Philosophical, and Physical Inquiries into the Foundations of Quantum Mechanics.* New York: Plenum Press.

Kandhadai, A. and Valentini, A. (2020). Mechanism for nonlocal information flow from black holes. *International Journal of Modern Physics A*, **35**: 2050031.

Keller, J.B. (1953). Bohm's interpretation of the quantum theory in terms of 'hidden' variables. *Physical Review*, **89**: 1040.

Kirsten, C. and Treder, H.J. (1979). *Albert Einstein in Berlin 1913–1933.* Berlin: Akademie-Verlag.

Kochen, S. and Specker, E.P. (1967). The problem of hidden variables in quantum mechanics. *Journal of Mathematics and Mechanics*, **17**: 59.

Kostelecký, V.A. and Mewes, M. (2002). Signals for Lorentz violation in electrodynamics. *Physical Review D*, **66**: 056005.

Liddle, A.R. and Lyth, D.H. (2000). *Cosmological Inflation and Large-Scale Structure.* Cambridge: Cambridge University Press.

Lustosa, F.B., Pinto-Neto, N. and Valentini, A. (2023). Evolution of quantum nonequilibrium for coupled harmonic oscillators. *Proceedings of the Royal Society A*, **479**: 20220411.

Misner, C.W., Thorne, K.S. and Wheeler, J.A. (1973). *Gravitation.* San Francisco: W. H. Freeman.

Murdoch, D.R. (1987). *Niels Bohr's Philosophy of Physics.* Cambridge: Cambridge University Press.

Pais, A. (1979). Einstein and the quantum theory. *Reviews of Modern Physics*, **51**: 863.

Pais, A. (1982). *Subtle is the Lord: the Science and the Life of Albert Einstein.* Oxford: Oxford University Press.

Pauli, W. (1953). Remarques sur le problème des paramètres cachés dans la mécanique quantique et sur la théorie de l'onde pilote. In George, A. (ed.) *Louis de Broglie: Physicien et Penseur.* Paris: Albin Michel.

Peter, P. and Uzan, J.-P. (2009). *Primordial Cosmology.* Oxford: Oxford University Press.

Pinto-Neto, N. (2005). The Bohm interpretation of quantum cosmology. *Foundations of Physics*, **35**: 577.

Pinto-Neto, N. (2021). The de Broglie–Bohm quantum theory and its application to quantum cosmology. *Universe*, **7**: 134.

Pinto-Neto, N. and Fabris, J.C. (2013). Quantum cosmology from the de Broglie–Bohm perspective. *Classical and Quantum Gravity*, **30**: 143001.

Pinto-Neto, N. and Sergio Santini, E. (2002). The consistency of causal quantum geometrodynamics and quantum field theory. *General Relativity and Gravitation*, **34**: 505.

Pusey, M.F., Barrett, J. and Rudolph, T. (2012). On the reality of the quantum state. *Nature Physics*, **8**: 475.

Shimony, A. (1984). Controllable and uncontrollable nonlocality. In Kamefuchi, S. *et al.* (eds.) *Foundations of Quantum Mechanics in the Light of New Technology.* Tokyo: Physical Society of Japan.

Struyve, W. and Valentini, A. (2009). de Broglie–Bohm guidance equations for arbitrary Hamiltonians. *Journal of Physics A*, **42**: 035301.

Towler, M.D., Russell, N.J. and Valentini, A. (2012). Time scales for dynamical relaxation to the Born rule. *Proceedings of the Royal Society A*, **468**: 990.

Tumulka, R. (2007). The 'unromantic pictures' of quantum theory. *Journal of Physics A*, **40**: 3245.

Tumulka, R. (2021). Bohmian mechanics. In Knox, E. and Wilson, A. (eds.) *The Routledge Companion to the Philosophy of Physics.* New York: Routledge.

Tumulka, R. (2022). *Foundations of Quantum Mechanics.* Cham: Springer.

Underwood, N.G. (2018). Extreme quantum nonequilibrium, nodes, vorticity, drift and relaxation retarding states. *Journal of Physics A*, **51**: 055301.

Underwood, N.G. and Valentini, A. (2015). Quantum field theory of relic nonequilibrium systems. *Physical Review D*, **92**: 063531.

Underwood, N.G. and Valentini, A. (2020). Anomalous spectral lines and relic quantum nonequilibrium. *Physical Review D*, **101**: 043004.

Valentini, A. (1991a). Signal-locality, uncertainty, and the subquantum H-theorem. I. *Physics Letters A*, **156**: 5.

Valentini, A. (1991b). Signal-locality, uncertainty, and the subquantum H-theorem. II. *Physics Letters A*, **158**: 1.

Valentini, A. (1992). On the pilot-wave theory of classical, quantum and subquantum physics. PhD thesis, International School for Advanced Studies, Trieste, Italy. [http://hdl.handle.net/20.500.11767/4334]

Valentini, A. (1996). Pilot-wave theory of fields, gravitation and cosmology. In Cushing, J.T., Fine, A. and Goldstein, S. (eds.) *Bohmian Mechanics and Quantum theory: An Appraisal.* Dordrecht: Kluwer.

Valentini, A. (1997). On Galilean and Lorentz invariance in pilot-wave dynamics. *Physics Letters A*, **228**: 215.

Valentini, A. (2001). Hidden variables, statistical mechanics and the early universe. In Bricmont, J., Dürr, D., Galavotti, M.C., Ghirardi, G.C., Petruccione, F. and Zanghì, N. (eds.) *Chance in Physics: Foundations and Perspectives.* Berlin: Springer.

Valentini, A. (2002a). Signal-locality in hidden-variables theories. *Physics Letters A*, **297**: 273.

Valentini, A. (2002b). Subquantum information and computation. *Pramana—Journal of Physics*, **59**: 269.

Valentini, A. (2004). Black holes, information loss, and hidden variables. arXiv:hep-th/0407032.

Valentini, A. (2007). Astrophysical and cosmological tests of quantum theory. *Journal of Physics A*, **40**: 3285.

Valentini, A. (2008). Hidden variables and the large-scale structure of spacetime. In Craig, W. L. and Smith, Q. (eds.) *Einstein, Relativity and Absolute Simultaneity.* London: Routledge.

Valentini, A. (2010a). Inflationary cosmology as a probe of primordial quantum mechanics. *Physical Review D*, **82**: 063513.

Valentini, A. (2010b). de Broglie–Bohm pilot-wave theory: many-worlds in denial?. In Saunders, S., Barrett, J., Kent, A. and Wallace, D. (eds.) *Many Worlds? Everett, Quantum Theory, and Reality.* Oxford: Oxford University Press.

Valentini, A. (2020). Foundations of statistical mechanics and the status of the Born rule in de Broglie–Bohm pilot-wave theory. In Allori, V. (ed.) *Statistical Mechan-*

ics and Scientific Explanation: Determinism, Indeterminism and Laws of Nature. Singapore: World Scientific.

Valentini, A. (2021). Quantum gravity and quantum probability. arXiv:2104.07966.

Valentini, A. (2023). Beyond the Born rule in quantum gravity. *Foundations of Physics*, **53**: 6.

Valentini, A. (2024). de Broglie–Bohm quantum mechanics. In *Encyclopedia of Mathematical Physics, 2nd edition.* Amsterdam: Elsevier.

Valentini, A. (2025a). de Broglie–Bohm pilot-wave theory. In *Oxford Research Encyclopedia of Physics.* Oxford: Oxford University Press. [https://oxfordre.com/physics]

Valentini, A. (2025b). *Beyond the Quantum: a Quest for the Origin and Hidden Meaning of Quantum Mechanics.* Oxford: Oxford University Press.

Valentini, A. (2025c). *Introduction to Quantum Foundations and Pilot-Wave Theory.* Oxford: Oxford University Press.

Valentini, A. and Westman, H. (2005). Dynamical origin of quantum probabilities. *Proceedings of the Royal Society A*, **461**: 253.

Vigier, J.P. (1985). Nonlocal quantum potential interpretation of relativistic actions at a distance in many-body problems. In Tarozzi, G. van der Merwe, A. (eds.) *Open Questions in Quantum Physics: Invited Papers on the Foundations of Microphysics.* Dordrecht: Reidel.

Vitenti, S., Peter, P. and Valentini, A. (2019). Modeling the large-scale power deficit with smooth and discontinuous primordial spectra. *Physical Review D*, **100**: 043506.

Weinberg, S. (1995). *The Quantum Theory of Fields: Volume I, Foundations.* Cambridge: Cambridge University Press.

Wigner, E.P. (1980). Thirty years of knowing Einstein. In Woolf. H. (ed.) *Some Strangeness in the Proportion: a Centennial Symposium to Celebrate the Achievements of Albert Einstein.* Reading, Massachusetts: Addison-Wesley.

Zeh, H.D. (1999). Why Bohm's Quantum Theory?. *Foundations of Physics Letters*, **12**: 197.

Zurek, W.H. (1998). Decoherence, chaos, quantum-classical correspondence, and the algorithmic arrow of time. *Physica Scripta*, **T76**: 186.

Zurek, W.H. (2003). Decoherence, einselection, and the quantum origins of the classical. *Reviews of Modern Physics*, **75**: 715.

Zurek, W.H., Habib, S. and Paz, J.P. (1993). Coherent states via decoherence. *Physical Review Letters*, **70**: 1187.

3

The de Broglie–Bohm theory is and is not a hidden variable theory

Jean Bricmont

3.1 Introduction

The title of this paper is evidently self-contradictory; but, as an American president might have said, what do you *mean* by 'hidden variable theory'? One of the goals of this chapter is to answer this question and to try to clarify any confusion surrounding this expression.

For example, in the wake of the awarding of the 2022 Nobel Prize in Physics to John Clauser, Alain Aspect, and Anton Zeilinger, for verifying the violation of Bell's inequalities, it has been frequently asserted that those violations imply the impossibility of hidden variables theories (which could be alternatives to ordinary quantum mechanics)[1] or maybe only of local hidden variable theories. The latter qualification is often associated with the idea that non-local hidden variable theories are too weird to consider.

Yet it is known since 1952, through the work of David Bohm, and in fact since 1927, through the work of Louis de Broglie, that a 'hidden variable' alternative to ordinary quantum mechanics exists. But why is it not excluded by the violation of Bell's inequalities? Is it because it is non-local? But, if yes, is that a defect or a quality of the de Broglie–Bohm theory?

3.2 What are 'hidden variables'?

By definition, the expression 'hidden variables' refers to any variable that would describe the physical state of a system beyond its quantum state.

But a moment of reflection shows that, if we accept this definition, the world is full of hidden variables. To illustrate this idea, consider the famous example of Schrödinger's cat. Since this example is well known, let us simply recall that, at the end of an experiment where one measures the property of a particle that can take two values, the measuring device, or the cat if we couple the device to the cat through a poison capsule, is (leaving aside normalization factors)

[1] For example, in its press release announcing the 2022 Nobel Prize in Physics, the Royal Swedish Academy writes: 'This means that quantum mechanics cannot be replaced by a theory that uses hidden variables', see https://www.nobelprize.org/prizes/physics/2022/press-release/.

$$\Psi_{\text{cat alive}} + \Psi_{\text{cat dead}}. \tag{3.1}$$

And that wave function, argued Schrödinger, cannot be a complete description of the cat, which is clearly either alive or dead but not both. The way out of this problem from the point of view of ordinary quantum mechanics is to introduce the collapse postulate: when one looks at the cat, one sees whether she is alive or dead and, depending on what one sees, one reduces the wave function of the cat (and of the particle that was measured and is thus coupled to the state of the cat) to either $\Psi_{\text{cat alive}}$ *or* $\Psi_{\text{cat dead}}$.

Since this is a *deus ex machina* from the point of view of the linear Schrödinger evolution, justifying it is often considered as the main problem in foundations of quantum mechanics.

Saying that the cat is alive or dead (depending on what we see) amounts to the introduction of a 'hidden variable' since the property of being dead or alive is not part of the quantum state (3.1).

But actually there is a deeper problem: neither $\Psi_{\text{cat alive}}$ nor $\Psi_{\text{cat dead}}$ are cats: they are functions defined on a high dimensional space $\mathbb{R}^{3N}$, with N being the number of particles composing the cat (putting aside the fact that there are different kinds of particles and they may also have a spin) while cats are located in $\mathbb{R}^3$. And it is not clear what it means to say that $\Psi_{\text{cat alive}}$ or $\Psi_{\text{cat dead}}$ are descriptions of cats, let alone 'complete descriptions' of them.[2] Like all wave functions, they allow us to predict results of measurements conducted on the objects that they 'describe' but nothing else.

Most of the people mentally identify cats and wave functions of cats. But the cat, simply by being located in the usual space $\mathbb{R}^3$, has properties that are not part of the wave functions $\Psi_{\text{cat alive}}$ or $\Psi_{\text{cat dead}}$.

The cat has a shape, a size, a weight, etc. that are, according to our definition, hidden variables. This does not mean that one could not, in principle, calculate those properties from a detailed knowledge of the wave function of the cat, but that these properties refer to objects in three dimensions and not to functions defined on $\mathbb{R}^{3N}$.

In the very useful terminology of John Bell, the cat (or any other macroscopic object) is a *local beable*. The word beable was created in opposition to the word observable, to refer to what exists in the world and not simply to what is observed (by us, humans), and the world local refers to the fact that those beables are located in the ordinary space $\mathbb{R}^3$.

3.3 Naive statistical interpretation

If we admit that there exist 'hidden variables' in the macroscopic world (variables that, in fact, exist everywhere and are not hidden at all but are the only things we ever see), why do we not try to assume that they also exist in the microscopic world?

[2]This idea is emphasized in Tumulka (2018), Section 5.1.

This leads us to the most natural interpretation of quantum mechanics, which is probably the one in the mind of most of the 'no worry about quantum mechanics' physicists (and also probably the one of Einstein), namely the 'statistical' one.[3]

According to that interpretation, a quantum state Ψ does not represent an individual system but an *ensemble* of systems.

And for any 'observable' represented by an operator A, there is a well-defined value $v(A)$ that a measurement of A for an individual quantum system would reveal (and not create, since it pre-exists to any measurement). It should be emphasized that, since the value $v(A)$ pre-exists to any measurement, it does not depend on the measuring apparatus and is thus an intrinsic property of the quantum system.

That value $v(A)$ is again a 'hidden variable', because it is a property of the system not included in the quantum state. In other words, there would be a 'hidden variable' $v(A)$ for each operator A. The map v is called a *value-map*.

In this interpretation, we assume that, if

$$\Psi = \sum_n c_n \Psi_n, \tag{3.2}$$

where Ψ_n's are the eigenvectors of A, with eigenvalues λ_n, then the frequency with which one obtains the result $v(A) = \lambda_n$ in an ensemble defined by Ψ is $|c_n|^2$ so that Born's rule holds.

In that view, there is no problem with the collapse or reduction of the quantum state: one updates one's probabilities given some new information; if we learn that the value associated with the physical quantity represented by the operator A is $v(A) = \lambda_n$, then we update our quantum state so that it becomes the corresponding eigenvector Ψ_n.

All would be well if this interpretation was tenable; unfortunately, it is not, for a purely mathematical reason. Indeed, it is a well-known property of quantum mechanics that if the operators A and B commute ($[A, B] = AB - BA = 0$), then they are simultaneously measurable, and the values $v(A)$, $v(B)$, etc. revealed by those measurements would have to satisfy

$$v(AB) = v(A)v(B) \tag{3.3}$$

and

$$v(A + B) = v(A) + v(B). \tag{3.4}$$

However, the following theorem renders the naive statistical interpretation untenable.

[3] For explicit statistical interpretations of quantum mechanics (but not formulated in the same way as here), see Ballentine (1970) and Blokhintsev (1968).

Theorem 3.1 *(No hidden variable theorem)*

Let $\mathcal{A}$ be the set of self-adjoint operators on a Hilbert space $\mathcal{H}$ with $\dim \mathcal{H}$ at least equal to 4.

Then, there does not exist a map $v : \mathcal{A} \to \mathbb{R}$ such that:

1. $\forall A \in \mathcal{A}$,
$$v(A) \text{ is an eigenvalue of } A.$$
2. $\forall A, B \in \mathcal{A}$ with $[A, B] = AB - BA = 0$, (3.3) or (3.4) is satisfied.

Remark: Theorem 3.1 is devised by John Bell (1966) and by Kochen and Specker (1967). For a simple proof provided by David Mermin, see Mermin (1993).[4]

To prove the theorem, we do not need to assume that the map is defined on all operators in $\mathcal{A}$. For example, the Kochen and Specker proof uses the squares of spin matrices, S_x, S_y, S_z, for spin associated with any three-dimensional set of orthogonal vectors x, y, z in $\mathbb{R}^3$. They have the following properties:

1. The eigenvalues of S_x^2, S_y^2, and S_z^2 are 0 and 1.
2. $[S_x^2, S_y^2] = [S_y^2, S_z^2] = [S_z^2, S_x^2] = 0$.
3. $S_x^2 + S_y^2 + S_z^2 = 2$.

From these conditions, assumptions 1 above and (3.4), it follows that the triplet $(v(S_x^2), v(S_y^2), v(S_z^2))$ must be either $(1, 1, 0)$ or $(1, 0, 1)$ or $(0, 1, 1)$.

But that must hold *for every* set of three-dimensional orthogonal vectors x, y, z in $\mathbb{R}^3$. Kochen and Specker were able to exhibit a finite number of such sets so that the above assumption on the values taken by $(v(S_x^2), v(S_y^2), v(S_z^2))$ leads to a contradiction.[5]

The obvious conclusion of this theorem is that one cannot have a statistical distribution of maps that do not exist.

A more general conclusion is that one cannot assume that measuring devices really measure intrinsic properties of physical systems, that pre-exist to their 'measurement'.

Measuring devices must have an active role. This was actually an intuition of Bohr:

> [...] the *impossibility of any sharp distinction between the behavior of atomic objects and the interaction with the measuring instruments which serve to define the conditions under which the phenomena appear* (Niels Bohr (1949), p. 210, quoted in Bell (2004), p. 2 (italics in the original).

[4] As a historical note, one may mention that a similar theorem was proved by von Neumann in his foundational book (von Neumann 1955), assuming equation (3.4) for all operators A and B, even when they do not commute, i.e. when they are not simultaneously measurable. If one makes that assumption, then, take $A = \sigma_x$, $B = \sigma_y$, two Pauli matrices; one easily checks that $v(A) = \pm 1$, $v(B) = \pm 1$, hence $v(A) + v(B)$ equals 0 or ± 2, while $v(A + B) = \pm\sqrt{2}$, which contradicts (3.4) (this argument originates from Bell (2004), p. 164).

[5] In their original argument, Kochen and Specker used 117 such sets, but that number was reduced to 33 by Peres, see Peres (1990, 1991). Note in passing that the Kochen–Specker argument only requires the Hilbert space $\mathcal{H}$ to have a dimension at least equal to 3. But Mermin's proof, that requires a Hilbert space of dimension at least equal to 4, is simpler.

But how do these devices interact with the physical system? In ordinary quantum mechanics, there is no answer to that question because the 'measurements' lead to collapses of the quantum state, which is a *deux ex machina* from the point of view of the Schrödinger evolution. We need a more detailed theory that explains what happens during 'measurements'.

Let us now turn to a statistical and 'hidden variable' theory that provides exactly such an explanation.

3.4 The de Broglie–Bohm theory

The theory of de Broglie (1927) (see his contribution to the 1927 Solvay Conference in Bacciagaluppi and Valentini 2009) and Bohm (1952), popularized by Bell (2004) and Dürr, Goldstein and Zanghì (2012)[6],

1. is a theory of 'hidden variables' (although they are not at all hidden),
2. that accounts for all the phenomena predicted by ordinary quantum mechanics,
3. that is not contradicted by no hidden variable theorems, and
4. that explains why measurements do not in general measure pre-existing properties of a system (in other words, it explains why measuring devices have an 'active role').

To start, let us think of the double-slit experiment, see Fig. 3.1:

How can electrons be both particles and waves? The answer of de Broglie–Bohm is that they are particles *guided* by waves.

In the de Broglie–Bohm's theory, the state of system is a pair (X, Ψ), where $X = (X_1, \dots, X_N)$ denotes the actual positions of all the particles in the system under consideration, that exist, whether we measure them or we 'look' at them or not.

Here, $\Psi = \Psi(x_1, \dots, x_N)$ is the usual wave function, $(x_1, \dots, x_N)$ denoting the arguments of the function Ψ. X are the 'hidden variables' in this theory; this is obviously a misnomer since particle positions are the only things that we ever directly observe (think of the double-slit experiment for example).

The dynamics of the de Broglie–Bohm's theory is as follows[7]; both objects Ψ and X evolve in time:

1. The wave function, at all times, and whether one measures something or not, evolves according to Schrödinger's equation.

[6] There are many expositions of the de Broglie–Bohm theory, see, e.g. Albert (1992) and Tumulka (2004) for elementary introductions, and Bacciagaluppi and Valentini (2009), Bohm and Hiley (1993), Bricmont (2016), Dürr and Teufel (2009), Dürr, Goldstein and Zanghì (2012), Dürr and Lazarovici (2020), Goldstein (2013), Holland (1993), Maudlin (2019), Norsen (2017) and Tumulka (2022) for more advanced ones. There are also pedagogical videos made by students in Munich, available at: `https://cast.itunes.uni-muenchen.de/vod/playlists/URqb5J7RBr.html`.

[7] For simplicity, we limit ourselves to the non-relativistic case and without spin. Extensions to relativistic equations exist (see, e.g. Tumulka (2022) Section 7.6) and we introduce the spin in Section 3.5 but without equations.

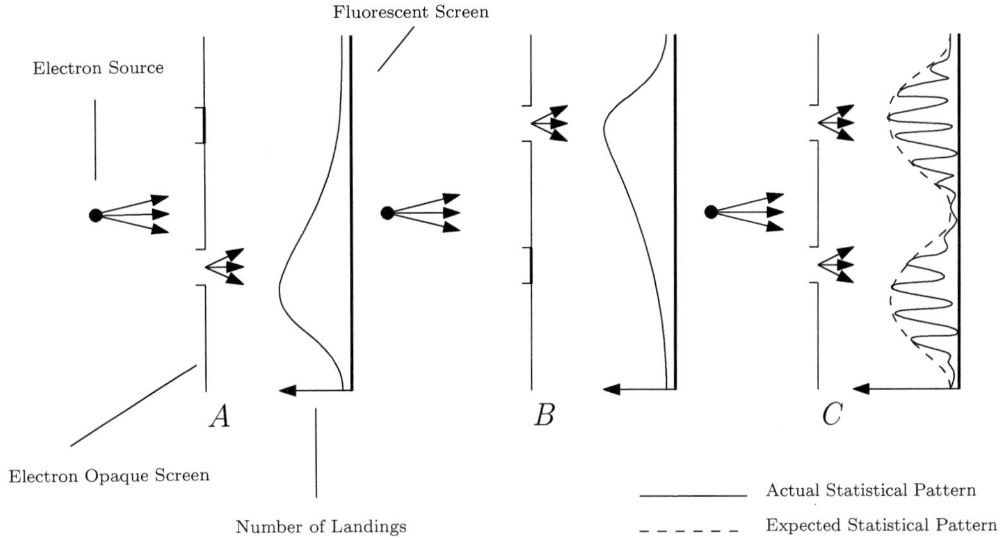

Fig. 3.1: A schematic description of the distribution of particles on the second wall on the right of the figure, depending on whether one or two slits are open in the first wall.

The wave function never collapses for a closed system.[8]

2. The evolution of the positions is guided by the wave function. Writing $\Psi = Re^{iS}$:

$$\frac{d}{dt}X_k(t) = \nabla_k S(X_1(t), \ldots, X_N(t)) \tag{3.5}$$

for $k = 1, \ldots, N$, where $X_1(t), \ldots, X_N(t)$ are the actual positions of the particles at time t.[9]

A classical way to illustrate the dynamics of that theory is given by the numerical solution of the double-slit experiment shown in Fig. 3.2:

The motion *in vacuum*, behind the slits, is highly *non-classical*, i.e. not rectilinear. Note that one can determine *a posteriori* through which hole the particle went. Also note the presence of a nodal line: it turns out that, by symmetry of Ψ, the velocity is tangent to the middle line; thus, particles cannot cross it.

This situation was nicely described by John Bell:

Is it not clear from the smallness of the scintillation on the screen that we have to do with a particle? And is it not clear, from the diffraction and interference patterns, that the motion of

[8]However, in situations that correspond to measurements in ordinary quantum mechanics, there is an effective collapse of the wave function, see Dürr *et al.* (1992), Section 5.

[9]This way of writing the guiding equation was used not only by Bell but also by de Broglie. Bohm, however, wrote the guiding equation with a second derivative in time, making it more similar to Newton's equation. But, in order to do that, he had to add to the classical potential another potential, called the 'quantum potential'. The nature of that potential has prompted many discussions. However, in the formulation used here, that potential plays no role.

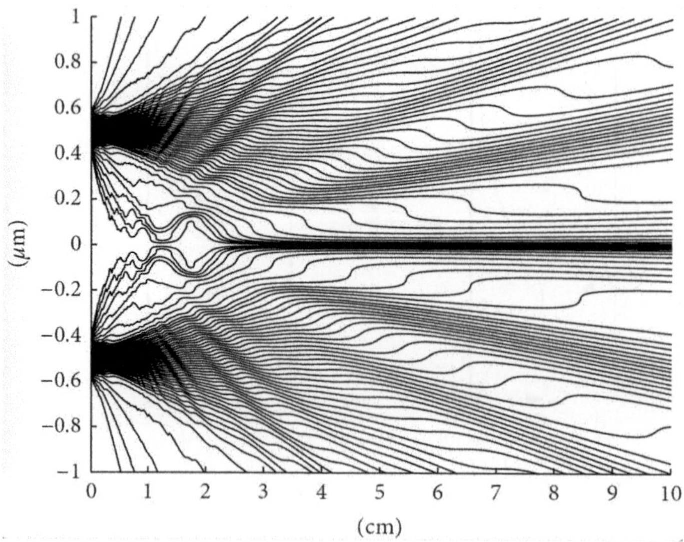

Fig. 3.2: de Broglie–Bohm trajectories computed for the double-slit experiment. Each line represents a particle trajectory. Reproduced with the kind permission of Alexandre Gondran, from Gondran and Gondran (2005). This picture is licensed under the Creative Commons Attribution-Share Alike 4.0 International license.

the particle is directed by a wave? De Broglie showed in detail how the motion of a particle, passing through just one of two holes in the screen, could be influenced by waves propagating through both holes.

And so influenced that the particle does not go where the waves cancel out, but is attracted to where they cooperate. This idea seems to me so natural and simple, to resolve the wave-particle dilemma in such a clear and ordinary way, that it is a great mystery to me that it was so generally ignored (Bell 2004, p. 191).

A natural question is: how does the theory of de Broglie–Bohm account for the statistical predictions of quantum mechanics?

That is due to a property of the de Broglie–Bohm dynamics called *equivariance*, which is illustrated in Fig. 3.3:

Assume that the initial density of particles ρ_0 is (approximately) given by $\rho_0(X) = |\Psi(X,0)|^2$, see the left side of Fig. 3.3. Then, the empirical density of particles at later times ρ_t will satisfy $\rho_t(X) = |\Psi(X,t)|^2$, where $\Psi(X,t)$ is the solution of the Schrödinger equation and ρ_t comes from the guiding equation (3.5). This is illustrated on the right side of Fig. 3.3.

So, if we assume that $\rho_0 = |\psi_0|^2$ at some initial time, it holds at all times. The statistical predictions of quantum mechanics are recovered in the de Broglie–Bohm theory, at least as far as the positions of the particles are concerned. But they are also recovered for all other observables, through an analysis of 'measurements' in that theory, as shown in the next section.

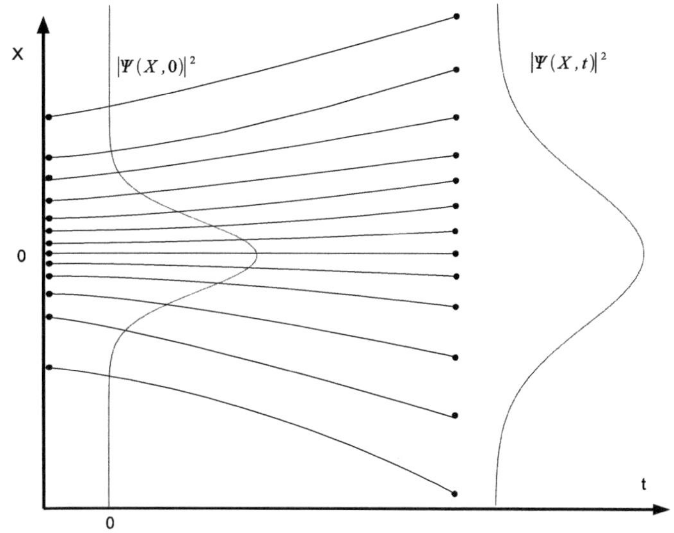

Fig. 3.3: Illustration of the property of equivariance of the $|\Psi(X,t)|^2$ distribution, in one dimension, for a Gaussian Ψ. Each dot represents the position of a particle, both at time 0 and time t, connected by trajectories.

The assumption that $\rho_0 = |\psi_0|^2$ is called quantum equilibrium. In any deterministic dynamical system, statistics of results depend on statistical assumptions on the initial conditions of the system (think of classical statistical mechanics as an example).

And, ultimately, any assumptions on the initial conditions bring us – through a version of the egg and the hen problem – to assumptions on the initial conditions of the universe, which is a complicated issue that we will not discuss here (see Dürr *et al.* (1992)). But, from a somewhat pragmatic point of view, one can put aside the egg and the hen problem, and simply assume a quantum equilibrium distribution of the particles' positions at the beginning of any physical situation that one analyses.

A natural question, to which we now turn, is why is the de Broglie–Bohm theory not refuted by the no hidden variable theorems, since it is a hidden variable theory?

3.5 The de Broglie–Bohm theory and the no hidden variable theorems

The answer to that question reveals the main quality of de Broglie–Bohm's theory, namely that it explains what happens during 'measurements'.

To illustrate this, we consider the following two examples:

First, a 'measurement' of something that does not exist: the spin. Then, a 'measurement' of something that does exist but is not measured by that 'measurement': the momentum.

To give an example of a 'measurement' of something that does not exist[10], consider

[10] This example is taken from Albert (1992), p. 153; see also Daumer *et al.* (1996).

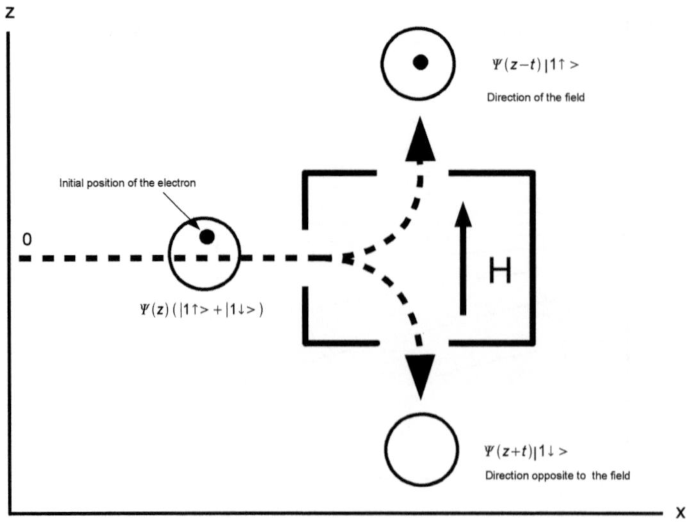

Fig. 3.4: An idealized spin measurement.

a Stern–Gerlach apparatus 'measuring' the spin,[11] see Fig. 3.4, where H is the magnetic field.

The $|1\uparrow\rangle$ part of the state always goes in the direction of the field, and the $|1\downarrow\rangle$ part always moves in the opposite direction.

If the particle is initially in the upper part of the support of the wave function (for a symmetric wave function), it always moves upwards. That is because there is here, as in Fig. 3.2, a nodal line in the middle of Fig. 3.4 that the particles cannot cross due to the symmetry of the wave function.

Now, repeat the same experiment, but with the direction of the gradient of the field reversed (see Fig. 3.5), and let us assume that the particle starts with *exactly the same wave function and the same position as before.*

If the particle is initially in the upper part of the support of the wave function, it again moves upward because of the presence, as in Fig. 3.2, of a nodal line in the middle of the figure that particles cannot cross. But moving upwards means now going in the direction *opposite* to the gradient of the field (since the latter is reversed).

So, the particle whose spin was 'up', will 'have' its spin 'down', although one 'measures' exactly the same quantity (the spin in the vertical direction), with *exactly the same initial conditions (for both the wave function and the position of the particle)*, but with two different arrangements of the apparatus. So, by changing only that arrangement, we obtain different results for the same initial conditions of the particle.

[11] In Section 3.4, we did not write down the guiding equations for particles whose quantum state is a spinor; see e.g. Dürr and Teufel (2009), Section 8.4, for the de Broglie–Bohm theory of those particles.

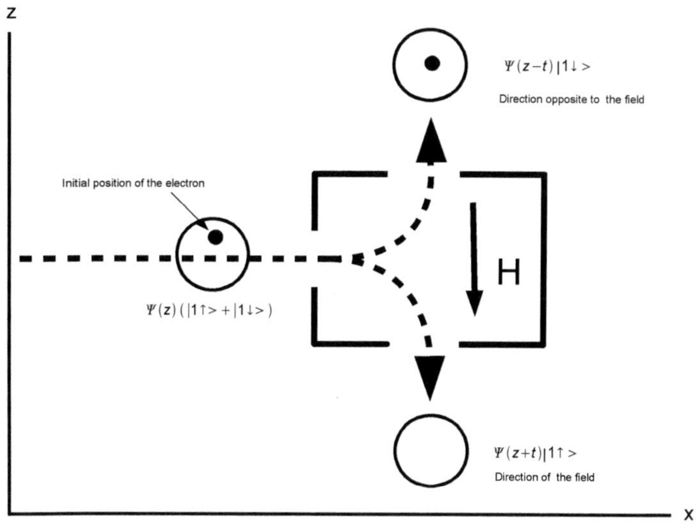

Fig. 3.5: An idealized spin measurement, with the orientation of the field reversed compared to Fig. 3.4.

This is a simple illustration of the active role of the measuring device. Another way to say this would be: *'spin is not real'*. Of course, what is real is that the quantum state has a spinor part. But what this expression means is that there is no pre-existing value of the spin, intrinsic to the particle, that would be revealed by such experiments.

Or, more precisely: *quantum 'measurements' are interactions that do not measure something intrinsic to the particle.*

To give now an example of a 'measurement' of something that does exist but is not measured by that 'measurement', consider a particle in a box.

If the wave function is real (for example, the ground state), then the particle is at rest.[12] Indeed, write (for one particle) $\Psi = Re^{iS}$ and

$$\frac{d}{dt}X(t) = \nabla S(X(t)). \tag{3.6}$$

If $S = 0$, $\frac{d}{dt}X(t) = 0$.

So, we know its velocity: it is zero! Note that we can also, if we want, measure its position to arbitrary accuracy. It seems that this violates Heisenberg's inequality. But, deducing the velocity from the theory is not what 'measuring the velocity' means in quantum mechanics.

To do that 'measurement', one needs to open the box and let the particle move freely; measure its position at some later time and divide the distance travelled by the elapsed amount of time.

[12] For spinless particles, if one includes the spin, particles are not at rest, see Das (2025).

This gives us the 'measured velocity', which is not zero and whose statistical distribution satisfies Heisenberg's inequality, but that has nothing to do with the 'true' velocity of the particle before the 'measurement', which, as we saw, was zero.[13]

All this justifies the idea that 'measurements' are genuine interactions between a system and an apparatus, which was actually an intuition of Bohr mentioned at the end of Section 3.3. But now this follows from the equations of the theory and not from some philosophical *a priori*.

John Bell also emphasized the active role of measurements:

> A second charge [against the word measurement] is that the word comes loaded with meaning from everyday life, meaning which is entirely inappropriate in the quantum context. When it is said that something is 'measured' it is difficult not to think of the result as referring to some pre-existing property of the object in question. This is to disregard Bohr's insistence that in quantum phenomena the apparatus as well as the system is essentially involved (Bell 1990).

Since 'measurements' are interactions that do not reveal pre-existing properties of the system, the de Broglie–Bohm theory is not proved incorrect by the no hidden variable theorems (see also Daumer *et al.* 1996 for a discussion of this fact). Another way to understand this is to observe that results of 'measurements' in the de Broglie–Bohm theory are 'contextual', namely they depend on the arrangements of the apparatuses used in those 'measurements". But, by assumption, the map v whose existence is shown to be impossible by Theorem 3.1 is supposed to assign intrinsic properties to quantum systems that a measurement would simply reveal and thus, do not depend on the properties of the measuring devices. So, there is simply no such map in the de Broglie–Bohm theory.

The de Broglie–Bohm theory is a statistical theory (the distribution of the particles' positions is 'random'), but a consistent one, because it does not associate pre-existing values to 'observables' other than positions.

One can (rather easily) show that if one assumes quantum equilibrium for the distribution of the particles' positions, then Born's rule is satisfied for the 'measurements' of all 'observables', see Dürr *et al.* (1992, 2004) for a general proof of this statement.

3.6 The no hidden variable theorems and non-locality

A standard objection against the de Broglie–Bohm theory is that it is non-local: indeed, in (3.5), the velocity of a given particle at time t depends on the positions of all other particles of the system, at that same time t, no matter how far apart the particles may be. So that if one acts on a given particle so as to change its position, then it instantaneously affects the velocities of all the other particles in the system. For a concrete illustration of how this non-locality expresses itself, see e.g. Bricmont (2016), Section 5.2.1.

We now argue that this non-local character is a quality rather than a defect of the de Broglie–Bohm theory, because Nature itself is non-local, as shown by a combination of an argument due to Einstein, Podolsky and Rosen (EPR) (1935) and another one due to Bell (1964), see also Goldstein *et al.* (2011). We also show how to simplify Bell's argument using the no hidden variables theorems. We start with the EPR argument,

[13] For the calculation of this result, see e.g. Bricmont (2016), Section 5.D.2.

in a slightly modified form and using the Bohm version of their argument, see Bohm (1951), instead of the original one.

3.6.1 The EPR dilemma

Consider a pair of particles, one of which is sent to Alice and another to Bob, (who are situated far apart), and whose quantum state is an entangled one:

$$|\Psi\rangle = \frac{1}{\sqrt{2}}(|A\uparrow\rangle|B\downarrow\rangle - |A\downarrow\rangle|B\uparrow\rangle), \tag{3.7}$$

with A being the particle sent to Alice and B the one sent to Bob. That state, according to ordinary quantum mechanics, means that the spin measured by Alice, in any given direction, will have equal probability to be up or down, but is perfectly correlated with the one of Bob: if the spin is up for Alice (the state $|A\uparrow\rangle$), it will be down for Bob (the state $|B\downarrow\rangle$) and vice versa.

Now comes the crucial question: how does one account for these perfect correlations?

There seems to be only two possibilities:

1. Either each particle carries with it instructions according to which, upon measurement, its spin will be up or down and the instructions of both particles are correlated.

2. Or the measurement of the spin on Alice's side instantaneously influences the result on Bob's side (or vice versa), and this influence guarantees that the result of Bob's spin measurement is correlated with that of Alice.

This is what we call *EPR's Dilemma*. But, EPR did not consider it as a dilemma because they only considered the first possibility as acceptable since the second possibility amounts to admitting the existence of instantaneous actions at a distance that EPR could not even envision, because the existence of those actions seemed to contradict the theory of relativity.

The instructions mentioned in the first possibility are called 'hidden variables' and assuming their existence means that the description of the physical situation by the quantum state is incomplete (which is what EPR thought to have proven).

To be historically correct, this is EPR's (logical) conclusion (without speaking of a dilemma), but not quite their argument[14] which was more complicated and unnecessarily so.

3.6.2 What did Bell show?

Bell's result is rather well known but often misunderstood. Bell showed that if we go back to EPR's dilemma:

1. Either there exist 'hidden variables', namely pre-determined results of experiments (variables not included in the quantum state),

[14] Moreover, this is Bohm's version of their argument; EPR considered positions and momenta of correlated particles instead of the spin variables.

2. Or there exists action at a distance in Nature.

Then, the first assumption alone leads to a contradiction with quantum predictions (very well verified by now) when different spin components are measured on the particle on Alice's side and that on Bob's side.[15] For a simple proof, see Dürr *et al.* (2005).

So, only the second possibility remains: there exist instantaneous actions at a distance in Nature. That is certainly the 'craziest' of the two possibilities, but the only one that is tenable given EPR's dilemma, Bell's argument and the empirical verifications of the violation of his inequalities.

Note that there is no assumption in Bell's argument about:

1. The existence of hidden variables,
2. Realism,
3. Or determinism.

Indeed, 'the existence of hidden variables' is deduced through the EPR dilemma from the assumption of locality, and the last two expressions, whenever they are precisely defined, are similar to the assumption of the existence of hidden variables.

Actually, Bell very clearly stated what he showed (here 'impasse' refers to non-locality and 'analysers' refers the apparatuses):

> Let me summarize once again the logic that leads to the impasse. The EPRB correlations are such that the result of the experiment on one side immediately foretells that on the other, whenever the analyzers happen to be parallel.
>
> If we do not accept the intervention on one side as a causal influence on the other, we seem obliged to admit that the results on both sides are determined in advance anyway, independently of the intervention on the other side, by signals from the source and by the local magnet setting.
>
> But this has implications for non-parallel settings which conflict with those of quantum mechanics. So we *cannot* dismiss intervention on one side as a causal influence on the other (Bell 2004, pp. 149–150, italics in the original).

3.6.3 Misunderstandings of Bell

However, Bell's result has been almost universally considered as a no hidden variable result, hence a proof that Bohr was right (against EPR) and that quantum mechanics is 'complete'.[16]

In a narrow sense, this is true, if one forgets EPR's dilemma, which of course Bell never forgot. If one does not forget this dilemma, Bell's result is a non-locality result (namely proving the existence of instantaneous actions at a distance) rather than a no hidden variable one.

[15] Actual experiments are performed with photons and their polarizations, but both situations are conceptually similar.

[16] Actually, Bohr's answer to the EPR paper (Bohr 1935) is very unclear. For a critical analysis of that paper, see Bell 2004, pp. 155–156.

3.6.4 Schrödinger's generalization of the EPR argument

Bell's argument relied on the statistics of measurements of the spin, or the polarization, in different directions for the particle on Alice's side and Bob's side. But those statistics had to be checked. Hence, the 2022 Nobel Prize.

Could one prove non-locality using only the perfect correlations? Yes, using Schrödinger's ideas in his 1935 paper (Schrödinger 1935).[17] The main innovation of Schrödinger was to extend the previous EPR-type argument to *all* observables, at least given suitable quantum states, namely:

Maximally entangled states
Consider a finite-dimensional (complex) Hilbert space $\mathcal{H}$ of dimension N.
A unit vector Ψ in $\mathcal{H} \otimes \mathcal{H}$ is *maximally entangled* if it is of the form

$$\Psi = \frac{1}{\sqrt{N}} \sum_{n=1}^{N} \psi_n \otimes \phi_n, \tag{3.8}$$

where ψ_n and ϕ_n are orthonormal bases in $\mathcal{H}$.

Since we are interested in quantum mechanics, we refer to those vectors as *maximally entangled states* and we associate, by convention, each space in the tensor product with a 'physical system', namely we consider the set $\{\psi_n\}_{n=1}^{N}$ as a basis of states for physical system 1 (associated with Alice when measurements are made on that system) and the set $\{\phi_n\}_{n=1}^{N}$ as a basis of states for physical system 2 (associated with Bob when measurements are made on that system).

Now, given a maximally entangled state, one can associate with each operator of the form $O \otimes \mathbf{1}$ (meaning that it acts non-trivially only on particle 1 since here $\mathbf{1}$ denotes the identity operator on $\mathcal{H}$) an operator of the form $\mathbf{1} \otimes \tilde{O}$ (meaning that it acts non-trivially only on particle 2), so that:

If ψ_n are eigenstates of O, with eigenvalues λ_n,

$$O\psi_n = \lambda_n \psi_n, \tag{3.9}$$

then the states ϕ_n are eigenstates of $\tilde{O}$, also with eigenvalues λ_n:

$$\tilde{O}\phi_n = \lambda_n \phi_n. \tag{3.10}$$

For the proof of this statement, see Bricmont *et al.* (2020a).

A simple example of a maximally entangled state is the one introduced in our discussion of EPR (3.7) (this shows that this notion of maximally entangled state is a proper extension of the state (3.7)).

[17] For more details about the reasonings provided later, see Bricmont *et al.* (2020a) or Hemmick (1996) and Hemmick and Shakur (2012). See also Bricmont *et al.* (2020b) for an extension of this result using the original EPR variables, positions and momenta, and see Bricmont *et al.* (2022) for a pedagogical summary.

In the previous notation, (3.8), one has

$$\psi_1 = |\uparrow\rangle,$$

$$\psi_2 = -|\downarrow\rangle,$$

$$\phi_1 = |\downarrow\rangle,$$

$$\phi_2 = |\uparrow\rangle.$$

If one takes

$$O = \begin{pmatrix} 1 & 0 \\ 0 & -1 \end{pmatrix} \tag{3.11}$$

which corresponds to the spin operator for system 1 and has eigenvectors ψ_1 with eigenvalue 1 and ψ_2 with eigenvalue -1, it turns out that the operator $\tilde{O}$ is of the form:

$$\tilde{O} = \begin{pmatrix} -1 & 0 \\ 0 & 1 \end{pmatrix} = -O, \tag{3.12}$$

which means that the spin operators for systems 1 and 2 are perfectly anti-correlated, since $\tilde{O}$ is minus the spin operator for system 2; this of course was the situation considered by EPR.

Let us now generalize the EPR dilemma by applying the previous result to spatially separated physical systems. Suppose that we have a pair of physical systems, whose states belong to the same finite-dimensional Hilbert space $\mathcal{H}$. And suppose that the quantum state Ψ of the pair is maximally entangled.

So, the quantum state Ψ of the pair is of the form (3.8). Any 'observable' acting on system 1 is represented by a self-adjoint operator O, which has therefore a basis of eigenvectors. It turns out that we may choose, without loss of generality, the set $\{\psi_n\}_{n=1}^{N}$ as being the eigenstates of O. Let λ_n be the corresponding eigenvalues.

If one measures the corresponding observable $\tilde{O}$ on system 2, the result will be one of the eigenvalues λ_n, each having equal probability $\frac{1}{N}$. If the result is λ_k, the (collapsed) state of the system after the measurement, will be $\psi_k \otimes \phi_k$.

Then, the measurement of observable O, on system 1, will necessarily yield the value λ_k.

So, we have again a situation (like for EPR) of:

Perfect correlations.

In any maximally entangled quantum state, there is, for each operator O acting on system 1, an operator $\tilde{O}$ acting on system 2, such that, if one measures the physical quantity represented by operator $\tilde{O}$ on system 2 and the result is the eigenvalue λ_k of $\tilde{O}$, then, measuring the physical quantity represented by operator O on system 1 will yield with certainty the same eigenvalue λ_k (and vice versa).

Let us go back to EPR's dilemma. In order to explain those perfect correlations, we have only two possibilities:

1. Either the results on both sides are pre-determined before any measurement (there exist 'hidden variables') and are perfectly correlated.
2. Or a measurement on one side instantaneously affects (or creates) the result on the other side: actions at a distance!

Let us call this *Schrödinger's dilemma.* And let us make the first possibility more explicit:

For any operator O acting on system 1, there is a value $v(O)$ that a measurement of O would reveal and not create since it must pre-exist to any measurement. And such a pre-existing value $v(O)$ must exist *for every* O acting on system 1.

To summarize, we have shown:

Theorem 3.2 *(Schrödinger's theorem)*

Let $\mathcal{A}$ be the set of self-adjoint operators on the component Hilbert space $\mathcal{H}$ of a physical system in a maximally entangled state Ψ. Then, assuming locality and the perfect correlations predicted by quantum mechanics, there exists a value-map $v : \mathcal{A} \to \mathbb{R}$, i.e. a map that assigns a value $v(O)$ to any experiment associated with what is called in quantum mechanics a 'measurement of an observable O'.

The conclusion of Schrödinger's theorem 3.2 and that of theorem 3.1 on the non-existence of 'hidden variables' or of value-maps clearly contradict each other.[18] So, the assumptions of at least one of them must be false. To derive Schrödinger's theorem, we assume only the perfect correlations and locality. The theorem on the non-existence of value-maps is purely mathematical.

The perfect correlations are an immediate consequence of quantum mechanics. The only remaining assumption is locality. Hence we can deduce:

Theorem 3.3 *(Non-locality 'theorem')*

The locality assumption is false.

Let us summarize the argument. Given a maximally entangled state, for every operator acting on system 1, there is a corresponding operator acting on system 2 such that the results of the measurements of both operators are perfectly correlated. There are only two possibilities:

– The results of those measurements are pre-determined before any measurement.

– Or non-locality (instantaneous actions at a distance).

The first possibility implies the existence of a value-map. But this is impossible because of the Bell–Kochen–Specker no hidden variables theorem 3.1. The only remaining possibility is non-locality.

[18]The reason we need this general discussion about maximally entangled states, is that the theorem 3.1 on the non-existence of 'hidden variables' holds only in spaces of four dimensions or more (actually three dimensions or more would suffice if we use only assumption (3.4), see Kochen and Specker 1967). So it would not be sufficient to consider only the maximally entangled state (3.7) since there the Hilbert space denoted by $\mathcal{H}$ is two-dimensional.

Of course, this is not a purely mathematical theorem since it depends on a physical assumption: the perfect correlations associated with maximally entangled state.

Nevertheless, the only things to check experimentally are those perfect correlations. There is no need for the Clauser–Aspect–Zeilinger experiments on correlations between 'measurements' of spins or polarizations in different directions for Alice and Bob.

How to reconcile this non-locality with relativity is a difficult open question, see Maudlin (2011) and Allori (2025) for more detailed discussions.

Finally, note that a famous 19th century British scientist expressed the EPR–Bell–Schrödinger proof of non-locality quite concisely:

When you have eliminated the impossible, whatever remains, however improbable, must be the truth (Sherlock Holmes).

3.7 Conclusions

The de Broglie–Bohm theory is not yet another 'interpretation' of quantum mechanics. It is not a different theory than quantum mechanics. It is simply the (rational) completion of quantum mechanics, which is manifestly incomplete since it does not speak of what happens outside of laboratories.

And ordinary quantum mechanics is simply de Broglie–Bohm theory applied to what happens in laboratories. Once this is understood, there is no need to run around in search of 'alternatives' to quantum mechanics or of 'interpretations' of it.

References

Albert, D. (1992). *Quantum Mechanics and Experience*, Cambridge, Massachusetts: Harvard University Press.

Allori, V. (2025). Relativistic pilot-wave theories as the rational completion of quantum mechanics and relativity. In Oldofredi, A. (ed.) *Guiding Waves in Quantum Mechanics: One Hundred Years of de Broglie–Bohm Pilot-Wave Theory*. Oxford: Oxford University Press.

Bacciagaluppi, G. and Valentini, A. (2009). *Quantum Mechanics at the Crossroads. Reconsidering the 1927 Solvay Conference*. Cambridge: Cambridge University Press.

Ballentine, L. E. (1970). The Statistical Interpretation of Quantum Mechanics. *Reviews of Modern Physics*, **42**: 358–381.

Bell J.S. (1964). On the Einstein–Podolsky–Rosen paradox. *Physics*, **1**: 195–200. Reprinted as Chap. 2 in Bell (2004).

Bell, J.S. (1996). On the problem of hidden variables in quantum mechanics. *Reviews of Modern Physics*, **38**: 447–452. Reprinted as Chap. 1 in Bell (2004).

Bell, J.S. (2004). *Speakable and Unspeakable in Quantum Mechanics. Collected Papers on Quantum Philosophy*, 2nd edn, with an introduction by Alain Aspect. Cambridge: Cambridge University Press, 1st edn. 1987.

Bell, J.S. (1990). Against measurement. *Physics World*, **3**: 33–40. Reprinted as Chap. 23 in Bell (2004).

Blokhintsev, D. I. (1968). *The Philosophy of Quantum Mechanics*, Dordrecht: Reidel. Original Russian edition, 1965.

Bohm, D. (1951). *Quantum Theory*. New York: Dover Publications, new edition 1989; Englewood Cliffs, NJ: Prentice Hall,1st edn.

Bohm, D. (1952). A suggested interpretation of the quantum theory in terms of "hidden" variables. I & II. *Physical Review*, **85**(2): 166–193.

Bohm, D. and Hiley, B.J. (1993). *The Undivided Universe. An Ontological Interpretation of Quantum Theory*. London: Routledge.

Bohr, N. (1935). Can quantum-mechanical description of physical reality be considered complete?. *Physical Review*, **48**: 696–702.

Bohr, N. (1949). Discussion with Einstein on epistemological problems in atomic physics. In Schilpp, P.A. (ed.) *Albert Einstein, Philosopher-Scientist*, pp. 201–241. Evanston, IL: The Library of Living Philosophers.

Bricmont, J. (2016). *Making Sense of Quantum Mechanics*. Berlin: Springer.

Bricmont, J. Goldstein, S. and Hemmick, D. (2020a). Schrödinger's paradox and proofs of non-locality using only perfect correlations. *Journal of Statistical Physics*, **180**: 74–91.

Bricmont, J. Goldstein, S. and Hemmick, D. (2020b). EPR-Bell-Schrödinger Proof of non-locality Using Position and Momentum. In Allori, V., Bassi, A., Dürr, D. and Zanghì, N. (eds.) *Do Wave Functions Jump?Perspectives of the Work of GianCarlo Ghirardi*, pp. 5–33. Cham: Springer.

Bricmont, J. Goldstein, S. and Hemmick, D. (2022). From EPR, Schrödinger paradox to non-locality, based on perfect correlation. *Foundations of Physics*, **52**: 53.

Das, S. (2025). Spin-aware movement of electrons and time-of-flight momentum spectroscopy. In Oldofredi, A. (ed.) *Guiding Waves in Quantum Mechanics: One Hundred Years of de Broglie–Bohm Pilot-Wave Theory*. Oxford: Oxford University Press.

Daumer, M., Dürr, D., Goldstein, S. and Zanghì, N. (1996). Naive realism about operators. *Erkenntnis*, **45**: 379–397.

Dürr, D., Goldstein, S. and Zanghì, N. (1992). Quantum equilibrium and the origin of absolute uncertainty. *Journal of Statistical Physics*, **67**: 843–907.

Dürr, D., Goldstein, S. and Zanghì, N. (2004) Quantum equilibrium and the role of operators as observables in quantum theory. *Journal of Statistical Physics*, **116**: 959–1055. Reprinted as Chap. 3 in Dürr *et al.* (2013).

Dürr, D., Goldstein, S., Tumulka, R. and Zanghì, N. (2005). John Bell and Bell's theorem. In Borchert, D.M. (ed.), *Encyclopedia of Philosophy*. USA: Macmillan Reference.

Dürr, D., Goldstein, S. and Zanghì, N. (2013). *Quantum Physics Without Quantum Philosophy*. Berlin, Heidelberg: Springer.

Dürr, D. and Teufel, S. (2009). *Bohmian Mechanics. The Physics and Mathematics of Quantum Theory*. Berlin, Heidelberg: Springer.

Dürr, D. and Lazarovici, D. (2020). *Understanding Quantum Mechanics. The World According to Modern Quantum Foundations*. Cham (Switzerland): Springer.

Einstein, A., Podolsky, B. and Rosen, N. (1935). Can quantum-mechanical description of physical reality be considered complete?. *Physical Review*, **47**: 777–780.

Goldstein, S., Norsen, T., Tausk, D.V. and Zanghì, N. (2011). Bell's theorem. *Scholarpedia*, **6**(10): 8378.

Goldstein, S. (2013). Bohmian mechanics. In Zalta, E. (ed.) *The Stanford Encyclopedia of Philosophy*, (Spring 2013 Edition). Available on `plato.stanford.edu/archives/spr2013/entries/qm-bohm/`.

Gondran, M. and Gondran, A. (2005). Numerical simulation of the double slit experiment with ultracold atoms. *American Journal of Physics*, **73**: 507–515.

Hemmick, D.L. (1996). *Hidden variables and non-locality in quantum mechanics*. Doctoral thesis, Rutgers University, October, 1996. Available on https://sites.google.com/site/dlhquantum/doctoral-thesis and arXiv:quant-ph/0412011v1.

Hemmick, D.L. and Shakur, A.M. (2012). *Bell's Theorem and Quantum Realism. Reassessment in Light of the Schrödinger Paradox*. Heidelberg: Springer.

Holland, P. (1993). *The Quantum Theory of Motion. An Account of the de Broglie–Bohm Causal Interpretation of Quantum Mechanics*. Cambridge: Cambridge University Press.

Kochen, S. and Specker, E.P. (1967). The problem of hidden variables in quantum mechanics. *Journal of Mathematics and Mechanics*, **17**: 59–87.

Maudlin, T. (2011). *Quantum non-locality and Relativity*, Cambridge: Blackwell. 1st edn, 1994.

Maudlin, T. (2019). *Philosophy of Physics: Quantum Theory*. Princeton: Princeton University Press.

Mermin, D. (1993). Hidden variables and the two theorems of John Bell. *Reviews of Modern Physics*, **65**: 803–815.

Norsen, T. (2017). *Foundations of Quantum Mechanics: An Exploration of the Physical Meaning of Quantum Theory*. Cham (Switzerland): Springer.

Peres, A. (1990). Incompatible results of quantum measurements. *Physics Letters A*, **151**: 107–108.

Peres, A. (1991). Two simple proofs of the Kochen–Specker theorem. *Journal of Physics A: Mathematical and General*, **24**: L175–L178.

Philippidis, C., Dewdney, C. and Hiley, B.J. (1979). Quantum interference and the quantum potential. *Il Nuovo Cimento B*, **52**: 15–28.

Schrödinger, E. (1935). Die gegenwärtige Situation in der Quantenmechanik. *Naturwissenschaften*, **23**: 807–812; 823–828; 844–849. English translation: The present situation in quantum mechanics, translated by Trimmer, J.D. in *Proceedings of the American Philosophical Society*, **124**, pp. 323–338 (1980). Reprinted in Wheeler and Zurek (1983), pp. 152–167.

Tumulka, R. (2004). Understanding Bohmian mechanics – A dialogue. *American Journal of Physics*, **72**: 1220–1226.

Tumulka, R. (2018). Paradoxes and Primitive Ontology in Collapse Theories of Quantum Mechanics. In Gao, S. (ed.) *Collapse of the Wave Function*, pp. 134–153. Cambridge: Cambridge University Press.

Tumulka, R. (2022). *Foundations of Quantum Mechanics*, Lecture Notes in Physics. Cham (Switzerland): Springer.

von Neumann, J. (1955). *Mathematical Foundations of Quantum Mechanics*, Princeton: Princeton University Press. First edition in German *Mathematische Grundlagen der Quantenmechanik* (1932).

Wheeler, J.A. and Zurek, W.H. (eds) (1983). *Quantum Theory and Measurement*. Princeton: Princeton University Press.

Part II

Historical Perspectives

4

On the vicissitudes of Bohm's 1952 interpretation of quantum mechanics

Olival Freire Junior

4.1 Introduction

When David Bohm published an interpretation of quantum mechanics in terms of hidden variables in 1952, he could not predict the complex fate it would have: disagreement of priorities with Louis de Broglie, a poor reception among most physicists, an enthusiasm among a few – particularly in France, including an alliance with de Broglie – frustration for the negative responses, its abandonment, and a revival of interest. After Bohm passed away, the interpretation gained its own life with different and competing approaches and supporters. According to his long-standing friend, the American physicist Melba Philipps, 'It is too bad, very sad indeed, that he did not live to see how his reputation has shot up recently. His interpretation of quantum mechanics is becoming respected not only by philosophers of science but also by "straight" physicists'.[1] Nowadays, indeed, Bohm's interpretation is one among other possible interpretations of quantum mechanics. It has been further developed into what we call Bohmian mechanics, including its application to applied physics and its relativistic extensions and quantum field theories[2], used as a probe in cosmology and as a theoretical framework in quantum cosmology, and it continues to be developed along lines closer to Bohm's and de Broglie's original suggestions.

In this chapter, Section 4.2 is a short biographical summary of David Bohm. Then, in Section 4.3, I focus on the initial poor reception of Bohm's proposal, as well as his sidelining in the late 1950s, which I claim was related to the prejudices among physicists concerning the hidden-variable issues and foundational issues at large. In Section 4.4, I analyse its later revival in the mid-1970s, which happened without any new evidence corroborating it, and claim that this was related to a new intellectual environment concerning research on the foundations of quantum mechanics. These are not the only questions of historical interest concerning Bohm's work on quantum mechanics. Examples of interest I am not delving into here are the role of Marxism

[1] Phillips' letter is in Freire Junior 2015, p. 63. The full correspondence among Bohm and his American friends in the 1950s (Phillips, Miriam Yevick, and Hanna Loewy) are published and commented in Talbot (2017).

[2] For an recent extensions of the Bohmian approach, see the contributions of Allori, Tumulka, Vassallo & Naranjo to this volume; for applications of the pilot-wave theory, see chapters of Das, Lombardi and Fortin, and Oianguren-Asua, Solé, Destefani, and Oriols.

in Bohm's shift to the hidden-variable interpretation as well as the strength of his commitment to deterministic descriptions at the same time.[3] In Section 4.5, I highlight that even nowadays there is a persistent misunderstanding regarding what kind of hidden variables may be compatible with quantum mechanics, as the recent discussions around the 2022 Physics Nobel Prize reveal. In the same section, I conclude by dscussing some lessons from this history for the practice of science; they are related to the role played by bias even in the most advanced fields of scientific research, and the need to overcome them.

4.2 Short biographical review

The singular most prominent feature of Bohm's professional career was his ongoing quest for understanding the quantum. It was not his only achievement as his contributions to the study of plasma as collective phenomena are also landmarks in the 20th-century history of physics. These studies were conducted during his doctoral studies at Berkeley as part of the war times effort and were continued in the late 1940s and early 1950s at Princeton, where he began his teaching services with his students Eugene Gross and David Pines. In addition, Bohm was a prolific writer with contributions to the philosophy of science.[4]

All these works, however, fade in comparison to the half-century of ongoing work on the understanding of quantum theory. His first attempts to cope with the quantum issues were condensed in the 1951 textbook *Quantum Theory* (Bohm 1951), where he presented a unique view combining Niels Bohr's complementarity principle with a realistic view of the quantum world. In addition, he analysed the Einstein–Podolsky–Rosen (EPR) argument for the incompleteness of quantum theory, claiming quantum mechanics was right; presented the case of a pair of electrons and their spins, exhibiting the singlet states, which would be later instrumental for John Bell's reasoning on the EPR argument; and showed how quantum states could be used to deal with measurement treating both systems and devices as quantum systems. Later on, when the EPR argument was carried to the labs, many authors called them EPR-B experiments in recognition of Bohm's contributions.

The same year the *Quantum Theory* textbook was published, Bohm changed his view on quantum mechanics dramatically. Motivations for this were diverse and included conversation with Albert Einstein on various issues concerning quantum mechanics. He suggested a new approach, far from what he called the usual interpretation. In Bohm's suggested interpretation, all quantum systems continuously evolve in space and time. More than merely suggesting the interpretation, Bohm built a model to describe particles with well-defined trajectories, through analogies between the quantum theory and the classical Hamilton–Jacobi equation. This model implied the introduction of a so far unknown potential, linking instantaneously distant particles. Furthermore, when this model was applied both to the quantum systems and measuring devices, the experimental outcomes did not reveal pre-existing variables rather

[3] See, for instance, Forstner (2008) and Van Strien (2020) and her contribution in this volume.

[4] On the impact and citations of Bohm's plasma works, see Freire Junior (2019), pp. 39, 51, 229. For a biography of Bohm, see Freire Junior (2019). On Bohm and McCarthyism, see Mullet (2008). On plasma work's inspirations, see Kojevnikov (2002).

the results were obtained through the interaction of the systems and the measurement apparatus. Bohm called the new potential 'quantum potential' and proposed the 'hidden-variables' interpretation as this model embraced more variables than those accepted by quantum mechanics. Later, following the first debates on this interpretation, Bohm accepted to call it the 'causal interpretation'. The most important of Bohm's achievements was that using this model he was able to reproduce all the predictions from quantum mechanics in the non-relativistic domain; which meant physicists now had a consistent alternative interpretation to the already known Niels Bohr's principle of complementarity.

When this interpretation was presented, it was constrained by certain vicissitudes involving Bohm's life and ideas. He suffered from McCarthyism, an anti-communist campaign in the US, and lost his position at Princeton University. From one of the most promising American theoretical physicists of his generation, he went into lifelong exile, working successively in Brazil, Israel, and later the United Kingdom, never settling back in the United States. This personal upheaval was coupled with the mixed reception of his ideas. He was warned by Einstein and Pauli of the similarity between his ideas and Louis de Broglie's 1927 proposal of a pilot wave guiding the quantum systems. Bohm's theory, however, was more developed than de Broglie's early ideas, for instance dealing with the measurement process and the EPR thought experiment, and what could have been a dispute about priorities became a long-standing alliance. A second obstacle, the poor initial reception of his proposal among his fellow physicists, would never disappear, and on this topic we have commented later.

Through the 1950s, Bohm continued to defend the hidden-variable interpretation but did not limit his quest for the understanding of the quantum. Indeed, with his student Yakir Aharonov, he used some experimental results obtained by Wu and Sakhnov, in a different context, to analyse the EPR argument, thus paving the way that would lead to the experiments with Bell's theorem (Silva 2022). Furthermore, they analysed an idealized situation of an electrical charge near a solenoid and suggested what is now a well-established quantum effect, the Aharonov–Bohm effect. This work culminated in the paper 'Significance of electromagnetic potentials in quantum theory', which has been by far Bohm's most cited scientific essay.

Bohm's life and ideas underwent significant changes in the transition from the 1950s to the 1960s. In the wake of the Soviet invasion of Hungary and the denunciation of Stalin's crimes, he severed his ideological ties with Communism. Then, he turned to a rapprochement with the Indian thinker Jiddu Krishnamurti. In addition, he realized that the US government had withdrawn his American citizenship after getting his Brazilian citizenship. Bohm had applied for the latter because the American consulate in São Paulo had confiscated his passport, telling him he could have it back in case he returned to the US, which he was afraid of in the mid-1950s. Brazilian citizenship and passport had been the manner he found to travel abroad (Freire Junior 2019, pp. 77, 100). Only in the late 1980s, as a result of a judiciary process, he was able to regain it. Thus, 20th-century vicissitudes, such as McCarthyism, the Cold War, and ideological splits, heavily constrained his life.

As for his suggested interpretation of quantum mechanics in terms of hidden variables, Bohm became discouraged by its poor reception and abandoned it. However, he never gave up attempts to understand the meaning of quantum physics. He continued this quest in collaboration, or in dialogue, with scientists such as Basil Hiley, Jeffrey Bub, Roger Penrose, John Bell, Karl Popper, Mario Bunge, Peter Holland, and many others. In the late 1970s, and thanks to the work of two of Bohm and Hiley's students, Chris Dewdney and Chris Philippidis, he returned to the interpretation contained in his 1952 papers. He spent the last decade of his life, in collaboration with Basil Hiley, dedicated to extending his early interpretation to the relativistic domain and to quantum fields, setting it in the wider frame of the foundational issues in quantum mechanics. The level he was able to reach with this work is summarized in the volume *The Undivided Universe* (Bohm and Hiley 1993), which was already finished when he died in 1992. This book has the meaningful title of *An Ontological Interpretation of Quantum Theory*, which reflected Bohm's lasting commitment to scientific realism and the quest for an ontology of the quantum world, a term already present in his 1951 textbook. Still, there was no reference to whether this world could be deterministic or probabilistic, while in his early presentation of the hidden-variables interpretation it was frequently referred to by himself as the causal interpretation. Indeed, Bohm and Hiley (1993), while defending the ontological view, wrote that "causal" and "hidden variables" are terms "too restrictive", ... "nor is this sort of theory necessarily causal", and "the question of determinism is therefore a secondary one, while the primary question is whether we can have an adequate conception of the reality of a quantum system, be this causal or be it stochastic or be it of any other nature". Thus, Bohm spent most of his scientific life acquiring a better understanding of quantum mechanics, a goal to which he dedicated almost all his efforts.

Contrary to the difficult reception of Bohm's pilot-wave theory, nowadays, his ideas on quantum mechanics have been developed and used in fields as diverse as cosmology, quantum engineering and quantum chaos to mention a few.

4.3 The early poor reception of hidden-variable interpretation

Bohm's paper with the suggestion of interpreting quantum mechanics in terms of hidden variables received more than 3,600 citations; thus, by any standard, it was a most well-received scientific paper. However, when you break down these numbers year after year, another perspective appears. In the first quarter of its existence, there were few citations. If you read the papers citing it in the early 1950s, most criticize Bohm's proposal. This poor early reception requires an explanation.

We have certain second-hand accounts or recollections of the manner some fellow scientists received Bohm's ideas.[5] However, the archival evidence is robust enough to give a vivid impression of this without depending on oral recollections. The most vocal and influential criticisms came from Wolfgang Pauli and Léon Rosenfeld. They resonated on Heisenberg's and Max Born's criticisms and influenced physicists everywhere.[6] Pauli had known and criticized Bohm's proposal since its first version,

[5] As an example of what I am calling second-hand recollection, see Max Dresden's recollections (as told by Peat and Cushing) of Oppenheimer saying, 'if we cannot disprove Bohm, then we must agree to ignore him', as reported by Becker (2018), pp. 157–159.

[6] All the following quotations are in Freire Junior (2019), pp. 81–97, where there are also reactions from Einstein, Schrödinger, and von Neumann, among others, and a detailed analysis of the early

criticisms which were however cleared up by Bohm. Later, Pauli focused his criticisms on issues of symmetry and predictability of new phenomena. In the published papers, Pauli wrote, as the causal interpretation does not have 'any effects on observable phenomena, neither directly nor indirectly [...] the artificial asymmetry introduced in the treatment of the two variables of a canonically conjugated pair characterizes this form of theory as artificial metaphysics'. And yet, '[if the] new parameters could give rise to empirically visible effects [...] they will be in disagreement with the general character of our experiences, [and] in this case this type of theory loses its physical sense'. In private letters Pauli was more acerbic. To Markus Fierz, he commented that Catholics and Communists depended on determinism to buttress their eschatological faiths, the former in the heaven to come and the latter in paradise on earth; references directed at Louis de Broglie and at Bohm and Vigier, respectively. He also warned Giuseppe Occhialini about 'Bohm in São Paulo and his 'causal" quantum theory', as Occhialini had worked in Brazil and continued scientific collaboration there.

Rosenfeld's criticisms were more philosophical, even ideological, because despite their shared Marxist views, their interpretation of quantum mechanics distinguished the two from each other. Rosenfeld denied the very existence of a controversy on the interpretation of quantum physics, writing to Bohm, "I certainly shall not enter into any controversy with you or anybody else on the subject of complementarity, for the simple reason that there is not the slightest controversial point about it". Rosenfeld saw the causal interpretation as a metaphysical regression, writing, "determinism has not escaped this fate [becoming an obstacle to progress]; the physicist who still clings to it, who shuts his eyes to the evidence of complementarity, exchanges (whether he likes it or not) the rational attitude of the scientist for that of the metaphysician".

Still according to Rosenfeld, every good Marxist should understand that 'the latter [metaphysician], as Engels aptly describes him, considers things "in isolation, the one after the other and the one without the other", as if they were "fixed, rigid, given once for all" '. Rosenfeld believed that complementarity was a dialectical achievement that had to be defended not only against Bohm's criticisms but also against Soviet critics who blamed it for introducing idealism into physics. Similar to Pauli, but more strongly and emphatically, Rosenfeld wrote to a number of colleagues around the world trying to convince them to keep their distance from Bohm's ideas. He went so far as to ask *Nature*'s editors not to publish a paper by Bohm entitled *A causal and continuous interpretation of the quantum theory*.

The echoes of this harsh reaction arrived in the mid-1960s and to the ears of younger physicists; the case in point being John Bell because he had been influenced by Bohm's work. On 2 December 1966, when he had just published his now well-known theorem on local hidden variables and quantum mechanics, he received the following letter from Léon Rosenfeld: 'I need not tell you that I regard your hunting hidden parameters as a waste of your talent; I don't know, either, whether you should be glad or sorry for that'. Later, when John Clauser was making one of the first experimental tests of Bell's theorem, a potential employer, the head of the Physics Department at San Jose, wrote to Bernard d'Espagnat asking if what Clauser was doing was 'real

reception of Bohm's interpretation. For a detailed analysis of Pauli's and Rosenfeld's reactions, see Oldofredi (2023).

physics'. The irony of this story is that those experiments were the basis, 50 years later, for the 2022 Physics Nobel Prize.[7]

John Bell was one of the first to realize that physicists' attitudes were neither adhesion to the complementarity view nor rejection of hidden variables. Rather, and worse, it was based on accommodation, considering that there was no good physics to be done on this topic as it has already been dealt with by the founding fathers of quantum mechanics. In 1966, he and Michael Nauenberg wrote, "[w]e emphasize not only that our view [that quantum mechanics is, at the best, incomplete] is that of a minority but also that current interest in such questions is small. The typical physicist feels that they have long been answered, and that he will fully understand just how if ever he can spare 20 minutes to think about it (Bell and Nauenberg 1966)". The poor initial reception of Bohm's proposal therefore resulted from the prevailing views among physicists about the meaning and implications of his research for the future of physics.

4.4 The later revival

From the early 1970s on, interest in the debate on the foundations of quantum mechanics and, in particular, the subject of hidden variables began to grow. Bell published his theorem contrasting quantum predictions with the assumption of local realism, or rather, local hidden variables. The race for the first experiments dealing with this theorem began. The measurement problem prompted several physicists to begin to work on it. Hugh Everett's interpretation gained traction and work on quantum cosmology was enhanced, especially in the hands of Bryce DeWitt. Interestingly, it was not a story of theory plus experiments; in fact, the *air du temps* in interest in the foundations of quantum mechanics was changing. This change was noted by M. Hammerton, from the Medical Research Centre, Cambridge, UK, who encapsulated it in the following letter to the magazine Physics Today (quoted in Freire Junior 2015, p. 228):

> The very interesting contributions to the quantum mechanics debate in your April issue, and the paper by DeWitt which triggered them, exemplify the highly complex and subtle ways in which scientific opinion can change. When I was an undergraduate reading physics 20 years ago, [...] the Copenhagen line was "scientific", anything else was meaningless, mumbo-jumbo, or, at best, mistaken. Now the curious thing is that, as far as I am aware, there has been no major finding or theoretical insight that could be held to demolish or supersede this interpretation. Nevertheless, there is now considerable dissatisfaction with it, and a willingness to regard other points of view—for example, hidden variables—as being at least respectable.

Bohm followed all this change with huge interest. He and Basil Hiley had animated a seminar at Birkbeck College, where Bohm was working since the early 1960s, with the occasional participation of Roger Penrose and Karl Popper, where the implications of quantum mechanics were under debate. Jeffrey Bub's doctoral dissertation was developed as part of that conversation. At the beginning they did not realize the full meaning of the Bell theorem, but in the early 1970s, Bohm and Hiley arrived at the conclusion that non-locality was the main feature of quantum mechanics, a feature also present in Bohm's original hidden-variable interpretation.

[7] For the letter to Bell and to d'Espagnat, see Freire Junior 2015, pp. 243 and 271, respectively.

They derived quantum non-locality from the quantum potential included in that interpretation and compared their view with Niels Bohr's concept of wholeness from the complementarity principle. They highlighted the similarities between the two interpretations, stated theirs was more intuitive, and acknowledged difficulties present in their interpretation related to the compatibility between the quantum potential and relativity. The paper was submitted on 29 March 1974 and titled *On the Intuitive Understanding of Nonlocality as Implied by Quantum Theory* (Bohm and Hiley 1975). Bohm maintained his interest in the experiments related to Bell's theorem, and two years later Bohm and Hiley analysed one of these experiments, which had been carried out in London by Wilson, Lowe and Butt (Bohm and Hiley 1976). Towards the end of the 1970s decade, Bohm and Hiley were surprised by the work of two of their doctoral students, Chris Philippidis and Chris Dewdney. They had rediscovered Bohm's 1952 paper by themselves and challenged Hiley to show them what were the flaws in that paper. After a weekend studying the paper, Hiley arrived at the conclusion that there were no flaws, something the students had already suspected. Then, using a computer the students calculated the trajectories predicted by Bohm's original paper. The pictures with the paths and plots of the quantum potential have since become iconic in the debates on the foundations of quantum mechanics.[8] It was then that Bohm came back with vigour to the analysis of his original ideas, trying to connect them with the wider development that the study of foundations had undergone.

Bohm moved back to his early hidden-variable interpretation, which he had abandoned throughout the 1960s and 1970s due to its bad reception among his fellow physicists, but had some similarities with Louis de Broglie's own trajectory. In 1928, de Broglie had abandoned the theory he had suggested because he felt physicists did not support it. Only in 1952, following Bohm's work, he came back to his original ideas. In both cases, because of bad receptions by the other specialists in the field, original and exciting scientific ideas of Louis de Broglie and David Bohm's could not be developed.

David Bohm passed away on October 27, 1992. He could not follow the increasing broader interest in his interpretation. Influenced by John Bell's interest in Bohm's ideas, a trio of physicists—Detlef Dürr, Sheldon Goldstein, and Nino Zanghì—launched what they called Bohmian mechanics, a more direct derivation of Bohm's insights without introducing the quantum potential, which has since attracted attention among the physicists. Furthermore, Bohm's ideas in cosmology have been used by several physicists, such as Anthony Valentini, Nelson Pinto-Neto, and J. Acácio de Barros. If one adds the developments led by Bohm's long-time collaborator, Basil Hiley, extending Bohm's initial approach we can say Bohm's legacy is at the inception of a flourishing and diversified research on the foundations of quantum mechanics. It is a positive response for someone who was considered one of the great quantum dissidents (Freire 2015, 2019).

[8] For the episode of the students, see Freire Junior (2019), pp. 170–174. For a first-hand recollection, see Dewdney (2023).

4.5 Enduring misconceptions about Bohm's ideas and conclusion

While the negative attitude against the interpretation Bohm had suggested based on hidden variables no longer existed, there remained misconceptions. From time to time a physicist would say that experiments on Bell's theorem proved quantum entanglement and disproved hidden variables, which is a half-truth, as one should add ... disproved local hidden variables. This terminology is important as Bohm's original idea is as non-local as quantum mechanics; thus, it was not disproved by the long series of the experiments on Bell theorem.[9] Bohm's interpretation therefore survives as one of the possible interpretations of quantum mechanics.[10]

When in November 2022 the Swedish Academy announced the Physics Nobel Prize for John Clauser, Alain Aspect, and Anton Zeilinger for 'experiments with entangled photons, establishing the violation of Bell inequalities and pioneering quantum information science', a short historical retrospective was presented at the press conference. The names of Einstein and Bell were highlighted but no mention of Bohm was made. In contrast, the magazine *Physics Today* presented a more comprehensive description of the prize, without leaving Bohm's name aside (Hill 2022). Thus, while the history of Bohm's name and work remains an inconvenient truth among certain physicists, the meaning of his legacy has been disputed among professional practitioners. Memory, in science as well as in society, is subject to struggle and reconstruction.

References

Becker, A. (2018). *What Is Real. The Unfinished Quest for the Meaning of Quantum Physics.* New York: Basic Books.

Bell, J.S., Nauenberg, M. (1966). The moral aspect of quantum mechanics. In De Shalit, A., Feshbach, H., Van Hove, L. (eds.) *Preludes in Theoretical Physics*, pp. 279–286. Amsterdam: North Holland.

Bohm, D. (1951). *Quantum Theory.* New York: Prentice-Hall.

Bohm, D. (1952). A suggested interpretation of the quantum theory in terms of hidden variables. I & II. *Physical Review*, **85**(2): 166–193.

Bohm, D. and Hiley, B. J. (1975). Intuitive understanding of nonlocality as implied by quantum-theory. *Foundations of Physics*, **5**(1): 93–109.

Bohm, D. and Hiley, B. J. (1976). Nonlocality and Polarization Correlations of Annihilation Quanta. *Il Nuovo Cimento B*, **35**: 137–144.

Dewdney, C. (2023). Rekindling of de Broglie-Bohm Pilot Wave Theory in the Late Twentieth Century: A Personal Account. *Foundations of Physics*, **53**: 24.

Forstner, C. (2008). The early history of David Bohm's quantum mechanics through the perspective of Ludwik Fleck's thought-collectives. *Minerva*, **46**(2): 215–229.

[9] On these experiments, see Freire Junior (2015, 2022) and Kaiser (2022).

[10] Jean Bricmont's contribution, this volume, on no-go hidden-variable theorems, explains why Bohm's proposal is not affected by those experimental results.

Freire Junior, O. (2015). *The Quantum Dissidents: Rebuilding the Foundations of Quantum Mechanics (1950–1990)*. Berlin: Springer.

Freire Junior, O. (2019). *David Bohm - A Life Dedicated to Understanding the Quantum World*. Cham, Switzerland: Springer.

Freire Junior, O. (2022). Alain Aspect's experiments on Bell's theorem: a turning point in the history of the research on the foundations of quantum mechanics. *European Physical Journal D*, **76**: 248.

Hill, H.M. (2022). Physics Nobel honors foundational quantum entanglement experiments. *Physics Today*, **75**(12): 14–17.

Kaiser, D. (2022). 'Tackling loopholes in experimental tests of Bell's inequality'. In Freire Junior, O., Bacciagaluppi, G., Darrigol, O., Hartz, T., Joas, C. Kojevnikov, A. and Pessoa, A., Jr. (eds.) *The Oxford Handbook of the History of Quantum Interpretations*, pp. 339–370. New York: Oxford University Press.

Kojevnikov, A. (2002). David Bohm and collective movement. *Historical Studies in the Physical and Biological Sciences*, **33**(1): 161–192.

Mullet, S.K. (2008). *Little man: four junior physicists and the red scare experience.* Ph.D. dissertation, Harvard University.

Oldofredi, A. (2023). Orthodox or dissident? The evolution of Bohm's ontological reflections in the 1950s, *European Physical Journal H*, **48**: 12.

Silva, I. (2022). 'Chien-Shiung Wu's contributions to experimental philosophy'. In Freire Junior, O., Bacciagaluppi, G., Darrigol, O., Hartz, T., Joas, C. Kojevnikov, A. and Pessoa, A., Jr. (eds.) *The Oxford Handbook of the History of Quantum Interpretations*, pp. 734–754. New York: Oxford University Press.

Talbot, C.(2017). *David Bohm: Causality and Chance, Letters to Three Women.* Cham: Springer.

Van Strien, M. (2020). Bohm's interpretation of quantum mechanics and the notion of classicality. *Studies in History and Philosophy of Modern Physics*, **71**: 72–86.

5

Why Bohm was never a determinist

Marij van Strien

5.1 Introduction

If David Bohm is known for one thing, it is for developing a fully deterministic interpretation of quantum mechanics.[1] His interpretation of quantum mechanics has generally been understood as an attempt to restore determinism in physics, and he has been applauded for demonstrating that a common sense, deterministic account of quantum mechanics is possible, as well as criticized for being conservative and unwilling to accept the revolutionary features of modern physics (Van Strien 2020). In the 1950s, he was mostly criticized: for example, Rosenfeld argued that Bohm's theory was motivated by a 'metaphysical prejudice' of determinism (Rosenfeld 1958), and Pauli described it as a 'classical-deterministic myth' of atomic physics (Pauli to Fierz, 1952, in Von Meyenn 1996, p. 500).

However, a closer look at Bohm's writings from the 1950s shows that his views were not at all conservative (Van Strien 2020; Oldofredi 2023). Bohm emphasized the complexity of nature and the interconnectedness of everything, and from the beginning he emphasized the holistic and non-local aspects of his interpretation of quantum mechanics.[2] And as I want to demonstrate in this chapter, he was in fact never a determinist. It has already been argued in Del Santo (2019) and Van Strien (2020) that for Bohm restoring determinism was never the central aim; here I want to defend the stronger claim that Bohm was in fact always opposed to determinism.[3]

It will be useful to have a definition of determinism. Determinism, roughly, is the view that any natural process can be fully described through a specification of initial (and boundary) conditions and a set of laws of nature: the laws of nature always yield a unique solution, such that they fully specify future states. This implies that given the state of a perfectly isolated system at an instant, it is in principle possible to derive the state of the system at all later times (and generally also at all earlier times). Or: given the state of the universe at an instant, it is in principle possible to derive the state of the universe at all later (and earlier) times. This is an *in-principle* argument: it does not need to be possible in practice to give exact predictions, and in fact this is almost

[1] But see Landsman (2022).

[2] These consist in the fact that the momentum of a particle depends on the system as a whole, and the motion of particles is guided by a quantum potential that does not vanish at large distances.

[3] See also Del Santo and Krizek (2025), in which the same claim is made; this paper and the present paper were written simultaneously and independently of each other.

never possible. But the determinist view is that in some way a complete specification of states and laws of nature exists or is possible, such that later states can be derived from earlier states plus the laws of nature. I will argue that in this sense, Bohm was never a determinist.

Determinism relies on the assumption that a physical system is always in a definite state, which can be specified through a number of variables. Furthermore, determinism can only hold in a non-trivial way if you allow for a limited number of laws of nature, which hold absolutely and are of limited complexity; if you allow for an unlimited number of laws of nature or for laws of unlimited complexity, determinism holds trivially because then for every conceivable sequence of events, one can find laws of nature describing exactly this sequence. Finally, either it must be possible to consider a system in perfect isolation, or determinism can only hold for the universe as a whole.

Generally, a lot of confusion has been caused by the fact that 'causality' is often equated with 'determinism', but that it can have other meanings as well. The term 'causality' can be used for example to refer to the principle that nothing happens without a cause, that the same cause is always followed by the same effect, or that there is no action at a distance; and none of these principles is equivalent to determinism. As we will see, also for Bohm, 'determinism' and 'causality' did not have the same meaning, and therefore the fact that Bohm named his interpretation of quantum mechanics the causal interpretation does not necessarily mean that he was committed to determinism.

In the next two sections, I describe the development of Bohm's views on determinism and causality during the 1950s, and how they were embedded in Marxist philosophy. This account is based on Bohm's published work as well as on his correspondence during this period. I conclude with some brief notes on the later developments of Bohm's views.

5.2 Determinism versus causality

In 1951, Bohm published a textbook titled *Quantum Theory*, which provides a general introduction to quantum physics. The book was finished in the summer of 1950. Through writing this book, Bohm tried to make sense of quantum physics himself, but after finishing it he was not convinced that the standard interpretation, as he had presented it here, could be correct. This then motivated him to develop an alternative interpretation, which he managed to do already in the spring of 1951. One may therefore suspect that *Quantum Theory* does not yet present Bohm's own views, but merely his understanding of the consensus at the time. However, his treatment of quantum physics is not standard in every way: it is oriented towards Niels Bohr's account of quantum mechanics, but is also based on an underlying assumption of realism, which is visible for example in Bohm's ontological interpretation of Bohr's notion of complementarity (Freire Junior 2019). The book also reveals Bohm's views on determinism at the time. If Bohm merely wanted to present a standard account of quantum physics, he could just have written that the theory is indeterministic; but he made a point of arguing that determinism had never been plausible and that there is no reason to expect a physical theory to be deterministic (Bohm 1951, pp. 150–153). He argues that the rise of determinism in classical mechanics had been connected with the development of the science of mechanics and the growing importance of machines, which led to a mechanistic world view. But indeterminism is in fact closer to everyday experience since what we experience is that a cause 'produces a tendency to an effect" but "does not guarantee the effect' (Bohm 1951, p. 150). That Bohm was already critical of determinism when writing *Quantum Theory* can also be seen from a letter which he wrote to Miriam Yevick in 1952; here, Bohm writes that when he started working on the book, he tended towards Bohr's views because Bohr 'tried to explain the physical meaning of the theory' and also because

> ...there was an element of dialectics in Bohr's point of view which attracted me. It seemed progressive because it broke the old mechanist materialist determinism, which left no room for growth and development of something new. (Bohm to Yevick, 1952, in Talbot 2017, p. 235)

Thus, Bohm was definitely not a determinist when he wrote *Quantum Theory*. Moreover, the fact that Bohm saw Bohr's break with determinism as dialectic suggests that already at this point, there was a connection between Bohm's rejection of determinism and Marxist philosophy.

On 4 December 1950, at the height of the cold war and the McCarthy era, Bohm was indicted for contempt of Congress, because of his affiliations with communism and his refusal to testify for the House Un-American Activities Committee (HUAC) (Freire Junior 2019, pp. 56–62). This led to a suspension from his job at Princeton University. Bohm had been a member of the communist party for a brief period in 1942, and although he had quickly become disappointed with the party, he had remained a Marxist. During the months he was suspended, Bohm had a lot of free time, which he used to develop an alternative interpretation of quantum physics (Bohm to Schatzman, 1952, in Besson 2018, p. 335).

By July 1951, Bohm had sent his article, 'A suggested interpretation of the quantum theory in terms of "hidden" variables', to *Physical Review*, where it was published in early 1952. In Bohm's interpretation, which is similar to an earlier proposal by de Broglie, particles have a well-defined position and momentum at all times and thus a well-defined path, and their movement is guided by a quantum potential which is derived from the Schrödinger equation. This interpretation yields exactly the same predictions as the standard interpretation. It is indeed deterministic; however, the fact that Bohm proposed an interpretation that had the feature of being deterministic does not necessarily imply that he was committed to determinism at this point. He considered this interpretation to be preliminary: it showed that an alternative interpretation of quantum physics was possible, but it should still be developed further, and Bohm in particular wanted to modify it in such a way that new predictions could be derived from it (Bohm 1952, p. 179). And from the paper, it does not seem that restoring determinism is Bohm's main motivation. Bohm's main concern is with the intelligibility of nature, the possibility to describe and understand natural processes, and the possibility to provide a realist account of quantum physics. He writes that the interpretation he proposes 'provides a consistent alternative to the usual assumption that no objective and precisely definable description of reality is possible at the quantum level of accuracy' (Bohm 1952, p. 188). Bohm does object to the present quantum theory that 'it requires us to give up the possibility of even conceiving precisely what might determine the behavior of an individual system at the quantum level' (Bohm 1952, p. 168). According to the usual interpretation of quantum mechanics, the theory can only yield probabilities for measurement outcomes, and more than this is not possible; in contrast, Bohm wants to be able to say what determines the behaviour of an individual system. This is of course closely related to determinism, but does not necessarily have to take the form of a full-blown determinism, and it is not clear from the paper that it does. In the conclusion, Bohm argues that we should assume that the world is objectively real; furthermore, he proposes the hypothesis that the world 'can correctly be regarded as having a precisely describable and analysable structure of unlimited complexity' (Bohm 1952, p. 189). This idea of unlimited complexity already seems in tension with determinism, since it implies that we can never arrive at a deterministic theory that provides a complete description of physical reality.

Overall, Bohm's paper by itself does not provide conclusive evidence of whether he was committed to determinism at this time. However, additional evidence makes it very implausible that he was ever committed to determinism. We have already observed that he rejected determinism not long before writing the paper in spring 1951, namely in *Quantum Theory*, which he finished in summer 1950. Furthermore, in his correspondence we find explicit rejections of determinism shortly thereafter. In May 1951, Bohm was acquitted from the charges of contempt of Congress. However, his contract at Princeton University ended in June, and the university decided not to prolong his contract; although not officially stated, it was clear that this was for political reasons, and it was clear to Bohm that he would not be able to find a job at another university in the United States. In October 1951, Bohm therefore moved to Brazil, where he obtained a professorship at the University of São Paulo (Freire Junior 2019). Shortly before moving to Brazil, Bohm wrote a letter to the physicist Evry Schatzman in France, in which he discusses his interpretation of quantum physics. He added, 'there are probably an infinite number of levels of reality, so that a *complete* theory will never be obtained' (Bohm to Schatzman, in Besson 2018, p. 310).[4] From São Paulo, Bohm sent letters to friends in the US. His letters to three female friends, namely Hanna Loewy, the physicist Melba Phillips, and the mathematician Miriam Yevick, have been published in the volume *David Bohm: Causality and Chance, Letters to Three Women* (ed. Talbot (2017)). In these letters, Bohm frequently discusses the conception of an infinite number of levels of nature. Our description of nature at the quantum level is very different from that at the classical level, and Bohm was convinced that in order to obtain a fuller understanding of certain phenomena we will need to go to a sub-quantum level, which will again look very different. The elements at each level emerge from elements at a lower level, from which they are qualitatively different. Bohm thought that the total number of levels of nature must be assumed to be infinite; thus, nature is infinitely complex.

In November 1951, Bohm writes to Yevick: 'Because of the existence of an infinite number of levels, the deterministic laws of order at each level probably follow only as a result of conditions of chaos existing at lower levels' (Talbot 2017, p. 205). The laws of thermodynamics arise from the unordered motions of the molecules and atoms at a

[4]The letter is undated, but he writes that he is about to move to São Paulo, where he arrived on 10 October (Freire 2019, p. 63).

lower level; Bohm thinks that something similar holds for all laws of nature. In March 1952, shortly after the article with his deterministic interpretation of quantum mechanics was published, Bohm describes the idea of a mechanically determined universe as a *nightmare*:

> I think that the explicit recognition of a limitless number of levels would be a big step forward in science. Most of the errors of both the positivist and the 19th century 'mechanical' materialists spring from an implicit assumption that the laws of nature will some day finally be understood in terms of a limited number of hypotheses. From this comes the nightmare of a mechanically determined universe that follows an inevitable course. (Bohm to Yevick, March 1952, in Talbot 2017, p. 254)

A bit further on in the same letter, Bohm writes that with the notion of an infinite number of levels of nature, 'the nightmare of complete determinism is avoided' (*ibid.*).

This does not mean that Bohm argued in favour of indeterminism at this point. In fact, he writes that positivists and idealists have tried to avoid the nightmare of determinism by assuming that there can be spontaneous, uncaused events, but according to Bohm, this is equally unfounded. Bohm argues that both the assumption of absolute determinism and the assumption of indeterminism obstruct the progress of science: a determinist assumes that one already knows all causes at play, while an indeterminist assumes that there are no causes to be found, and therefore both views 'discourage a search for "deeper" causes than those already known' (*ibid.*).

Bohm chose the term 'causal interpretation' for his interpretation of quantum mechanics because it enables a causal account of quantum phenomena. In an article published in 1953, Bohm argues that the assumption that certain events can take place purely by chance is unscientific, since to assume that something is uncaused means giving up on trying to understand it:

> ... when we face a new and as yet unexplained phenomena, the most fruitful attitude is always to assume they have a cause, which we must discover. Even if there really were no cause, no error could come from the assumption that there was one. All that would happen would be that our efforts to find the cause would not be successful. But if, as is much more likely, there really is a cause, and we assume there is not, then we may be led to overlook important new factors that are needed in the theory. (Bohm 1953a, p. 284)

By assuming that events at the quantum level may be genuinely uncaused, 'the usual interpretation relegates real physical phenomena to a domain that is by definition forever beyond the possibility of scientific investigation' (*ibid.*). In fact, 'the domain of causality defines the domain of science itself' (*ibid.*). And in another paper published in 1953, Bohm argued that if uranium nuclei decay at different times, it is only reasonable to suppose that there must be a difference between them: 'the most elementary scientific procedure would suggest that if two objects are observed to act differently, this should be regarded as a posteriori evidence that there must in fact be some physical difference between them' (Bohm 1953b, p. 465). He argues that whereas according to the usual interpretation of quantum mechanics there is no difference between two uranium nuclei which decay at different times, according to the causal interpretation there indeed is a difference: the time of decay is determined by the exact positions of the particles within the nucleus, and these positions must thus be different for the two nuclei (Bohm 1953b, pp. 465–466). In this way, Bohm's interpretation indeed offers a causal account of radioactive decay.

Thus, Bohm argues that we must assume that every event has a cause and that it is in principle possible to find this cause. At the same time, he describes determinism as a 'nightmare'. How to understand this? Can one accept that everything has a cause without being a determinist?

Although causality has often been equated with determinism, Bohm in fact makes an explicit distinction. In his correspondence from the early 1950s, Bohm equates determinism with mechanical determinism, according to which nature consists of fixed basic elements that move according to fundamental laws of nature. With this conception of determinism, nature is conceived of as a machine, in which there is only quantitative change, which consists in the rearrangement of basic elements. Causality, in contrast, can describe qualitative as well as quantitative change. Bohm argues that there is an emergence of qualitatively new behaviour at higher levels of nature, and this can be described with causal laws (Bohm to Phillips, 1953, in Talbot 2017, p. 164). However, whereas Bohm considers determinism to be mechanical, one may object that non-mechanical conceptions of determinism are also possible.

But Bohm draws further distinctions between causality and determinism. Determinism is about the possibility to predict events, and if determinism holds, the future determines the past as well as the past determines the future. Causality is additionally about the possibility to change the course of events: by changing the causes, we can change the effects (Bohm to Yevick, 1952, in Talbot 2017, p. 255). Thereby, causality also involves a temporal element: we can change the causes and thereby change the effects, but not vice versa (Bohm 1953a, p. 285)). Bohm argues that causality is in fact incompatible with determinism: if determinism holds, it is not possible for us to actually change anything. If determinism holds, then rather than an event having a specific cause, it is determined by the entire earlier state of the universe. Therefore, Bohm writes that if we accept Laplacian determinism, 'causality itself loses all meaning' (*ibid.*).

According to Bohm, this problem is avoided because we do not have to accept Laplacian determinism. The latter only follows if we take the laws of Newtonian mechanics to be absolutely valid; but the history of science teaches us that we should never assume that we have found a final theory that holds absolutely. As soon as we accept that Newton's laws only hold within a limited domain and only with good approximation, it no longer follows from these laws that the universe as a whole is deterministic. Thus, Newtonian mechanics is deterministic, and the causal interpretation shows that a deterministic theory of quantum mechanics may also be possible; but neither of these hold absolutely, and neither of these can be taken to be a final theory. Therefore, neither of these imply a universal, Laplacian determinism (*ibid.*).

We may ask if Bohm's conception of levels of nature could nevertheless allow for an ontological conception of determinism of the universe as a whole: could one argue that, given the state of the universe as a whole, including all levels, at an instant, only one future development is possible? In his correspondence, Bohm argues that as long as we assume the number of levels of nature to be finite, we do arrive at the conclusion that nature as a whole must be deterministic. However, if there is an infinite number of levels of nature, no complete analysis is possible, since we cannot conceive of the state of the universe as a whole:

Although each level is causal, the totality of levels cannot ever be taken into account. Thus, as a matter of principle, we say that complete determinism could not even be conceived of, yet, each level can be determined. (Bohm to Yevick, 1952, in Talbot 2017, p. 254)[5]

With an infinite number of levels of nature, there can always be an emergence of genuinely new behaviour 'because the effects of the limitless number of lower levels can always surge up into a higher level (and vice versa)' (Bohm to Yevick, 1952, in Talbot 2017, p. 255). Furthermore, what happens at a certain level does not only depend on lower levels but can also depend on a higher level, and therefore we cannot obtain exact knowledge of what happens at a certain level by analysing only the lower levels:

...the higher levels will also always help determine the character of things that may exist at the lower levels. Thus, every level is in a sense, just as real as every other, since the "whole picture" cannot be deduced by starting at the "lowest level" and working upward. (Bohm to Yevick, 1952, in Talbot 2017, p. 246; see also p. 171)

Therefore, according to Bohm, 'if we have a finite number of causal levels, then the future is already contained logically in the present, but not if we have an infinite number' (Bohm to Yevick, 1952, in Talbot 2017, p. 255); 'It is the unlimited number of levels which give matter its 'non-mechanical' aspects' (Talbot 2017, p. 254).

Bohm's views imply that we can never arrive at a theory that is deterministic as well as fully accurate. A specific level of nature may be described by a deterministic theory; however, since levels are never fully independent from each other, this can only be an idealization. No complete description of nature is possible, not even in principle: it is principally impossible to fully specify the state of a system, and therefore, Bohm is not a determinist according to the definition given in the introduction. In this way, Bohm can argue that in principle, it is always possible to find a cause for any event, without implying that nature as a whole is deterministic.

There is a Marxist element in Bohm's conception of infinite levels of nature: Bohm himself directly connects it with dialectical materialism (Talbot 2017, p. 255).[6] In an interview, Bohm said:

I remember also, before leaving the United States, I picked up something in the Princeton library, some Soviet publication which mentioned Lenin saying that the electron was inexhaustible. (...). That sort of struck a chord because I said, 'Well, not only the electron, but everything, all matter is inexhaustible'.[7]

This refers to a passage from Lenin's book *Materialism and Empirio-criticism*, in which Lenin argues against positivism. Lenin writes: 'The electron is as *inexhaustible* as the atom, nature is infinite, but it infinitely *exists*' (Lenin 1947, p. 243). Also for

[5] In a letter to Charles Biederman in 1961, Bohm writes that we could say that every event is determined if we consider the totality of the world, but that this would be a trivial determinism (Bohm to Biederman, 1961, in Pylkkänen 1999, p. 164). This is related to the point that the concept of determinism is only meaningful if it is based on the idea that there is a limited number of laws of nature which hold absolutely.

[6] On Marxist elements in Bohm's early work, see also Kojevnikov (2002).

[7] Interview of David Bohm by Maurice Wilkins on 22 December 1986, Niels Bohr Library & Archives, American Institute of Physics, College Park, MD USA, www.aip.org/history-programs/niels-bohr-library/oral-histories/32977-6.

Lenin, this meant that our knowledge of nature can never be complete. Bohm's view of an infinite number of levels of nature was probably also influenced by Friedrich Engels, who also proposed a view according to which there are different levels of nature, with qualitatively new properties emerging at higher levels (Sheehan 1985, p. 34). Engels' materialism is not a crude materialism but rather organicist and emergentist: there is always an emergence of qualitatively new properties.

5.3 Determinism and randomness

An issue with Bohm's interpretation of quantum mechanics is how to account for the Born rule: the probability density for the position of particles must be equal to $P = |\Psi|^2$ (with Ψ being the wave function) in order to agree with experimental results. In the standard interpretation of quantum mechanics, this is simply postulated, but in Bohm's interpretation, particles have well-defined trajectories and equations of motion, and one has to account for the fact that the probability density is given by the above equation. It can be proven that once the probability density satisfies the Born rule, this is stable and will continue to hold, but how is this situation reached when you start with an arbitrary configuration of particles? Here, Bohm brought in the idea of random or chaotic movements. In a manuscript which Bohm sent to de Broglie in 1951, he argues that the Born rule can be explained through 'the chaotic character of the particle motion (resembling Brownian motion), which arises whenever the particle interacts with other systems' (Bohm, in Drezet and Stock 2021, p. 171). There are deterministic laws which determine the trajectories of particles, but there is also always some deviation from these trajectories due to chaotic motion resulting from the interactions with other systems, and after a while this results in a convergence of the probability density to $P = |\Psi|^2$. Also in (1952), Bohm points at the 'chaotically complicated character' of the coupling between the particle and the surrounding systems with which it interacts to explain the Born rule. And in (1953b), Bohm provides a demonstration of how an arbitrary initial probability distribution approaches $P = |\Psi|^2$ as a result of random collisions.

In 1953, the French physicist Jean-Pierre Vigier, an assistant of de Broglie, came to Brazil to work with Bohm for a few months. Their work resulted in a joint paper, in which they explain the claim that any initial probability density will develop to the probability density $P = |\Psi|^2$ through random fluctuations in the quantum potential. Bohm and Vigier use a hydrodynamic model and argue that there exist fluctuations in the quantum potential like in any other physically real fluid or field, and that these can originate from motions at a sub-quantum level. The fluctuations in the motion of particles can be compared to Brownian motion, the random motion of particles suspended in a fluid which results from their collision with molecules in the fluid (Bohm and Vigier 1954). Thus, Bohm and Vigier appeal to a lower level of nature in order to solve an issue in Bohm's interpretation of quantum mechanics.

This appeal to random motion is not necessarily incompatible with determinism. Though Brownian motion itself is random, it arises as a result of the motion of molecules in a fluid, and it is possible that these molecules move according to deterministic equations, although their motion is too complex to be tractable. In this way, Brownian motion may be seen as an example of how random behaviour can emerge from underlying determinism; and analogously, Bohm's and Vigier's claim that random fluctuations arise as a result of motions at a sub-quantum level does not necessarily exclude the possibility that there is a deterministic theory for the sub-quantum level. However, Bohm and Vigier do seem to understand their derivation of the Born rule as an introduction of randomness and as thereby going against determinism. In (1957, p. 116), Bohm notes about his work with Vigier: 'in our model we have not insisted on a purely causal theory, for we have also utilized the assumption of random fluctuations originating at a deeper level'. And in a letter to Popper in 1984, Bohm writes:

> It is true that I first used a determinist version of (...) quantum theory. But later, (...) a paper was written, in which we assumed that the movement of the particle was a stochastic process. Clearly that is not determinism. (Bohm to Popper, 1984, quoted in Del Santo 2019)

Thus, from the beginning, Bohm thought that random or chaotic processes played a role in his interpretation of quantum mechanics, and by 1954 he had supplemented the deterministic account given in his 1952 paper with an element of randomness arising from a lower level of nature, which for him meant that his interpretation of quantum mechanics was no longer deterministic.

5.4 Causality and chance

Although as we have seen, it seems that Bohm was never committed to a complete determinism, until around 1954 he held on to the idea that chance must always be taken to arise from underlying causality and seems to have considered causality as more fundamental. Furthermore, his correspondence shows that around 1952-1954, he was working on models in which deterministic laws give rise to chaos and randomness, and thought that 'all chaos comes from causality' (Bohm to Yevick, 1953, in Talbot 2017, p. 317). He also argued that probability distributions can be derived from causal laws. He developed these ideas in particular through his correspondence with Miriam Yevick, who later also published on this topic (Yevick 1957). Bohm found it problematic if we need two different accounts of probabilities: in statistical mechanics, we need to use probabilities because we cannot specify the exact state of the system for practical reasons, though probabilities would theoretically be eliminable if could give a full specification of the system, whereas in the standard account of quantum mechanics, there are probabilities that are fundamental and irreducible (Bohm 1953b, p. 466). Bohm thought that an advantage of his theory of quantum mechanics was that with this theory probabilities could be interpreted in the same way as in statistical mechanics; thus, no second notion of probability is needed. Thus, even though Bohm rejected absolute determinism, he thought that there always needs to be a causal basis for chance and probability.

At some point in 1954, however, his views seem to have shifted: whereas before, he thought that causality was more fundamental than chance, he now argued that there is a dialectical relation between causality and chance and that both are equally objective (Talbot 2017, pp. 49–62). In this, he was influenced by the Brazilian physicist Mario Schönberg, with whom he worked in São Paulo, as well as the Argentine physicist Mario Bunge (Talbot 2017, p. 56). Both were also Marxists and interested in dialectical materialism. Soon after his arrival in Brazil, in November 1951, Bohm mentioned in his correspondence that he was reading Friedrich Engels' *Dialectics of*

Nature in Portuguese, 'borrowed from the collections of Schönberg' (Bohm to Yevick, 1951, in Talbot 2017, pp. 200–203) – apparently, Bohm used Engels' work for learning Portuguese. Engels' *Dialectics of Nature*, which remained unfinished and fragmentary and was published only after his death, contains remarks on the relation between chance and necessity. Here, Engels objects to the view that some things are necessary and other things are purely accidental: there is always an interplay of both. He also objects to the deterministic view according to which everything is necessary and nothing is accidental: for Engels, this is an 'empty phrase' because it is not verifiable in any way, and because in practice, chance always continues to play a role (Engels 1954, p. 219). Engels appeals to Hegel's dialectical views on causality and argues that causality arises from chance and chance from causality, and that what is necessary from one perspective becomes accidental when seen from another perspective and vice versa. Also Schönberg thought that there was a dialectical relation between causality and chance. Bohm had a difficult relationship with Schönberg but became more receptive to his views after a while (Peat 1997, p. 155; Talbot 2017, pp. 351, 404). In an interview, Bohm says that Schönberg 'helped to show me that I had been approaching the thing in a narrow way, by just looking at causality, without bringing into the opposite side of chance'.[8] And:

> I had been discussing causality and essentially through discussions with Schönberg had turned into more a dialectical direction. That is he was very interested in dialectic. He used to say that Lenin had said communists should read Hegel, and very few did.[9]

Bohm took up Schönberg's advice to read Hegel. He became very interested in the philosophy of Hegel, an interest that remained strong throughout his life (Kožnjak 2022; Peat 1997).

Also Mario Bunge considered causality and chance as dialectical opposites. In 1953, Bunge came to work with Bohm in São Paulo for six months. Two years before, he had published an article titled 'What is chance?', in which he argued that 'chance and necessity convert one into the other according to the general laws of dialectics' (Bunge 1951, p. 231). As science develops, we find that what seemed to be necessary in fact arises from chance events, and what seemed to be chance events in turn arise from necessity (Bunge 1951, p. 218). Bunge argued that the dilemma between determinism and indeterminism is a false one: these are both unfounded metaphysical positions. The article is explicitly written from a Marxist perspective, including quotes from Engels. Bohm found this article very interesting: it was part of his motivation to invite Bunge to São Paulo (Bunge 2016, p. 90).

By 1954, Bohm had adopted these views of a dialectic relation between causality and chance, as can be seen from his correspondence from this period. Whereas before he was concerned with showing how randomness can arise from causality, he now argued that causality and randomness are 'equally real, and both are inter-dependent and interconnected' (Bohm to Yevick, 1954, in Talbot 2017, p. 382). Causal and statistical

[8] Interview (Footnote 7).

[9] Interview of David Bohm by Maurice Wilkins on 3 October 1986, Niels Bohr Library & Archives, American Institute of Physics, College Park, MD USA, www.aip.org/history-programs/niels-bohr-library/oral-histories/32977-5

laws are both needed in our descriptions of nature and have an equally objective validity.

Thus, during his time in Brazil, Bohm's interest in dialectical materialism deepened. Forstner (2008) has analysed the changes in Bohm's thought in this period in terms of the concept of thought-collectives developed by Ludwik Fleck, arguing that after moving to Brazil, Bohm became detached from his former community of physicists in the United States and departed from the thought style of that community, and dialectical materialism became more central to his thought.

In early 1955, Bohm moved to Israel. In 1957, he published his book *Causality and Chance in Modern Physics*, about which he later said: 'I just about finished most of it when I got out of Brazil. I finished it finally in Israel, but the basic thing was done in Brazil'; 'I think that book was basically a result of Brazil. I think in Brazil the main thing that happened was that my ideas transformed a great deal'.[10] In *Causality and Chance in Modern Physics*, Bohm argues that causality and contingency are both abstractions that offer opposite views of the same object. 'Each view, then, limits the other, corrects the other, and through its relationship with the other enables us to form a better concept of what the object is' (Bohm 1957, p. 3). Bohm argued that the assumption of determinism and the assumption of fundamental chance are equally unsupported. Whenever one has deterministic laws of nature, a closer study may reveal that these laws arise through averaging over chance fluctuations; and similarly, whenever one has statistical laws of nature, which merely ascribe probabilities to future events, a more detailed study may reveal how these laws arise from underlying determinism. Thus, wherever one finds determinism, there can be underlying chance and vice versa.

Bohm objects to the standard interpretation of quantum mechanics that it takes chance to be fundamental and excludes the possibility of ever being able to describe the exact trajectories of particles. However, he also argues against a strictly determinist picture, since it relies on the assumption that our current theory is final and its laws hold absolutely. This, according to Bohm is unjustified: scientific theories always have a limited domain of validity, and even within this domain they do not hold absolutely, since different levels of reality are never fully isolated from each other. Since our laws of nature never hold absolutely and can never give a complete description of nature, they generally do not determine the future uniquely: 'Rather, they make possible only a one-to-many correspondence between cause and effect, in the sense that a specification of certain causes will in general limit the effect to a certain range of possibilities' (Bohm 1957, p. 16). In an isolated mechanical system, we can have something close to deterministic laws, but true determinism remains an idealization: no system is ever perfectly isolated, which means that there are always some external disturbances, and even in a perfectly isolated mechanical system,

> ...there would still exist disturbances coming from motions at the molecular level. Of course, one could in principle try to take these into account by applying the laws of motion to the molecules themselves, but then one would discover still further disturbances coming from the quantum-mechanical and other deeper-lying properties of matter. (Bohm 1957, p. 20)

[10] Interview (Footnote 9).

Bohm argues that the assumption that the indeterminacy of quantum mechanics is final is just as unjustified as the assumption that classical determinism holds absolutely – in both cases, the possibility that future developments will lead to changes in the theoretical framework is excluded (Bohm 1957, p. 102). Though Bohm speaks about an infinity of levels of nature, he notes towards the end of the book that as science develops further, we may find that this structure of levels breaks down at some point, and instead of describing nature at smaller and smaller scales we may 'disclose a still more general pattern of organization of things' (Bohm 1957, p. 140). But what Bohm is sure of is that nature is inexhaustibly rich and that science will never be finished.

Bohm's *Causality and Chance in Modern Physics* does not mention the names Engels, Marx, Lenin, Hegel, or the word 'dialectical', but it is very clearly based on dialectical materialism. In fact, during his time in Brazil he mentioned in his correspondence that he was working on a book on dialectical materialism, which is probably this book (Talbot 2017, p. 291). It is of course understandable that Bohm did not make the dialectical materialist basis of his work explicit, since this would not have helped the acceptance of his ideas. However, various elements of dialectical materialism can be recognized in *Causality and Chance in Modern Physics*. First, the idea that nature is inexhaustibly rich is, as we have seen, also expressed in Lenin's statement of the inexhaustibility of the atom and in Engels' views on the levels of nature. Furthermore, Bohm writes that everything is interconnected, and that everything is always in transformation and nothing remains constant; these ideas can be found in Engels' natural philosophy as well (Sheehan 1985, pp. 36–44). Bohm also writes that quantitative change leads to qualitative change, giving the example of phase transitions: if the temperature of a gas is lowered, the gas condenses and enters a liquid phase, and thereby new qualities appear (Bohm 1957, p. 52). This derives from Engels' formulation of Hegel's first law of dialectics, 'the law of the transformation of quantity into quality and vice versa'.[11] According to Engels, this is a natural law that is based on our study of nature, and Engels also gives phase transitions as an example. Finally, as we have seen, Bohm argues for a dialectic relation between causality and chance.

Freire Junior (2015, p. 52; 2019) has argued that in the early 1950s, Bohm was committed to determinism, and that this was connected to the fact that he was a communist during this period. He writes furthermore that in the second half of the 1950s, Bohm abandoned determinism, and that it is plausible that this development is connected to his break with communism in 1956, 'following the revelations of Nikita Khrushchev, the Soviet leader, at the 20th Congress of the Soviet Communist Party, about the crimes in the Stalin-era, and the Soviet invasion of Hungary', which deeply shocked Bohm (Freire Junior 2019, p. 105; and see Talbot 2017, pp. 67–75).

However, the opposite seems to be the case. Rather than motivating a determinist view, Bohm's interest in Marxist philosophy helped him develop an account of causality that was not deterministic, and then to arrive at the idea of a dialectic relation between causality and chance. His break with communism took place before the publication of

[11] The second law of dialectics, in Engels' formulation, is 'The law of the interpenetration of opposites': there are contradictory tendencies in everything. The third law of dialectics is 'The law of the negation of the negation', which means that everything carries within itself the conditions for its annihilation (Engels 1954).

Causality and Chance in Modern Physics, but after he had developed the central ideas presented therein: we have seen that according to Bohm, the book was mostly written when he arrived in Israel in 1955.

Marxism is often associated with determinism, and especially with the idea that there are laws determining the development of society. But there are different schools in Marxist philosophy. Freire himself has pointed out that there was a variety of Marxist views on quantum physics, and has in fact discussed a few cases of physicists who adhered to a dialectical materialist philosophy and rejected determinism, including Paul Langevin in France and Mituo Taketani in Japan (Freire Junior 1997).[12] A further example of a Marxist physicist rejecting determinism is the Russian physicist and historian of science Boris Hessen (Talbot and Pattison 2021). And in the 1950s, the Belgian physicist Léon Rosenfeld and the Soviet physicist Vladimir Fock argued against determinism from a Marxist perspective. They both argued that Bohr's notion of complementarity could be understood in a dialectic way (see Jacobsen 2007; Cross 1991; Freire Junior 1997; Martinez 2019). Rosenfeld and Fock were in fact among Bohm's strongest critics: they both considered Bohm's interpretation a step back to an outdated mechanistic determinism (although, as I have argued, this perception was not entirely correct). We have already seen that both Bunge and Schönberg also rejected determinism from a dialectical materialist point of view. Vigier, who was a member of the communist party in France, initially rejected Bohm's conception of an infinite number of levels of nature, but by 1954 he had become convinced of the idea and argued that it was in agreement with dialectical materialism (Besson 2018, p. 165). He argued that for this reason, Bohm's interpretation should not be understood as a return to Laplacian mechanical determinism (Vigier 1956, p. 125), and in a lecture in 1957, speaking about the approach of Bohm and himself, he said: 'We do not wish to return to Laplacian determinism and to reduce all laws to causal laws' (Vigier in Körner (ed.) 1957, p. 77). In 1961 Vigier published an article titled 'Théorie des Niveaux et Dialectique de la Nature' in the communist journal *La Pensée*, in which he expressed ideas closely resembling those of Bohm in *Causality and Chance in Modern Physics*, but with the dialectical materialist foundations made explicit (Vigier 1961).

So even if, especially in the Soviet Union, there were also Marxist physicists who sought for a deterministic physics, it seems that in many (if not most) cases, an adherence to Marxist philosophy was instead connected to a rejection of determinism in physics. In this regard, it is not surprising that Marxist philosophy in fact helped Bohm develop non-determinist views.

5.5 Later years

After the late 1950s, Bohm rarely mentioned dialectical materialism again, and this is probably indeed connected with his disillusionment with communism. Hegel, however, remained a strong influence throughout his life: Bohm endlessly read and reread Hegel (Kožnjak 2022). The dialectic elements of his view on causality and chance are also expressed in his correspondence with the artist Charles Biederman in the early 1960s. Here, Bohm argues that chance and contingency have as much objective existence as

[12] On Marxist approaches to quantum mechanics, see, e.g. Cross (1991), Freire Junior (1995).

causality, and writes, 'Every time we have a regularity, we must have some limit to this regularity. (...) Wherever there is law, there must be lawlessness' (Bohm to Biederman, 1960; in Pylkkänen 1999, p. 34). Therefore, 'the future is not fully determined by the past' and there can be genuine freedom (Bohm to Biederman, 1960; in Pylkkänen 1999, p. 14).

Already in the 1950s, Bohm remarked that his 1952 interpretation of quantum mechanics was preliminary and in some respects unsatisfactory, but he thought that it showed the possibility in principle of alternative interpretations of quantum mechanics and that it was a promising starting point for further development (Bohm 1957, p. 127). By the early 1960s, Bohm had abandoned his causal interpretation in search of a more radically new theory of quantum physics. However, there seems to be little evidence that this should be understood as a result of a change in his views on causality; rather, he seems to have been disappointed by the reception of his theory and by the lack of concrete results. He did move further away from mechanism in this period, arguing that our conceptions of space and time may also have to be revised.

In the 1970s, Bohm took up his (1952) interpretation of quantum mechanics again, after some of his students became interested in it. In *The Undivided Universe* (1993), Bohm and his co-author Hiley provide a detailed account of Bohm's interpretation of quantum mechanics, although with some differences from Bohm's work from the 1950s. In the introduction, they write that they prefer the term 'ontological interpretation' over 'causal interpretation'; however, I think it is not clear that this indicates a weakening of Bohm's commitment to causality. Bohm and Hiley write that the theory they propose does not have to be deterministic: 'The question of determinism is therefore a secondary one, while the primary question is whether we can have an adequate conception of the reality of a quantum system, be this causal or be it stochastic or be it of any other nature' (Bohm and Hiley 1993, p. 2). When they discuss causality and determinism later in the book, the views expressed are similar to those in Bohm's *Causality and Chance in Modern Physics*: whenever there is determinism, there can be underlying chance and the other way around, and 'ultimately our overall world view is neither absolutely deterministic nor absolutely indeterministic' (Bohm and Hiley 1993, p. 324).

5.6 Conclusion

To summarize, in *Quantum Theory* (finished in 1950), Bohm wrote that there is no reason to expect a physical theory to be deterministic. In the period 1951–1953, he argued in correspondence that we must assume that everything has a cause, and that chance reduces to causality, but he explicitly distinguished causality from determinism and rejected a fully deterministic account of nature, even referring to determinism as a 'nightmare'. Although the interpretation of quantum mechanics he proposed in 1952 was indeed deterministic, it was not motivated by a commitment to determinism, and Bohm in fact suggested from the beginning that an element of chaos or randomness had to be added in order to account for the Born rule. From ca. 1954, Bohm argued that there is a dialectic relation between causality and chance: causality and chance always transform into each other and both are equally objective.

Nevertheless, Bohm's theory of quantum mechanics has almost universally been received as an attempt to restore determinism in physics. This makes him part of a more general pattern. Beller (1999) has pointed out that there has been a tendency to interpret all opposition to the standard interpretation of quantum mechanics as a metaphysical attachment to classical determinism, and that in this way, opposition has been portrayed as conservative and dogmatic (Beller 1999, p. 281). Other examples of this tendency are Schrödinger and Einstein. Schrödinger's objections to the standard interpretation of quantum mechanics have often been interpreted as a refusal to accept indeterminism, despite the fact that Schrödinger already argued that we should give up on determinism before the introduction of an explicitly indeterministic theory of quantum mechanics in 1925–26 and despite the fact that Schrödinger's objections were in fact different (Ben-Menahem 1989). Also Einstein's concerns with quantum physics have often been thought to stem from an attachment to outdated classical determinism: his objections are often summarized by his statement that 'God does not play dice'. However, also in the case of Einstein, it has been argued that indeterminism was actually never his main concern and that his actual concerns with quantum mechanics went deeper (see, e.g. Fine 1986; Howard 1993; Paty 1995).

Like Schrödinger and Einstein, Bohm was primarily after an understanding of quantum mechanics. Bohm was not satisfied with the standard account of quantum mechanics because it implied that there is a hard limit to the possibility to describe and understand natural processes. This is related to causality, but it does not mean that he was committed to determinism.

References

Beller, M. (1999). *Quantum Dialogue: The Making of a Revolution.* Chicago: University of Chicago Press.

Ben-Menahem, Y. (1989). Struggling with causality: Schrödinger's case. *Studies in History and Philosophy of Science*, **20**(3): 307–334.

Besson, V. (2018). *L'interprétation causale de la mécanique quantique: biographie d'un programme de recherche minoritaire (1951–1964).* Doctoral dissertation, Université de Lyon; Universidade Federal da Bahia.

Bohm, D. (1951). *Quantum Theory.* New York: Prentice Hall.

Bohm, D. (1952). A suggested interpretation of the quantum theory in terms of "hidden" variables. I & II. *Physical Review*, **85**(2): 166–193.

Bohm, D. (1953a). Comments on an Article of Takabayasi conserning the Formulation of Quantum Mechanics with Classical Pictures. *Progress of Theoretical Physics*, **9**(3): 273–287.

Bohm, D. (1953b). Proof that Probability Density Approaches $|\psi|^2$ in Causal Interpretation of the Quantum Theory. *Physical Review*, **89**(2): 458.

Bohm, D. (1957). *Causality and Chance in Modern Physics.* London: Routledge & Kegan Paul.

Bohm, D. and Hiley, B. (1993). *The Undivided Universe: An Ontological Interpretation of Quantum Theory.* London: Routledge.

Bohm, D. and Vigier, J.P. (1954). Model of the causal interpretation of quantum theory in terms of a fluid with irregular fluctuations. *Physical Review*, **96**(1): 208–216.

Bunge, M. (1951). What is chance? *Science & Society*, **15**(3): 209–231.

Bunge, M. (2016). *Between Two Worlds: Memoirs of a Philosopher-Scientist.* Switzerland: Springer.

Cross, A. (1991). The crisis in physics: Dialectical materialism and quantum theory. *Social Studies of Science*, **21**: 735–759.

Del Santo, F. (2019). Striving for Realism, not for Determinism: Historical Misconceptions on Einstein and Bohm. Backpage of *APS News*, May 2019.

Del Santo, F. and Krizek, G.C. (2025). Against the 'nightmare of a mechanically determined universe': Why Bohm was never a Bohmian. In Oldofredi, A. (ed.) *Guiding Waves in Quantum Mechanics: One Hundred Years of de Broglie-Bohm Pilot-Wave Theory.* Oxford: Oxford University Press.

Drezet, A. and Stock, B. (2021). A causal and continuous interpretation of the quantum theory: About an original manuscript by David Bohm sent to Louis de Broglie in 1951. *Annales de la Fondation Louis de Broglie*, **46**: 169–195.

Engels, F. (1954). *Dialectics of Nature.* Moscow: Progress Publishers.

Fine, A. (1986). *The Shaky Game: Einstein, Realism and the Quantum Theory.* Chicago: University of Chicago Press.

Forstner, C. (2008). The early history of David Bohm's quantum mechanics through the perspective of Ludwik Fleck's thought-collectives. *Minerva*, **46**: 215–229.

Freire Junior, O. (1997). Quantum controversy and Marxism. *Historia Scientiarum: International Journal of the History of Science Society of Japan*, **7**(2): 137–152.

Freire Junior, O. (2015). *The Quantum Dissidents: Rebuilding the Foundations of Quantum Mechanics (1950-1990).* Berlin: Springer.

Freire Junior, O. (2019). *David Bohm: A Life Dedicated to Understanding the Quantum World.* Cham, Switzerland: Springer.

Howard, D. (1993). Was Einstein really a realist? *Perspectives on Science*, **1**: 204–251.

Jacobsen, A. (2007). Léon Rosenfeld's Marxist Defense of Complementarity. *Historical Studies in the Physical and Biological Sciences*, **37**: 3–34.

Kojevnikov, A. (2002). David Bohm and collective movement. *Historical Studies in the Physical & Biological Sciences*, **33**(1): 161–192.

Körner, S. (ed.) (1957). *Observation and Interpretation: A Symposium of Philosophers and Physicists.* London: Butterworths.

Kožnjak, B. (2022). Waterfalls, Societies, and Temperaments: Fragmentation and Wholeness in the Lives and Work of David Bohm and Georg Wilhelm Friedrich Hegel. *Synthesis Philosophica*, **37**(1): 89–128.

Landsman, K. (2022). Bohmian mechanics is not deterministic. *Foundations of Physics*, **52**(4): 73.

Lenin, V.I.I. (1947). *Materialism and Empirio-Criticism*. Moscow: Progress Publishers.

Martinez, J.P. (2019). Beyond Ideology: Epistemological Foundations of Vladimir Fock's approach to Quantum Theory. *Berichte zur Wissenschaftsgeschichte*, **42**(4): 400–423.

Oldofredi, A. (2023). Orthodox or dissident? The evolution of Bohm's ontological reflections in the 1950s. *European Physical Journal H*, **48**: 12.

Paty, M. (1995). The nature of Einstein's objections to the Copenhagen interpretation of quantum mechanics. *Foundations of Physics*, **25**(1): 183–204.

Peat, F.D. (1997). *Infinite Potential: The Life and Times of David Bohm*. Reading, MA: Addison-Wesley.

Pylkkänen, P. (1999). *Bohm-Biederman Correspondence: David Bohm and Charles Biederman. Vol. 1: Creativity and Science*. London and New York: Routledge.

Rosenfeld, L. (1958). Physics and metaphysics. *Nature*, **181**: 658.

Sheehan, H. (1985). *Marxism and the Philosophy of Science: A Critical History*. New Jersey: Humanities Press International.

Talbot, C. (ed.) (2017). *David Bohm: Causality and Chance, Letters to Three Women*. Cham, Switzerland: Springer.

Talbot, C. and Pattison, O. (eds.). (2021). *Boris Hessen: Physics and Philosophy in the Soviet Union, 1927–1931: Neglected Debates on Emergence and Reduction*. Cham, Switzerland: Springer Nature.

Van Strien, M. (2020). Bohm's theory of quantum mechanics and the notion of classicality. *Studies in History and Philosophy of Science Part B: Studies in History and Philosophy of Modern Physics*, **71**: 72–86.

Vigier, J.-P. (1956). *Structure des micro-objets dans l'interprétation causale de la théorie des quanta*. Paris: Gautier-Villars.

Vigier, J.-P. (1961). Théorie des niveaux et dialectique de la nature. *La Pensée: Revue du Rationalisme Moderne*, **99**: 36–48.

Von Meyenn, K. (ed.) (1996). *Wolfgang Pauli, Wissenschaftlicher Briefwechsel mit Bohr, Einstein, Heisenberg u.a. Band IV, Teil I: 1950–1952*. Berlin, Heidelberg: Springer.

Yevick, M.L. (1957). Probability and determinism. *American Journal of Physics*, **25**(8): 570–580.

6

Against the 'nightmare of a mechanically determined universe': Why Bohm was never a Bohmian

Flavio Del Santo and Gerd Christian Krizek

'Why on earth are they calling it Bohmian mechanics?
Haven't they read a word I have written?!'
David Bohm (reported by Basil Hiley)

6.1 Introduction

David Bohm (1917–1992) went down in history as the physicist who achieved the impossible by providing a deterministic interpretation of quantum mechanics Bohm (1952a, 1952b).[1] Acclaimed or blamed therefore as a champion of determinism, he was (and still is) regarded by many as a cure against the claims of the Copenhagen school that quantum mechanics necessarily requires a completely novel way of looking at the world.[2] According to this narrative, Bohm restored the seemingly lost comfort of mechanistic determinism, which had characterized physics for centuries, and his work seems therefore animated by a certain intellectual conservatism (see, e.g. Cushing 1994).[3]

Here, we show that his aim was unequivocally not to return to an antiquated pre-quantum paradigm. Bohm's views on philosophy of physics have instead been explicitly aimed, with continuity throughout his whole career, at demolishing certain established views that he perceived as limiting and dogmatic. As discussed later, one of these was the concept of *mechanism*, a form of reductionism that Bohm regarded as the

[1] Bohm himself referred to his interpretation as 'alternative interpretation' (Bohm 1952a, 1952b, 1957), as 'causal interpretation' (Bohm 1953b; Bohm and Vigier 1954), and as 'quantum potential interpretation'. In the literature it is referred to as 'Ontological interpretation' (Bohm and Hiley 1993; Pylkkänen 2016), 'De Broglie–Bohm causal interpretation' (Holland 1995), or 'De Broglie–Bohm Pilot-Wave Theory', 'Bohmian Mechanics' (Dürr and Teufel 2009; Maudlin 2019), or 'Bohm theory' (Hiley 2001; Esfeld *et al.* 2013). The variety of terminologies reflects different stances and views of Bohm's collaborators and successors which substantially deviate in some cases from Bohm's own ideas and whose discussion would go beyond the scope of this work.

[2] There is not a single, consistent Copenhagen interpretation (Howard 1993). However, one can identify fundamental principles in the views of Bohr, Heisenberg, Born, etc., such as the principle of complementarity and the correspondence principle.

[3] See also Van Strien (2025), in which similar claims are made; that paper and the present paper were written simultaneously and independently of each other.

> assumption that the great diversity of things that appear in all of our experience, every day as well as scientific, can all be reduced completely and perfectly to nothing more than consequences of the operation of an absolute and final set of purely quantitative laws determining the behaviour of a few kinds of basic entities or variables (Bohm 1957, p. 37).

In this effort, Laplacian determinism, a concept somewhat loosely defined, was regarded by Bohm as the first and foremost expression of mechanism, and he thus searched for alternatives throughout his whole life.

As noted by Nobel laureate Roger Penrose, "there can be few physicists who have delved into the philosophical implications of their subject as has David Bohm" (Hiley 1997). It is indeed possible to identify at least three fundamental tenets in David Bohm's philosophy of physics, namely (i) *realism*, (ii) *causality*, and (iii) *anti-mechanism*. Here we do not deal with Bohm's realism that has already been the subject of numerous studies, and it is undisputed that Bohm was committed to (some form of) realism (see, e.g. Del Santo 2019b; Van Strien 2020; Freire Junior 2019; Oldofredi 2023, and references therein). On the other hand, we focus on the latter two tenets, which have been astonishingly misunderstood in most of the vast literature devoted to Bohm's thought and his intellectual legacy. In particular, the term causality has been commonly assumed to be a synonym of determinism; a mistake unfortunately still present in the literature, at least in physics, to date. Furthermore, Bohm always opposed mechanism, which, we stress again, has its most striking example (but not the only one) in determinism.

The main aim of this chapter is to clarify some of Bohm's original philosophical stances by demolishing certain established misconceptions around his commitment to determinism, which we cannot emphasize enough, was *never* present in his thought. It is a peculiar case that a scholar to whom so many historical and philosophical studies have been devoted has been so misrepresented. Bohm's sustained rejection of determinism was only partly acknowledged in Beller (1999) and new important evidences made available thanks to the publication of a collection of letters in Talbot (2017). Moreover, one of us (F.D.S.) already pointed out in Del Santo (2019b) that Bohm's commitment to determinism was secondary to his commitment to realism. The same thesis was then put forward in Van Strien (2020). Here, we show that Bohm's position was more radical than this: not only was determinism not his philosophical priority, but he actually always opposed it.

In Section 6.2, we recollect the standard narrative about Bohm's ideas. Albeit with some variations, indeed, there seems to be a consensus about the fact that Bohm's main philosophical concern was to retrieve determinism in modern physics (at least at a certain stage of his working life).

We strongly counter, in Section 6.3, this standard narrative with a more accurate account of the actual philosophical views of David Bohm, focusing on his take on causality and (non)determinism. We show that one of Bohm's main commitments was always anti-mechanism, a position that he had understood very early to be incompatible with determinism. This is what actually led him to initially (partly) support the indeterministic doctrine of Bohr and Heisenberg, which, however, he abandoned when he realized that randomness is another, for him unacceptable, form of mechanism. Hence, his commitment to determinism – stemming from his celebrated alternative interpretation – is only ostensible. Bohm's anti-mechanistic position led him to develop a dialectic philosophical view of an unlimited number of levels of description of reality that can be neither deterministic nor fully random, but still allow either of these descriptions to exist at different levels.

Here, we mainly focus on the period of the 1950s because it is in that decade that Bohm allegedly underwent a change from being a supporter of Bohr to becoming a determinist and then supposedly abandoned this debate altogether as his commitment to Marxism faded away. To avoid further misinterpretations on our part, we favour quoting as much as possible from Bohm's original writings rather than presenting our own summaries and analyses.[4]

6.2 The standard narrative: Bohm's alleged commitment to determinism

During World War II, the practices of physics underwent a drastic change. The foundational debate that had characterized the early days of quantum physics gave way to a pragmatic approach, the so-called "shut up and calculate" attitude, oriented towards applications often of a military nature (Kaiser 2011). It was only a handful of physicists (and a few philosophers) who started reviving the uneasiness towards the Copenhagen interpretation (see Kaiser 2011; Freire Junior 2015; Baracca *et al.* 2017; Besson 2018; Del Santo 2019a). Among them, David Bohm was a link between the old generation of critics – such as Albert Einstein, who played an active role in his intellectual life, Erwin Schrödinger, or (the early) Louis de Broglie –and the new underground culture concerned with quantum foundations to come.

After completing his PhD with Robert Oppenheimer at Berkeley in the 1940s and a position at the Physics Department at Princeton University, in 1951, Bohm fell victim of the witch-hunt of McCarthyism because of his adherence to Marxism. This led him to a life of exile: firstly to Brazil, then to Israel, and finally to the United Kingdom, where he spent the rest of his life (see Peat (1997) and Freire Junior (2019) for biographies of Bohm).

During his PhD, Bohm got acquainted with the ideas of Bohr (see Freire Junior 2019, p. 1, and Peat 1997, Chapter 4), the initiator of the so-called Copenhagen interpretation of quantum physics. The latter maintains that quantum mechanics necessarily leads to abandoning certain fundamental precepts of classical physics, among them determinism. Instead the Copenhagen interpretation embraces the genuine probabilistic nature of quantum phenomena.

Bohm went so deep in his reflections about quantum theory and its foundations that, in 1951, he published the textbook *Quantum Theory* (Bohm 1951), fully in the spirit of the Copenhagen interpretation. Shortly after the publication, indeed, Bohm himself stated about his book: 'a clear presentation of Bohr's point of view (the first clear, if I may boast a little)' (Letter from Bohm to Miriam Yevick; Letter 66, Folder C117, dated: 23 January 1952, in Talbot (2017), p. 235.).

[4]Moreover, in the interest of conciseness, we had to cut down the amount of Bohm's relevant quotations, which can however be found in an extended version of this chapter Del Santo and Krizek (2023).

However, in the very same year, Bohm submitted, on 5 July, a seminal work (published in two parts Bohm (1952a, 1952b)) wherein he presented the first consistent alternative interpretation of the quantum formalism. He introduced the initial position of quantum particles as a 'hidden variable' that, if known, would lead to *deterministic* trajectories similar to the familiar ones of classical mechanics (but guided by a genuinely additional quantum part in the potential).

So far, these are mere historical facts. Based on these, however, a standard narrative about David Bohm has crystallized, which can be summarized as follows: *In the span of around a year, Bohm had a dramatic shift in his philosophical agenda moving one of his tenets from indeterminism to determinism.* This narrative is not only popularized among physicists in the sort of working history that hovers in the community, but has been advocated by most historians, too. This is however not surprising, since admittedly it *prima facie* seems a rational account of the facts. A more thorough historical reconstruction, proposed among other works in the recent comprehensive biography of Bohm by Olival Freire Junior, states a nuanced story (Freire Junior 2019). First of all, it points out that already in Bohm (1951), Bohm had placed some hints of his uneasiness with Copenhagen, such as endorsing ontological realistic assumptions (see Freire Junior 2019, pp. 48–51). Moreover, historians tend to add a third phase in which Bohm supposedly distanced himself again from determinism at the end of the 1950s, concurrently with his dropping of Marxism. This double shift, also in relation to Marxism, was already strongly emphasized by Pylkkänen (1999), and also Freire, although more cautiously, endorses a similar position (Freire Junior 2019, p. 123). At any rate, the main point of the standard narrative is essentially also present in these more informed accounts.

The historical question that naturally arises then is: *why* did Bohm go through such a drastic and abrupt change from an adherent of the school of Copenhagen (in particular the thought of Bohr), i.e. a doctrine explicitly advocating the failure of determinism, to a novel deterministic interpretation? (And, possibly, why did he give up determinism again a few years later?). That is, what caused the sudden 'conversion' of Bohm from an open supporter of indeterminism to a staunch determinist (and perhaps back)?

Numerous studies have tried to answer this question (Jammer 1974; Peat 1997; Beller 1999; Pylkkänen 1999; Forstner 2008; Talbot 2017; Freire Junior 2019), apparently quite successfully despite a few minor details that are still the subject of historical debate. But what if the question was the wrong one in the first place? What if determinism has never been a desideratum for Bohm, rather, this change was not about his world view, but simply it was reflecting different phases of Bohm's experimentation in his attempt to achieve a physical theory that would satisfy his main philosophical tenets? In Section 6.3, we, in fact, defend this thesis. That is, that Bohm always upheld an anti-mechanistic view that was clearly incompatible with determinism alone. Before doing that, in the remainder of this section, we will continue summarizing the standard narrative, or rather, its reply to the main question it poses.

There is an almost absolute consensus on the fact that the two elements that played the major role in Bohm's turn towards determinism have been, on the one hand, his encounter with Einstein, and, on the other hand, his Marxist views. This twofold explanation is by now well established among historians, who mostly debate about the extent of one or the other influences (possibly, concurrently with Bohm's political prosecution; see Forstner 2008).

It is indeed well known that Einstein had opposed Bohr's views since the early days of quantum theory and his attempt to maintain determinism, summarized by the motto "God does not play dice", has entered the popular culture. However, while Einstein was invariably troubled by the abandonment of realism (and possibly of locality and localizability) implied by Bohr and his school, there are quite incontrovertible evidences that determinism was not Einstein's main philosophical concern (see Del Santo 2019b and references therein), and even less so in his late years. In the light of this, we can conjecture that the impact that Einstein had on Bohm at the time of their encounter at Princeton in the early 1950s was probably that of casting doubt on the Copenhagen interpretation and suggesting that one could search for an alternative. However, it does not seem likely that he directly pushed Bohm towards determinism, let alone hidden variable models that he never supported (see Del Santo 2019b).

As for whether and to what extent Marxism has been a guiding principle for Bohm in developing his deterministic hidden variable interpretation, the question is subtler. This has been considered in detail by Forstner (2005, 2008), and partly by Peat (1997), Freire Junior (2019), and Talbot (2017). Bohm certainly agreed with the ontology supported by Marx and Engels, namely, a materialistic philosophy which 'says that the sole reality is the natural world, and this world is made up solely of matter' and 'material things are not dependent for their existence or nature on any mind or minds', thus implying realism (from A. W. Wood, cited in Talbot (2017), p. 24). Moreover, Marx and Engels put together this materialistic view and the dialectic of Hegel, which turned into the main guiding philosophy of Marxism, i.e., *dialectical materialism.* While dialectical materialism applied in a scientific context primarily deals with the nature of the world, it is in the Marxist analysis of the progress of history and society, *historical materialism*, that one finds determinism as a main characteristic. In fact, for Marx it is the mode of production and the struggle between social classes that *necessarily* determine historical change.

As explained by Freire Junior (2019), it is objectively difficult to know to which Marxist writings Bohm had access to and therefore which parts of that philosophy had a concrete impact on his scientific and philosophical views. However, we will see in Section 6.3 that it is the dialectic aspect (and partly the materialist one, for what concerns realism) of Marxism that seems to have played the major role in the views about philosophy of science that guided Bohm, rather than the deterministic character of historical materialism.[5] As a matter of fact, Bohm was already a Marxist when he published his book Bohm (1951) in which he endorsed the view of Bohr, and we indeed show that his interest in Bohr actually stemmed, at least partly, from Marxism. This should be considered as Bohm's first attempt to get away from a mechanistic philosophy in a dialectic (i.e. Marxist) spirit.

Historians are not the only ones who have misconceived Bohm's point of view. The idea that Bohm's first and foremost concern was that of restoring determinism at

[5] It is known about Rosenfeld that he was a Marxist and a supporter of Bohr.

any cost was surely always widespread among physicists too. Starting with the contemporaries who were supportive of him – like Einstein, Louis de Broglie, and several Marxist physicists, in particular Jean-Pierre Vigier – and closely followed by his critics like Pauli and Rosenfeld, they all emphasized Bohm's commitment to determinism: the former as a merit and the latter as a untenable conservative attitude (see Freire Junior 2019, Chapters 4.2-4.5, for the early reactions on Bohm's hidden variable model).[6] As a matter of fact, due to his hidden variable model, Bohm started being regarded as a staunch determinist.

6.3 An alternative narrative: Bohm against mechanistic determinism

6.3.1 Indeterminism in Bohm's book *Quantum Theory* (1951) and beyond

As we have previously recalled, the first work of Bohm in which he manifestly deals with foundational questions is his 1951 book on quantum theory (Bohm 1951). It is generally known, as we have discussed, that this book takes an approach close to the orthodox view of Copenhagen. Note that in doing so, Bohm was not blindly following the mainstream, but rather he was actively looking for ways to provide quantum mechanics with solid and understandable physical foundations, against the widespread pragmatic acceptance of an operationally interpreted abstract formalism.

He therefore saw in the thoughts of Bohr an attractive philosophy because it was provided with two main features: the principle of complementarity, and irreducible probability (i.e. non-determinism). In the former he saw elements of dialectics, which we claim was Bohm's main influence from Marxism. In fact, this is a first attempt, that Bohm was to develop in greater detail in the following years (see below), to apply the ideas of Engels who, in his *Dialectics of Nature*, 'is especially opposed to attempts at mechanical reductionism' (Talbot 2017). In the context of quantum physics, the interaction between two qualitatively different descriptions (the classical and the quantum ones) determines reality. Therefore, they form something qualitatively new, not according to necessity. This also satisfied Bohm's anti-reductionist convictions because the classical world ought to lie outside of the quantum domain as a primitive and cannot be in general fully reduced to a quantum description. As for the acceptance of objective chance (i.e. potentialities), he saw in this the most natural possibility to abandoning the view of mechanistic determinism.

Moreover, soon after the publication of the book, he explained to his friend, the mathematician Miriam Yevick, why he got interested in Bohr:

[6] Incidentally, it should be recalled that Bohm's interpretation did not receive the praise that he expected and that he might have deserved. Even Einstein, who supported Bohm in his career and considered him a very talented physicist, stated that Bohm's way of restoring determinism 'seems too cheap' (see Del Santo 2019b). There are several hypotheses about why this has been the case, related to the *Zeitgeist* of post-war physics, Bohm's political views, the authority of the Copenhagen school, etc. (See Freire Junior 2015, 2019; Peat 1997; Cushing 1994). It was only in more recent years that the so-called Bohmian mechanics found new momentum in a sub-community of scholars interested in foundations of quantum physics (see Dürr and Teufel 2009; Dürr *et al.* 2013; Maudlin 2019). Also Bohm's close collaborators rediscovered Bohm's original interpretation and encouraged further works closer to Bohm's non-mechanistic ideas (see Baracca *et al.* 1975; Philippidis *et al.* 1979; Bohm and Hiley 1993).

All I knew was that there was one school, which utterly repelled me, in which one was supposed to introduce abstract mathematical postulates, and be satisfied if the calculations agreed with experiment. Against this, Bohr's school seemed to be a big improvement, because at least he tried to explain the physical meaning of the theory. Moreover, there was an element of dialectics in Bohr's point of view which attracted me. It seemed progressive because it broke the old mechanist materialist determinism, which left no room for growth and development of something new (Bohm to Miriam Yevick; Letter 65, Folder C117, dated: 7 January 1952, Talbot 2017, p. 227).

Note that at the time when he wrote this letter, Bohm was a staunch Marxist and most remarkably had completed his work on deterministic hidden variables half a year ago[7], and yet he was evidently criticizing *mechanistic materialist determinism.*

For what concerns its content, Bohm's book is an excellent technical manual of quantum mechanics and, although it endorses the view of the Copenhagen school, it is already possible to pin down where the main philosophical concerns of its author lie: causality is already his main focus together with his refusal of mechanism. However, at this stage, he explicitly endorses indeterminism as a way out of mechanism, a view that was soon to change when he realized that indeterminism can also be mechanistic.

In the previous section, we have recalled that Freire Junior (2019) already noted that a first element that distances Bohm from the Copenhagen school, is that in his 1951 book he looks for a realist account of nature. Another main difference with Copenhagen becomes manifest for what concerns causality. While for Heisenberg 'quantum mechanics proves the invalidity of the law of causality'[8] (Heisenberg 1927), for Bohm causality was an absolutely indispensable tenet. However, in his book it is clearly mentioned that while maintaining causality he wants to escape determinism. Hence, a first major distinction, certainly not well-understood at that time (and alas not even today in most of physics circles), is the conceptual difference between causality and determinism. In his 1951 book, Bohm actually already advocates a causally non-deterministic nature of physical laws, in terms of tendencies (this is closely related to Popper's view in terms of *propensities;* see Section 3.5 of the extended version of this Chpater Del Santo and Krizek (2023)):

we wish to call attention to the fact that, even in very early times, two alternative general types of causal laws appeared. One of these involved the notion of complete determinism; the other involved the notion of causes as determining general tendencies but not determining the behavior of a system completely (Bohm 1951, Chapter 8, Section 'Completely Deterministic vs. Causal Laws as Tendencies').

Bohm goes as far as to brilliantly show that actually the determinism of classical physics makes the concept of causality redundant ('classical theory leads to a point of view that is prescriptive and not causal', Bohm 1951, Chapter 8, Section 'Classical Theory Prescriptive and not Causal'). Hence, he found a way out of the effective lack of causality in a completely deterministic theory in terms of the tendencies or potentialities entailed by (the Copenhagen interpretation of) quantum physics:

With the advent of quantum theory, the idea of complete determinism was shown to be wrong and was replaced by the idea that causes determine only a statistical trend, so that a

[7] The first of the two hidden variable papers has been received on 15 July 1951 (Bohm 1952a).

[8] The original German phrase reads: '*so wird durch die Quantenmechanik die Ungtültigkeit des Kausalgesetzes definitiv festgestellt.*

given cause must be thought of as producing only a tendency toward an effect. [...] (Bohm 1951, Chapter 8, Section 'New Properties of Quantum Concepts: Approximate and Statistical Causality').

However, he soon realized that Bohr's philosophy was not as satisfactorily as he previously had sensed because it indeed contained a dialectical approach but not as much of materialism as he would have wanted:

After I had written the book, I finally began to grasp the full meaning of the theory, and could see that it leads inevitably to a form of (dialectical) idealism. But this was not so clear when I started, because of the general confusion in the literature (Bohm to Miriam Yevick; Letter 65, Folder C117, dated: 7 January 1952, Talbot 2017, p. 227).

And again:

I notice that you call me "a disciple of Einstein". This is not very accurate. Actually I was a strong "Bohrian" and wrote my book under the assumption (later proved wrong) that the principle of Complementarity was a materialist point of view. It certainly is very dialectical, but I did not see at that time that it is not materialist (Bohm to Schatzman; Letter A1.15, dated: 7 September 1952, Besson 2018, p. 335).

6.3.2 Against determinism, despite hidden variables (1952)

Exactly in the same period when the book Bohm (1951) was appearing, Bohm was formulating his alternative, deterministic interpretation in terms of hidden variables. Given his clear motivation recalled in the previous section, why did he do that? Bohm must have found himself in a strange position, when he managed to conceive a consistent model based on hidden variables that restored determinism. He clearly wanted to prove something that was considered impossible by the founding fathers of theory, in particular John von Neumann who had allegedly proven that a hidden variable completion of quantum mechanics was in principle impossible.[9] Moreover, Bohm wanted to prove that Bohr's and Heisenberg's view were not necessarily the ultimate description of reality.

According to Hiley, indeed, Bohm 'was not a deterministic man, he used causality. [...] He was not bound to it [determinism]. David Bohm always used to say to me: "I am making a proposal". So, all this people think he had rigid views. He didn't have rigid views. He was always making proposals, because he thought he never fully got to the bottom of quantum mechanics' (Del Santo and Krizek 2019).

In fact, although Bohm stresses in his papers that the '"hidden" variables determine the precise results of each individual measurement process' (Bohm 1952b), repeatedly acknowledging very clearly the deterministic character of his model, he certainly never adopted a fundamental ontology merely made of particles plus their deterministic dynamics guided by the wave function. This is something that some of his followers, the so-called Bohmians, have instead assumed, namely, considering Bohm's proposal as the ultimate description of reality, much against the view of Bohm himself.

Indeed, at least since 1951, most likely when he was still in Princeton (see Talbot 2017, Footnote 48, p. 31), Bohm started developing a new philosophy based on the concept of having different levels of description, each of which can be either deterministic or indeterministic, but each of them giving only a partial account of reality. His

[9] On the history of von Neumann's impossibility proof, see Dieks (2017).

ontology was thus made of the wholeness of the different levels of qualitatively different entities. However, he postulated the number of levels to be infinite, thereby making it fundamentally impossible to have mechanism, and in particular determinism:

> Because of the existence of an infinite number of levels, the deterministic laws of order at each level probably follow only as a result of conditions of chaos existing at lower levels. If the lower-level conditions of chaos could be altered, then the very framework of description of the higher level laws would also have to be altered. Thus, we are led to a more dynamic concept of the laws of nature (Bohm to Miriam Yevick; Letter 58, Folder C116, dated: 23 November [1951],Talbot 2017, p. 205).

Note that this idea, while being further refined, remained essentially unchanged throughout Bohm's transition from the period of his 1951 book to his hidden variable proposal, and reached its main expression in the volume *Causality and Chance* Bohm (1957) published in 1957 (see Section 6.3.4).[10] For instance, after he had already completed his hidden variable interpretation, he wrote to Yevick:

> The "things" at each level, are made up of smaller "elements" at a more fundamental level, and it is the motion of these more fundamental elements (not usually directly visible to us, except with the aid of elaborate scientific research) which causes the appearance and disappearance of the "things" existing at a higher level. These more fundamental "elements" however, cannot be permanent, but must be made up of still more fundamental "elements" and so on ad infinitum (Bohm to Miriam Yevick; Letter 65, Folder C117, dated: 7 January 1952, Talbot 2017, p. 227).

Bohm also points out his position on the need for infinite levels to this collaborator Schatzman in a letter from 1952:

> It is most likely that not even the substratum particles could be indestructible and unanalysable. Instead, there is probably another substratum below this (of a qualitatively different kind most probably) and so on ad infinitum. Thus, we should have an infinite series of qualitatively different levels of laws. Any finite number of levels can always be understood by humanity, but never <u>all</u> of them (Besson 2018, p. 351).

And soon after his letter to Miriam Yevick in January, he wrote what is one of the most important quotations from the whole collection of known writings of David Bohm, because it unambiguously states that he could not accept mechanic determinism, even in the period when he was promoting his hidden variable model:

> Most of the errors of both the positivist and the 19th century "mechanical" materialists spring from an implicit assumption that the laws of nature will some day finally be understood in terms of a limited number of hypotheses. From this comes the nightmare of a mechanically determined universe that follows an inevitable course. To avoid this nightmare, positivists and idealists have given up causality and assumed a "spontaneous" (i.e., uncaused) element in physical processes.
>
> The concept of a limitless number of levels [...] provides a motive power for continual development & growth. Moreover, the nightmare of complete determinism is avoided. Although each level is causal, the totality of levels cannot ever be taken into account. Thus, as a matter of principle, we say that complete determinism could not even be conceived of, yet, each level can be determined. Here, we part company with the believers in "spontaneity" for we say that what appears to be spontaneous is caused by factors, in principle, knowable, but now hidden to us. But to be able to say this without implying complete determinism, we must assume an

[10] A first sketch of the idea has been published in Bohm (1953a).

unlimited number of levels (Bohm to Miriam Yevick; Letter 73, Folder C118, dated: Received 31 March [1952], Talbot 2017, pp. 254–55).

It is now clear that Bohm did not undergo a conversion from indeterminism (à la Copenhagen) to determinism (with hidden variables) as the standard narrative implies. He actually stayed faithful to his tenets of realism and causality, and his shift was merely that of realizing that Bohr's approach was not enough to achieve what he had in mind. It should be further noted that his contributions to plasma physics, in collaboration with his former student David Pines from 1951 to 1957, also encompass elements of his holistic ideology (Bohm and Pines 1951; Pines and Bohm 1952; Bohm and Pines 1953). So, it seems that his philosophical theory of the infinite levels was conceived to 'cure' his own model from the 'nightmare' of determinism. One should also remark that this idea of unlimited levels is very much in the spirit of dialectics, and indeed this is the most Marxist trait in Bohm's work. As pointed out by Talbot, such a connection is perhaps less abstract that one could think, drawing directly from the work of Engels: 'especially in the Dialectics of Nature, Engels introduces the idea of levels, or what he calls "forms of motion". [...] Engels is especially opposed to attempts at mechanical reductionism, which "blots out the specific character" and "qualitative difference" of non-mechanistic forms of motion' (Talbot 2017, p. 25). For Bohm, this dialectic view of nature is a way to maintain a non-trivial form of causality, intended as the possibility of creating non-necessary new things, contrarily to the mechanistic view. In a letter to his friend – the American physicist Melba Phillips – Bohm specified this connection in detail:

[T]he concept of the infinity of levels shows that there need exist in nature no such thing as a basic element which never changes. [...] Thus, causality is a broader concept than that of mechanical determinism. [...] A "mechanistic" attitude toward science however, tends to limit the growth of our concepts in an arbitrary and dogmatically conceived way. Such a mechanistic attitude refers not only, however, to the mechanistic determinists, but also to the "mechanistic indeterminists", who insist that in the quantum of action, we have reached an ultimate, indivisible, and unanalyzable entity, which will never be found to have a structure understandable in terms of a deeper level (Bohm to Melba Phillips. Letter 43, Folder C48, dated: 13 October 1953, Talbot 2017, p. 164).

In the following years, Bohm kept developing his philosophy of the infinite levels, sharpening the distinction between causality and deterministic mechanism, and advocating the former in strong opposition to the latter. For Bohm, causality is the possibility of creating new qualitative entities in a non-trivial sense, i.e. without being able to reduce everything to a finite collection of basic elements that cannot change and are subject to fix laws:

Now, at first sight, it may seem that we could eliminate the large-scale level by analyzing it in terms of its basic molecular motions. And if there were a finite number of levels, this would be true. But if there are an infinite number, then each level stands on a footing that is, in the long run, as basic as that of any other. For every level has below it a deeper one. Indeed, matter can be regarded as made up of the totality of all levels. Each level makes its own specific contribution to the totality (Bohm to Melba Phillips. Letter 46, Folder C48, dated: 15 March 1954, Talbot 2017, p. 170).

6.3.3 An indeterministic causal model by Bohm and Vigier (1954)

So far, the evidence that Bohm was against determinism even during the years in which he devised and promoted his hidden variable model is limited to private correspondence. However, in 1954, Bohm published a paper with Vigier – *Model of the causal interpretation of quantum theory in terms of a fluid with irregular fluctuations* (Bohm and Vigier 1954) – that is a first attempt to put into practice the ideas of a model of causal interpretation which is however fundamentally non-deterministic, due to different levels of description. In fact, therein Bohm and Vigier postulate a field that is described by a fluid of density $|\psi|^2$, which is then able to recover the standard quantum mechanics

> by introducing the hypothesis of a very irregular and effectively random fluctuation in the motions of the fluid. [...] Such random fluctuations are evidently consistent within the framework of the causal interpretation of the quantum theory. Thus, there are always random perturbations of any quantum mechanical system which arise outside that system (Bohm and Vigier 1954).

They indeed clarify that "the causal interpretation of the quantum theory permits an unlimited number of new physical models" and that their suggested 'model is an extension of the causal interpretation of the quantum theory already proposed, which provides a more concrete physical image of the meaning of our postulates than has been available before, and which suggests new properties of matter that may exist at deeper levels' (Bohm and Vigier 1954). Here, causal means the possibility of explaining the theory in terms of a sub-quantum level (the fluid) that accounts for the higher quantum level. Note that, contrarily to the first hidden variable theory (Bohm 1952a, 1952b), this model is based on fundamental random fluctuations, thereby dispelling even more the doubt that Bohm was a committed determinist: 'In the model that we have proposed here, however, the statistical fluctuation in the results of such [quantum] measurements are shown to be ascribable consistently to an assumed deeper level of irregular motion'. It is interesting to note that while the postulated fluctuations of the fluid are considered to be (at this level of description) genuinely indeterministic, Bohm and Vigier think of these fluctuations as having a certain structure in terms of potentialities: 'The fact that the mean density remains equal to $|\psi|^2$, despite the effects of the random fluctuations, implies then that a systematic tendency must exist for fluid elements to move towards regions of high mean fluid density'. The ontological basis of this new indeterministic model and how it relates to Bohm's philosophy of the infinite levels is explained by Bohm in correspondence with Einstein:

> The general idea is that at a level more fundamental than that of quantum mechanics, there is a field which satisfies causal laws. This field is, however, in a state of statistical fluctuations. These fluctuations are somehow described by the Ψ field (Bohm to Einstein; Letter 16, page 5 Folder C14, dated: 3 February 1954, Bohm 1954a, p. 5).

> My own point of view is that below the quantum theory there exists a sub quantum-mechanical level of continuous and causally determined motion, and that the quantum theory is related to the sub-quantum mechanical level, more or less as ordinary Brownian motion is related to the atomic level. In other words, events at the atomic level are contingent on the (in general irregular) motions of some as yet unknown but qualitatively new kind of entity, existing below the atomic level. As a result, the relationships between things, that can be defined at the atomic level will be characterized by the laws of chance, since they will

be determined only in terms of some quasi-ergodic type of motion of new kinds of entities existing at the lower level (Bohm to Einstein; Letter 21, Folder C15, dated: 14 November 1954, Bohm 1954b).

Einstein's replies may seem surprising to those who still believe that he was also a committed determinist at any cost, because they show once more that he was dissatisfied with Bohm's first (deterministic) hidden variable model: 'I am glad that you are deeply immersed seeking an objective description of the phenomena and that you feel that the task is much more difficult as you felt hitherto' (Einstein to Bohm; Letter 17, Folder C14, 10 February 1954, Bohm 1954a). And again: 'In the last years several attempts have been made to complete quantum theory as you have also attempted. But it seems to me we are still quite remote from a satisfactory solution of the problem' (Einstein to Bohm; Letter 20, Folder C15, dated: 28 October 1954, Bohm 1954b).

Bohm did not further develop this approach, which he most likely perceived as a proposed first step towards his philosophy of levels of description, but he came back to a stochastic causal interpretation, also with Hiley, in the 1980s (cf. Bohm 1988; Bohm and Hiley 1989).

6.3.4 Causality and Chance in Modern Physics (1957)

It is around the same period that Bohm started thinking not only that either a deterministic or an indeterministic description was possible in every level of an infinite series, but that both individual and statistical laws are necessary for a causal interpretation:

> The picture which I propose is this: The totality of causal laws includes both statistical and individual laws. We start with this totality as our basic reality. [...] The fundamental reality is that of matter in being and in process of change, or of becoming, as it may more accurately be called (Bohm to Miriam Yevick. Letter 121, Folder C124, dated: 10 September 1954, Talbot 2017, p. 419–22).

These dialectic ideas grew into a book, *Causality and Chance*, which Bohm published in 1957 (Bohm 1957). Therein, Bohm identifies two types of causal laws (both considered fundamental): simple causal laws that connect past and future one-to-one (i.e. deterministic) and more general ones that are one-to-many, (i.e. that do not lead to a unique evolution but only to an array of possibilities):

> [L]et us note that the one-to-many character of a causal law has no essential relationship to a lack of knowledge on our part concerning the additional causal factors to which the more precise details of the effect can be traced. [...] In other words, a one-to-many law represents an objectively necessary causal connection, but in this case, what is necessary is that *the effect remain within certain bounds*; and not, as in simpler types of causal laws, that the effect be determined uniquely (Bohm 1957, p. 17).

And again, Bohm clarifies, as he always maintained (cf. Section 6.3.1), that causality is a more general concept than that of necessity (i.e. determinism):

> We see, then, that it is appropriate to speak about objectively valid laws of chance, which tell us about a side of nature that is not treated completely by the causal laws alone. Indeed, the laws of chance are just as necessary as the causal laws themselves. [Footnote:] Thus necessity is not to be identified with causality, but is instead a wide category (Bohm 1957, p. 23).

Furthermore, Bohm here again stresses the fact that objective chance should be interpreted as a potentiality, i.e., a property of the system and its causal conditions:

On the basis of the above considerations, we are then led to interpret the probability of, for example, a given result in the game of dice as an objective property associated with the dice that are being used and with the process by which they are thrown (Bohm 1957, p. 27).

Note that this example is exactly the same used by Popper (1959) when he introduced the propensities interpretation (see Section 3.5 of the extended version of this chapter Del Santo and Krizek (2023)), again showing the compatibility between Bohm and a world view based both on causality and indeterminism.

Beyond causality, a large part of Bohm's 1957 book is devoted to defend another of his main tenets, namely, anti-mechanism. However, while being still convinced that determinism is an unacceptable form of mechanism, there is a fundamental difference with respect to his book on quantum theory (Bohm 1951). Here, in fact, Bohm does not consider randomness alone as a way out of mechanism:

The question of what constitutes a mechanistic philosophy, therefore, cuts across the problems of determinism and indeterminism. For this reason, we shall call the philosophy described in this section by the name of "indeterministic mechanism" (Bohm (1957), pp. 62–63).

Bohm's criticism of mechanism (and thereby of determinism), does not spare his own hidden variable interpretation, which he considers again an unsatisfactory physical model, whose main feature, he stresses, is consistency:

While our theory can be extended formally in a logically consistent way by introducing the concept of a wave in a $3N$-dimensional space, it is evident that this procedure is not really acceptable in a physical theory, and should at least be regarded as an artifice that one uses provisionally until one obtains a better theory in which everything is expressed once more in ordinary three-dimensional space (Bohm 1957, p. 117).

Finally, in his *Causality and Chance*, Bohm for the first time systematically[11] defends publicly his philosophical view of the infinite levels of description as the main alternative to mechanism, be it deterministic or indeterministic. As noted already by Freire Junior (2019), this marks Bohm's entry in the philosophical debate and would allow him to engage with prominent philosophers of science, the like of Paul Feyerabend and Karl Popper (see further). However, these ideas of infinite levels were not appreciated by his more traditional Marxist followers, who saw in this the undermining of determinism: a positive feature for Bohm and an unacceptable price for them. This is the case of Évry Schatzman and Vigier who wrote to Bohm: "We may be wrong, but we do not agree at all with your ideas about the different levels of reality. It seems to us that it is a formal interpretation of the famous sentence of Lenin, in *Materialism and Empiriocriticism*, about the different levels of reality" (quoted in Freire Junior (2019), p. 108).

To conclude, in *Causality and Chance* Bohm synthesizes his main philosophical tenets that have been present in his writing since the beginning, but in a quite scattered way. Therein, Bohm defends, for the first time systematically, causality in its broadest sense, advocating the fundamental necessity of both individual and statistical laws, depending on the context. Moreover, he firmly rejects mechanism, not only in the form of determinism (which he did for many years already), but also in its indeterministic form. Finally, Bohm opposes mechanism with a dialectic philosophy of infinite levels of description that he had developed throughout the 1950s.

[11] Cf. Footnote 10.

For what concerns physics proper, in 1957, Bohm published with his student Yakir Aharonov a paper where he rejects his own 1952 model, not on the basis of determinism but on nonlocality: 'It must be admitted, however, that this quantum potential seems rather artificial in form [...] that it implies instantaneous interactions between distant particles, so that it is not consistent with the theory of relativity' (Bohm and Aharonov 1957). Bohm thus kept proposing his dialectical views of different levels, similar to the paper with Vigier (Bohm and Vigier 1954), looking for a a 'deeper subquantum-mechanical level' (Bohm and Aharonov 1957).

It is interesting to note that still at this stage, Bohm's views were completely misunderstood. Louis de Broglie, who wrote the foreword of his *Causality and Chance*, for instance, keeps attributing to Bohm the great merit of giving hope to those who look for a deterministic hidden variable explanation of quantum theory: 'It is possible that looking into the future to a deeper level of physical reality we will be able to interpret the laws of probability and quantum physics as being the statistical results of the development of completely determined values of variables which are at present hidden from us. It may be that the powerful means we are beginning to use to break up the structure of the nucleus and to make new particles appear will give us one day a direct knowledge which we do not now have of this deeper level' (Bohm 1957, p. x). This goes completely against what Bohm conveys in his book, making us wonder whether people like de Broglie were actually reading Bohm's works or they just imposed on him what they wished to hear.

Towards the end of the 1950s, Bohm abandoned Communism, following the revelations of Stalin's crimes by Nikita Khrushchev in 1956 (see Freire Junior 2019). As already recalled, this has been identified in the literature as the main motivation to abandon his commitment to determinism. But as we have shown, such an alleged commitment to determinism was never present in the first place and his dialectic attitude remained an important factor in his philosophy. However, probably due the frustration of being continuously misunderstood, Bohm's engagement with different models of the causal interpretation became sparser. Actually, since his moving to the United Kingdom, firstly in Bristol and then in London, he engaged more and more in the philosophical debate, becoming friend with Paul Feyerabend, Karl Popper, and Stephen Körner, and he kept his interpretational considerations away from his physics colleagues.

Hiley joined Bohm at Birkbeck College in London in 1961 and, as a matter of fact, they passed 'ten years without actually talking about the causal interpretation' (Del Santo and Krizek 2019). As recalled by Hiley (Del Santo and Krizek 2019), it was only in the 1970s that two of Bohm's students, Chris Dewdney and Chris Philippidis, 'rediscovered' the hidden variable papers Bohm (1952a, 1952b). Hiley finally thoroughly read Bohm's original papers and motivated the students to use the computer to calculate the trajectories of particles using Bohm's model. This marks the revival of Bohm's hidden variables (see also Freire Junior 2019, Chapter 6.1), a revival to whom Bohm, however, obviously did not participate.[12] Actually, when approached by Dewdney and Philippidis, 'Bohm himself [...] admitted that he had made a tactical error in his original presentation of the theory. The term *hidden variables*, he said, created the

[12] With the only exception of Bohm *et al.* (1985).

wrong impression, and the papers themselves were too rigid and deterministic' (Peat 1997), p. 266). Indeed, Bohm later explained the purpose of his original proposal in the following way:

> To show that it was wrong to throw out hidden variables because they could not be imagined, it was therefore sufficient to propose any logically consistent theory that explained the quantum mechanics, through hidden variables, no matter how abstract and hypothetical it might be. Thus, the existence of even a single consistent theory of this kind showed that whatever arguments one might continue to use against hidden variables, one could no longer use the argument that they are inconceivable. Of course, the specific theory that was proposed was not satisfactory for general physical reasons, but if one such theory is possible, then other and better theories may also be possible, and the natural implication of this argument is 'Why not try to find them?' (Bohm 1980, p. 104).

In the following decades, Bohm dedicated his work to a holistic approach that continued his ideas from the work on the causal interpretation of quantum theory. His late scientific programme was based on quantum field theory to approach the concept of the infinite levels he already pointed out in the early works. However, his philosophical ideas remained consistent to his early works in the refusal of mechanistic ideas (see also Bohm 1980, p. 223).

6.4 Discussion and conclusion

In this chapter, we have shown that Bohm was always against mechanism and therefore determinism. We have contradicted the historical narrative according to which one can identify an early period when Bohm was a supporter of Bohr, a later period when he was a committed determinist (influenced by Einstein and Marxism), and finally a period, after his break with Marxism, in which determinism quit being a main concern of his. On the contrary, Bohm's philosophical tenets have never changed throughout his whole life: he was always committed to develop a realistic, causal, non-mechanistic view of physics. This led him to develop a new dialectical philosophy composed of infinite levels of description that guided him in his work for the following decades. As such, Bohm would have never accepted determinism, at any stage of his life. In a slogan, *Bohm was never a Bohmian.*

Although the content of this chapter has mostly a historical scope, it may also affect the physicists and philosophers who have proclaimed themselves Bohmians. It is undeniably true that Bohm provided the first deterministic hidden variable model of quantum theory. And yet, we just want to stress that for him this was nothing more than a model, a proof of principle that it was possible to do what was considered fundamentally unattainable at the time. However, simultaneously, this was for him most unsatisfactory, for it betrayed one of his deepest convictions about nature, namely, that a basic ontology of particles moved around by deterministic laws cannot be the end of the story. Therefore, the many scholars who today support Bohmian mechanics at face value, giving to it an ontological role, should be aware that they are advocating a world view that stems from what its original proposer considered a mere model that could not satisfy the basic standards of acceptability for a physical theory (except internal consistency). Now, while this is obviously a logically acceptable position, they should be aware that they are going directly against the fundamental views of Bohm, and therefore cannot appeal to his authority. This separation between the original thoughts of Bohm and those who adopted his model was so striking that soon before his death when he became aware of Sheldon Goldstein and Detlef Dürr's work on his ideas, Bohm bitterly confessed to his main collaborator Basil Hiley: 'why on earth are they calling it Bohmian mechanics? Haven't they read a word I have written?' (Del Santo and Krizek 2019). So, concerning determinism, Bohm finds himself in a position comparable (fortunately with less ethical implications) to that of Einstein with respect to the atomic bomb: it is a historical fact that it was Einstein who suggested to US president Franklin Roosevelt to research on nuclear weapons to preempt Nazi Germany to achieve the same threat. However, for his whole life – before and after – Einstein was a committed pacifist. Similarly, it is a historical fact that Bohm developed a deterministic interpretation of quantum theory. However, for his whole life – before and after – he was a committed anti-determinist. Invoking Bohm to defend deterministic views of physics is like invoking Einstein to promote nuclear weapons.

Acknowledgements

F.D.S. acknowledges support from FWF (Austrian Science Fund) through an Erwin Schrödinger Fellowship (Project J 4699s).

References

Baracca, A., Bergia, S. and Del Santo, F. (2017). The origins of the research on the foundations of quantum mechanics (and other critical activities) in Italy during the 1970s. *Studies in History and Philosophy of Science Part B: Studies in History and Philosophy of Modern Physics*, **57**: 66.

Baracca, A., Bohm, D., Hiley, B.J., and Stuart, A.E.G. (1975). On some new notions concerning locality and nonlocality in the quantum theory. *Il Nuovo Cimento B (1971–1996)*, **28**(2): 453.

Beller, M. (1999). *Quantum Dialogue: The Making of a Revolution.* Chicago: University of Chicago Press.

Besson, V. (2018). *L'interprétation causale de la mécanique quantique: biographie d'un programme de recherche minoritaire (1951–1964).* PhD thesis, Université de Lyon; Universidade Federal da Bahia.

Bohm, D. (1951). *Quantum theory.* New York: Prentice-Hall, Inc.

Bohm, D. (1952a). A suggested interpretation of the quantum theory in terms of "hidden" variables. I. *Physical Review*, **85**(2): 166–179.

Bohm, D. (1952b). A suggested interpretation of the quantum theory in terms of "hidden" variables. II. *Physical Review*, **85**(2): 180–193.

Bohm, D. (1953a). Comments on an article of Takabayasi conserning the formulation of quantum mechanics with classical pictures. *Progress of Theoretical Physics*, **9**(3): 273–287.

Bohm, D. (1953b). Proof that probability density approaches $|\psi|^2$ in causal interpretation of the quantum theory. *Physical Review*, **89**(2): 458.

Bohm, D. (1954a). *Correspondence. Papers and correspondence of David Joseph Bohm 1917–1992, letter C14*. Birkbeck Library Archives and Special Collections, University of London. GB 1832 BOHM/C.

Bohm, D. (1954b). *Correspondence. Papers and correspondence of David Joseph Bohm 1917–1992, letter C15*. Birkbeck Library Archives and Special Collections, University of London. GB 1832 BOHM/C.

Bohm, D. (1957). *Causality and Chance in Modern Physics*. University of Pennsylvania Press.

Bohm, D. (1980). *Wholeness and the Implicate Order*. London: Routledge.

Bohm, D. (1988). Non-locality in the stochastic interpretation of the quantum theory. In *Annales de l'IHP Physique théorique*, **49**(3):287.

Bohm, D. and Aharonov, Y. (1957). Discussion of experimental proof for the paradox of Einstein, Rosen, and Podolsky. *Physical Review*, **108**: 1070.

Bohm, D., Dewdney, C. and Hiley, B.J. (1985). A quantum potential approach to the wheeler delayed-choice experiment. *Nature*, **315**(6017): 294.

Bohm, D. and Hiley, B.J. (1993). *The Undivided Universe. An Ontological Interpretation of Quantum Theory*. London: Routledge.

Bohm, D. and Hiley, B.J. (1989). Non-locality and locality in the stochastic interpretation of quantum mechanics. *Physics Reports*, **172**(3): 93.

Bohm, D. and Pines, D. (1951). A collective description of electron interactions. I. Magnetic interactions. *Physical Review*, **82**: 625.

Bohm, D. and Pines, D. (1953). A collective description of electron interactions. III. Coulomb interactions in a degenerate electron gas. *Physical Review*, **92**: 609.

Bohm, D. and Vigier, J.-P. (1954). Model of the causal interpretation of quantum theory in terms of a fluid with irregular fluctuations. *Physical Review*, **96**(1): 208.

Cushing, J.T. (1994). *Quantum Mechanics: Historical Contingency and the Copenhagen Hegemony*. Chicago: University of Chicago Press.

Del Santo, F. (2019a). Karl Popper's forgotten role in the quantum debate at the edge between philosophy and physics in 1950s and 1960s. *Studies in History and Philosophy of Science Part B: Studies in History and Philosophy of Modern Physics*, **67**: 78.

Del Santo, F. (May 2019b). Striving for realism, not for determinism: Historical misconceptions on Einstein and Bohm. *APS News*.

Del Santo, F. and Krizek, G.C. (2019). *Interview to Basil Hiley*. Birkbeck College, London (UK).

Del Santo, F. and Krizek, G.C. (2023). Against the 'nightmare of a mechanically determined universe': Why Bohm was never a Bohmian [Extended preprint version]. *arXiv preprint arXiv:2307.05611*.

Dieks, D. (2017). Von Neumann's impossibility proof: Mathematics in the service of rhetorics. *Studies in History and Philosophy of Science Part B: Studies in History and Philosophy of Modern Physics*, **60**: 136.

Dürr, D., Goldstein, S. and Zanghì, N. (2013). *Quantum Physics without Quantum Philosophy.* Berlin: Springer.

Dürr, D. and Teufel, S. (2009). *Bohmian Mechanics: The Physics and Mathematics of Quantum Theory.* Berlin: Springer.

Esfeld, M., Hubert, M., Lazarovici, D. and Dürr, D. (2013). The ontology of Bohmian mechanics. *The British Journal for the Philosophy of Science*, **65**(4): 773–796.

Forstner, C. (2005). *Dialectical Materialism and the Construction of a New Quantum Theory: David Joseph Bohm, 1917–1992.* Berlin: Max-Planck-Institut für Wissenschaftsgeschichte.

Forstner, C. (2008). The early history of David Bohm's quantum mechanics through the perspective of Ludwik Fleck's thought-collectives. *Minerva*, **46**(2): 215.

Freire Junior, O. (2015). *The Quantum Dissidents: Rebuilding the Foundations of Quantum Mechanics (1950–1990).* Berlin: Springer.

Freire Junior, O. (2019). *David Bohm: A Life Dedicated to Understanding the Quantum World.* Cham, Switzerland: Springer.

Heisenberg, W. (1927). Über den anschaulichen Inhalt der quantentheoretischen Kinematik und Mechanik. *Zeitschrift Für Physik*, **43**: 172–198.

Hiley, B.J. (1997). David Bohm. 20 December 1917–27 October 1992. *Biographical Memoirs of Fellows of the Royal Society*, **43**: 107.

Hiley, B.J. (2001). Non-commutative geometry, the Bohm interpretation and the mind-matter relationship. In *AIP Conference Proceedings*, **573**: 77–88. American Institute of Physics.

Holland, P.R. (1995). *The Quantum Theory of Motion. An Account of the de Broglie-Bohm Causal Interpretation of Quantum Mechanics.* Cambridge: Cambridge University Press.

Howard, D. (1993). Was Einstein really a realist? *Perspectives on Science*, **1**(2): 204.

Jammer, M. (1974). *Philosophy of Quantum Mechanics. The Interpretations of Quantum Mechanics in Historical Perspective.* New York: Wiley.

Kaiser, D. (2011). *How the Hippies Saved Physics: Science, Counterculture, and the Quantum Revival.* New York: WW Norton & Company.

Maudlin, T. (2019). *Philosophy of Physics: Quantum Theory.* Princeton: Princeton University Press.

Oldofredi, A. (2023). Orthodox or dissident? The evolution of Bohm's ontological reflections in the 1950s. *European Physical Journal H*, **48**: 12.

Peat, D.F. (1997). Infinite potential: The life and times of David Bohm. Reading, MA: Addison-Wesley.

Philippidis, C., Dewdney, C. and Hiley, B.J. (1979). Quantum interference and the quantum potential. *Nuovo Cimento B*, **52**(1): 15.

Pines, D. and Bohm, D. (1952). A collective description of electron interactions. II. collective vs individual particle aspects of the interactions. *Physical Review*, **85**: 338.

Popper, K.R. (1959). The propensity interpretation of probability. *The British journal for the Philosophy of Science*, **10**(37): 25.

Pylkkänen, P. (1999). *Bohm–Biederman Correspondence: David Bohm and Charles Biederman: Vol. I.: Creativity and Science*. London: Routledge.

Pylkkänen, P. (2016). Quantum theory, active information and the mind-matter problem. In Dzhafarov, E., (ed.) *Contextuality from quantum physics to psychology*, pp. 325–334. Singapore: World Scientific.

Talbot, C. (2017). *David Bohm: Causality and Chance, Letters to Three Women.* Cham, Switzerland: Springer.

Van Strien, M. (2020). Bohm's theory of quantum mechanics and the notion of classicality. *Studies in History and Philosophy of Science Part B: Studies in History and Philosophy of Modern Physics*, **71**: 72.

Van Strien, M. (2025). Why Bohm was never a determinist. In Oldofredi, A. (ed.) *Guiding Waves in Quantum Mechanics: One Hundred Years of de Broglie-Bohm Pilot-Wave Theory*. Oxford: Oxford University Press.

Part III

Philosophical Discussions

7

Relativistic pilot-wave theories as the rational completion of quantum mechanics and relativity

Valia Allori

7.1 Introduction

In this chapter I wish to argue that the best way of reconciling quantum theory and relativity is to think of them both as incomplete. The 'spirits' of these two theories are in tension: the 'spirit' of quantum theory is non-locality, but it is developed in a theory without a fundamental space-time; while the 'spirit' of relativity is space-time in which interactions propagate locally in a Lorentz invariant framework. So, how should we proceed in our research towards a relativistic quantum theory?

I argue that quantum theory, *contra* to popular opinion, is incomplete because it lacks a microscopic spatiotemporal ontology to explain the phenomena constructively: macroscopic bodies are composed of the microscopic fundamental entities, and their properties are derived from the microscopic dynamics. The pilot-wave theory completes quantum mechanics by specifying a spatiotemporal ontology of particles. Nonetheless, its non-locality clashes with relativity. This is (part of) the reason why people have gained interest in alternative quantum theories, like for instance the many-worlds theory, whose non-locality is more controversial.

However, many have argued that the experimental violation of Bell's inequality has unquestionably shown that all quantum theories are non-local.[1] If so, the non-locality of the pilot-wave theory cannot count against it. Quantum non-locality indeed suggests a natural fix: a preferred frame. If so, relativity also turns out to be incomplete, as it lacks the (natural) spatiotemporal structure necessary to account for non-locality. I conclude by arguing that the simplest and most straightforward type of theory that best combines relativistic spatiotemporal structure and quantum non-locality is a relativistic pilot-wave theory, in which there is a microscopic spatiotemporal ontology to constructively explain the phenomena and there is a Lorentz invariant dynamical preferred spatiotemporal foliation implementing the non-local interaction. That is, relativistic pilot-wave theories are the rational completion of both relativity and quantum mechanics.

[1] However, see later, especially Footnote 10.

Here, a roadmap of the chapter is presented. In Section 7.2 I start with quantum theory. I briefly argue that the real problem with quantum theory is not the measurement problem but rather the absence of a spatiotemporal ontology. I then discuss locality, its desirability, and the charge, originally advanced by Einstein, that quantum theory is non-local. I conclude by showing how, given the experimental violation of Bell's inequality, the 'spirit' of quantum theory is non-locality: no theory which reproduces the quantum predictions can be local. In Section 7.3 I present the pilot-wave theory as a theory of particles that interact non-locally. Defined in this way, the theory straightforwardly has none of the previous problems, and, in accordance with the violation of Bell's inequality, it is non-local. I then turn in Section 7.4 to relativity, I explore the possible 'spirits' of relativity, over and above the Lorentz invariance of the laws, and I argue that it is fundamentally a theory about space-time. In Section 7.5, I analyse the various proposed relativistic quantum theories: first relativistic pilot-wave theories, and then relativistic spontaneous localization theories. They all have Lorentz invariant laws, and they are all non-local, but while the majority of relativistic pilot-wave theories add a preferred foliation to the relativistic spatiotemporal structure, relativistic spontaneous collapse theories are foliation free. Nonetheless, I show that, in addition to being counterintuitive and *ad hoc*, they threaten the notions of influence upon which the desirability of locality was based. In this way, relativistic spontaneous collapse theories show the cost of insisting about relativity being a complete theory. In Section 7.6 I argue that the best way of thinking about relativity given the non-locality of nature is as an incomplete theory with an important heuristic value: it is a principle theory that puts constraints *on theories*, rather than on phenomena.

7.2 Quantum theory and its problems

Quantum theory is unproblematic if understood from an antirealist perspective as it provides incredibly accurate and precise predictions. However, it is unclear what it states about the world. This is usually explained presenting the measurement problem, which is the problem of macroscopic superpositions.

7.2.1 Unobserved macroscopic superpositions

According to quantum mechanics as shown in physics books, the complete state of any physical system is given by the quantum state, represented by the wave function ψ, evolving according to the Schrödinger equation which is linear. As such, this theory is problematic because the sum of two solutions of a linear equations is still a solution, and this leads to unobserved macroscopic superpositions. This is the so-called problem of the Schrödinger cat, otherwise known as the measurement problem. The standard solution of this problem is given by the collapse rule: when a measurement is performed, the superposition randomly and instantaneously collapses in one of its terms. However, this raises questions about what is a measurement and why is it not another physical process? To avoid these questions, people have proposed more precise quantum theories, and the pilot-wave or de Broglie–Bohm theory is among them.[2]

[2]de Broglie (1924) and Bohm (1952).

7.2.2 Absence of a spatiotemporal ontology

Having said that, I think that a deeper problem with quantum theory is that it does not have a spatiotemporal ontology. The wave function in standard quantum theory could be naturally understood as a field, analogue to electromagnetic fields. However, unlike electromagnetic fields, the wave function is a field in a high-dimensional space. As such, the wave function cannot straightforwardly be seen as a wave oscillating in space.[3] Moreover, if one favours constructively explaining the behaviour of macroscopic objects as composed of microscopic entities, then one arguably needs to require that both macro- and micro-objects live in the same space, namely three-dimensional space.[4] In addition, it has been argued that theories in which the fundamental field describing matter are not in space-time are problematical, as they obscure the role of space-time, especially when discussing the compatibility with relativity.[5] As described in the last subsection, the pilot-wave theory does not have this problem because it is a theory of particles.

7.2.3 Non-locality as the 'spirit' of quantum theory

In addition, as emphasized by Einstein, quantum theory with collapse is non-local. He considered this non-locality to show that the theory is incomplete: if quantum theory were complete then it would be non-local, and non-locality is not acceptable. Locality is the idea that influences travel at finite velocity. It is always assumed that nature is local for two main reasons. Firstly, since causes precede effects in time, interaction between two objects takes time. Moreover, locality seems necessary to treat systems as effectively isolated: we assume that what happens in a certain spatial region is substantially influenced only by objects that are close by. Note that Newtonian mechanics violates locality: the gravitational and the electromagnetic forces act instantaneously on matter. For instance, lifting my arm now instantly influences the motion of Saturn. Moreover, relativity imposes the constraint that the interaction cannot propagate faster than light. However, the concept of field was later introduced to take care of the first problem: it is a mediator of the interaction, present in every point at each instant. Also in classical mechanics the intensity of the forces decreases as the inverse of the distance squared, so that the forces generated by distant objects can be neglected and we can treat distant objects as *effectively* isolated, bypassing the second problem.

[3]This problem was originally pointed out by de Broglie, Lorentz, Einstein, and Schrödinger. In 1927, de Broglie justified his introduction of particles because 'it seems a little paradoxical to construct a configuration space with the coordinates of points that do not exist' (Bacciagaluppi and Valentini 2009, p. 346). Lorentz wrote to Schrödinger: 'If I had to choose now between your wave mechanics and the matrix mechanics [referring to the quantum formalism proposed by Heisenberg earlier that year], I would give the preference to the former, because of its greater intuitive clarity, so long as one only has to deal with the three coordinates x, y, z' (Prizbram 1967). Similar concerns come from Einstein: 'The field in a many-dimensional coordinate space does not smell like something real' (Howard 1990). Schrödinger agreed: 'of course this use of the q-space (configuration space, *noa*) is to be seen only as a mathematical tool, [...] ultimately [...] the process to be described is one in space and time' (Bacciagaluppi and Valentini 2009, p. 447). The so-called wave function realists do not find these arguments compelling; nonetheless, they do not deny that space-time and a spatiotemporal ontology need to suitably emerge (see Ney 2021 and references therein).

[4]Allori (2013) and Maudlin (2019).

[5]Wallace and Timpson (2010) and, in different terms, by Myrvold (2015).

Einstein argued that the wave function collapse is non-local, making quantum theory unacceptable. He proposed various incompleteness arguments that rely on the assumption of locality, and which culminated in the famous EPR paper (1935). In the version discussed by Bohm (1951), consider two particles in a spin singlet state flying in opposite directions, where two experimenters are ready to measure their spin properties. According to standard quantum theory, an individual particle in such a pair does not possess a definite spin property in any direction until this property is measured. If quantum theory were complete, then a spin measurement along some direction on one particle would collapse the wave function, instantaneously determining the spin measurement outcome for the other particle, regardless of their mutual distance. However, this straightforwardly violates locality. Collapse is instantaneous as the action of the classical forces, but there is no three-dimensional field mediating the interaction, as the wave function is not in space-time. Moreover, unlike classical fields, the strength of the interaction is unaffected by distance. It is this last feature that prevents us from assuming quantum systems to be isolated: they are instead entangled.

Thus, EPR concluded, the only other way of accounting for these correlations is the assumption of 'pre-existing' values of spin properties along some direction which are revealed by spin measurements, but which quantum theory does not specify ('hidden variables'). That is, EPR thought that one could (and should!) locally complete quantum mechanics to make this non-locality go away. But EPR were empirically wrong. Bell (1964) constructed an EPR-like hidden-variable theory and could write a mathematical constraint on this theory's results (namely Bell's inequality), which does not hold for quantum theory.[6] One can then set up a crucial test, which was later performed, and its results were compatible with quantum mechanics,[7] thereby falsifying any EPR-like hidden-variable theory aimed at explaining the perfect correlations locally. In other words, the EPR correlations can only be explained by a non-local interaction: any theory matching the quantum predictions has to be non-local.[8] In light of this, it seems fair to say that, rather than a problem, non-locality is the true novel feature of quantum theory, and that it embodies its 'spirit'. Thus, quantum theory is a theory of matter moving in space-time, interacting non-locally.

7.3 The pilot-wave theory

The pilot-wave theory, which is about to be described in this section, has no problem of superpositions, as it has a spatiotemporal ontology, but it is non-local. These features would make it a perfectly suitable candidate for a fundamental physical theory according to any physicist in the pre-quantum era, if it were not for its non-locality. Nonetheless, now we know more: the experimental violation of Bell's inequality has shown that we cannot do better than this: non-locality is the spirit of quantum theory.

[6] Note that the derivation of Bell's inequality is more general than that; see notably Goldstein *et al.* (2011).

[7] Aspect (1982).

[8] For more on Bell's proof of non-locality, see Goldstein *et al.* (2011). See also Maudlin (1994) for a general discussion on the interplay between non-locality and relativity, as well as Dürr and Lazarovici (2020). Importantly, I am leaving aside approaches that reject the non-locality conclusion by questioning the hypothesis of statistical independence (superdeterminism) as well as retrocausal approaches.

Accordingly, some have advertised the pilot-wave theory as the rational completion of quantum mechanics:[9] it is the simplest, most explanatory spatiotemporal theory in which the interaction is straightforwardly non-local, as required by the empirical violation of Bell's inequality.

7.3.1 Ontology and nomology

The pilot-wave theory has a clear spatiotemporal ontology: it is a theory of microscopic point particles that can be understood as composing macroscopic objects, as in the classical case. However, contrary to classical mechanics, particles evolve according to an equation which constrains their velocities:

$$\frac{d}{dt}x(t) = \frac{\hbar}{m}\nabla S(x,t)$$

where $\hbar = h/2\pi$ is the reduced Planck constant, and S determines a guiding field for the particles, and is the phase of the wave function: $\psi(x,t) = R(x,t)e^{\frac{i}{\hbar}S(x,t)}$.

This ψ function evolves according to the Schrödinger equation:

$$\frac{\partial}{\partial t}\psi(x,t) = -\frac{\hbar^2}{2m}\frac{d^2}{dx^2}\psi(x,t) + V\psi(x,t).$$

For a system of N particles, the wave function is a function of their positions: $\psi = \psi(x_1, x_2, ..., x_n, t)$ namely a function in the configuration space of these particles. For the reasons explained in the previous section, the wave function describes the interaction between particles. That is, it is part of the nomology: the wave function and its evolution equation need to be specified to properly define the theory, just like the law of the forces are needed to properly define classical mechanics.[10]

7.3.2 Non-locality

The pilot-wave theory is explicitly non-local: for N particles, the wave function depends on the configuration of all particles at the same instant. This means that pushing one particle affects the behaviour of all the others, instantly, no matter how far apart. The effective collapse of the wave function is an explicit example of non-locality in the pilot-wave theory. In this theory, the wave function for all particles (namely, the wave function of the universe) $\Psi(x_1, x_2, ..., x_n)$ always evolves deterministically according to the Schrödinger equation, so no collapse ever happens. The evolution of sub-systems, like, say, one with three particles with positions x_1, x_2, x_3, is effectively determined by its conditional wave function. This is computed by fixing the position variables in $\Psi(x_1, x_2, ..., x_n)$ with the real positions (denoted using capital letters) of all the particles except x_1, x_2, x_3: $\psi(x_1, x_2, x_3) = \Psi(x_1, x_2, x_3, X_4, ..., X_N)$.[11] Under certain

[9] Bricmont (2016).

[10] For a recent proposal on how to think about the wave function along these lines, see Allori (2021). Alternative views on the nature of the wave function in the pilot-wave theory, include the ones discussed in Albert (1996), Norsen (2010), Esfeld *et al.* (2014), Suàrez (2015), and Hubert and Romano (2018).

[11] See Dürr, Goldstein, and Zanghì (1992).

circumstances, sub-systems may have a Schrödinger evolving conditional wave function, but not when the sub-system is being measured: after the interaction with the measuring device, the sub-system conditional wave function ψ *effectively* collapses into one of the terms of the superpositions of Ψ.[12] So, the pilot-wave theory explains the effective collapse of the wave function granting that the particles non-locally interact.

7.3.3 The role of hidden variables

The pilot-wave theory is a hidden-variable theory in the sense that it completes the description of the state of any physical system by adding to the wave function the specification of the particles configurations. Nonetheless, it is not a hidden-variable theory as EPR would have wanted. As reconstructed in Bricmont *et al.* (2022), EPR needed hidden variables such as spin properties to locally explain the perfect correlations. However, these properties had to be found to be the same in all possible measurements of them. That is, spin values had to be *non-contextual*: the value of a non-contextual property is revealed by the measurement, and it does not depend on the way in which it is measured. All properties we think of as genuine properties are non-contextual. The violation of Bell's inequality falsifies these theories, in which the hidden variables explain the correlations locally. Instead, the pilot-wave theory correctly violates Bell's inequality, just like quantum theory: in this theory, the correlations are explained by what Einstein thought unthinkable, namely non-locality. Note that in order for the pilot-wave theory to violate Bell's inequality, and thus make the same prediction of quantum theory, it cannot have any hidden variables *a-la* EPR (otherwise the theory would have been falsified). It means that in the pilot-wave theory spin properties are contextual: spin 'measurements' do not reveal the spin values pre-measurement, so they are not really measurements of anything.[13] Genuine measurements should be processes in which the system–detector interaction is negligible so that the outcome faithfully reproduces the value of the property before it is measured. Instead, a spin 'measurement' is a process in which the system–device interaction significantly changes the system's property value so that the outcome does not reflect any property of the system before the interaction. Accordingly, there is no spin property in the pilot-wave theory. The only real hidden variables that can be revealed by measurements are particles positions. Note that the particles positions are hidden variables which are not introduced to explain the observed correlations of EPR type of experiments; the non-locality of the interaction explains them. Rather, as anticipated above, particles positions are needed to ground the theory to space-time, providing quantum theory with a spatiotemporal ontology.[14]

7.4 The 'spirit' of relativity

Some argue that the type of non-locality in the pilot-wave theory makes this theory particularly unsuited for a relativistic extension. To properly assess these claims, one needs to better understand what a relativistic extension requires.

[12] For more on this, and on the conditional wave function in the pilot-wave theory, see, notably, Norsen (2016).

[13] See Bricmont (2024).

[14] For more on this, see Allori (2013).

Relativity was originally formulated in terms of two principles: the constancy of the velocity of light c in all inertial reference frames and the relativity principle, according to which physical laws should have the same form in all inertial frames. Together these two principles imply that the description of the phenomena in two inertial frames are connected by Lorentz transformations.

7.4.1 Relativity and locality

The first principle, about the velocity of light, comes from the observation that in electromagnetism the velocity of light is constant, independent of the reference frame. As anticipated, when combining this principle with the second principle, it follows that this constant is the maximum possible velocity. In this respect, this encodes the idea of locality that interaction travels, and it restricts it by stipulating that it never travels faster than light. Thus, no interaction can be instantaneous. As we have shown, locality is not a prerogative of relativity, but a more a generally desirable feature of physics: for instance, electrodynamics is local, as the interaction is mediated by the fields travelling at the velocity of light. But, as we also have shown, the violation of Bell's inequality makes locality empirically not an option.[15] So, given quantum non-locality, the idea of relativity as prescribing a limit to the velocity of interaction has to go. Aspect's experiments falsify this: some influences are instantaneous.

7.4.2 Relativity and Lorentz invariance

Turning to the second principle, the idea behind it is very general: it is the requirement that the description we provide of the phenomena is as independent as possible from our point of view. The principle of Galilean relativity is the expression of the principle of relativity within classical mechanics. It states that the form of the laws is the same in all (inertial) frames, which for classical mechanics translates as follows: laws in different frames are connected by Galilean transformations. Or laws are invariant under Galilean transformations. In special relativity, the relativity principle still holds: the description of the phenomena should be as observer independent as possible. However, when combined with the light principle, it leads to the Lorentz transformations, rather than the Galilei ones. Thus, the relativity principle is equivalent of stating that physical laws should be invariant under Lorentz transformations. Einstein wrote: 'The universal principle of the special theory of relativity is contained in the postulate: The laws of physics are invariant with respect to Lorentz transformations (for the transition from one inertial system to any other arbitrarily chosen inertial system)'.[16] Is something else required from a theory to be 'genuinely' relativistic? Is there a deeper 'spirit' of relativity, aside from Lorentz invariance?

7.4.3 Relativity as a theory of space-time structure

Another formulation of relativity, which does not explicitly use the two principles but is nonetheless equivalent, is in terms of Minkowski space-time. Since c is constant,

[15] For more on relativity and non-locality, see Maudlin (1994) and Dürr and Lazarovici (2020).
[16] Einstein (1949).

one can consider ct as an additional spatial coordinate in a newly combined 'space-time' manifold. This is Minkowski space-time, which is a four-dimensional space that unifies three-dimensional space and time. In it, each point has coordinates given by the quadruple $(x, y, z, ct) = (\mathbf{x}, ct)$. Two events in space-time, A and B, may be represented in coordinate systems S as well as coordinate system S', moving uniformly at velocity v with respect to S. S and S' are connected by Lorentz transformations, and it turns out that if the axes of S are orthogonal with one another, the ones of S' instead meet at acute or obtuse angles. Nonetheless, given the second principle, they are equivalent, and given the first principle, light travels in both along the diagonal lines.

As direct consequences of Lorentz transformations, one has a new transformation for velocities, effects like time dilation and length contraction. Moreover, one could show that if signals could travel faster than light one would encounter causal paradoxes in which a signal arrives before it is being sent. So, faster than light motion is impossible. In addition, it follows from the Lorentz transformations in terms of coordinate differences that two events that are simultaneous in S are not necessarily simultaneous in S'. Since all frames are equivalent, the notion of absolute simultaneity is lost. This is the so-called *relativity of simultaneity*. In Minkowski space-time the trajectory of an object moving at the speed of light is a 'light-like curve'. One therefore can imagine that for any point P, space-time divides into three regions delineated by the light-ray trajectories originating from and arriving at P. They form the surface of a double cone (the 'light cone') whose vertex is P. These three regions are the absolute past of P (the lower part of the light cone), the absolute future (the upper light cone), and the points elsewhere. An object in P can only come from the past light cone points, and it can only evolve into the future light cone points. Light travels on the surface of the light cone, while particles are confined to time-like worldlines, inside the light cone. Two points in the elsewhere region (outside the light cone) would have to travel faster than light to reach one another. Since nothing propagates faster than light, two such points, called space-like separated, cannot interact. Minkowski space-time has a metric, specifying the distance between spatiotemporal points, or events. This is given by the interval $I = \sqrt{(\mathbf{x}_2 - \mathbf{x}_1) - c^2(t_2 - t_1)^2}$ Accordingly the surfaces containing points at equal distance form P are hyperboloids. One can slice space-time into instantaneous snapshots called space-like hypersurfaces. Events on a given snapshot are simultaneous, namely they happen at the same time. However, as anticipated, observers in relative motion will disagree about the simultaneity of two events. It means that there are different possible ways of slicing space-time into instantaneous surfaces, one for each observer. These different 'slicings' are titled with respect to one another, ultimately because of the Minkowski metric, which has hyperboloids as same distance surfaces. The upshot is that no slicing is 'the' correct one: there are simultaneous events only with respect to a particular slicing. This is the relativity of simultaneity: there is no preferred spatiotemporal slicing.

Especially given its Minkowski formulation, relativity seems straightforwardly a theory of space-time. Because of this, any theory containing a non-spatiotemporal object, such as the wave function, seems already in tension with relativity. In other words, as anticipated, one cannot see the wave function as a spatiotemporal mediator field like one could do in the case of electrodynamics. In this reading, only a theory in which matter and its interaction are represented by spatiotemporal entities (emerging or fundamental) could qualify as a candidate for genuinely relativistic status. Nonetheless, the requirement of a theory to be framed in space-time is clearly not enough to qualify a theory as relativistic, as obviously one can imagine spatiotemporal theories that have nothing to do with relativity. A better idea is to think that a truly relativistic theory only uses relativistic space-time structure as encoded in Minkowski space-time, such as the light cone. Note that a global slicing of space-time, as the ones we discuss in the next section, would be a structure over and above what one finds in Minkowski space-time.

7.5 Relativistic non-local theories

So, to summarize the discussion of the previous sections, a quantum theory would be non-local, and it would be relativistic if it were Lorentz invariant. Moreover, ideally, the non-local interaction would have to happen via a Minkowski spatiotemporal structure. The following two sections briefly discuss several relativistic pilot-wave and spontaneous collapse theories and their features. It is worth noting that, while they are certainly notable proposals, none of them is (yet) fully capable of describing an interacting picture.[17]

7.5.1 Relativistic pilot-wave theories

Some think that the non-locality of the pilot-wave theory implies that the pilot-wave theory fundamentally violates Lorentz invariance. Accordingly, in such a theory, Lorentz invariance would be an emergent symmetry of the observational level.[18] Nonetheless, Lorentz invariant extensions of the pilot-wave theory have been proposed: some have a Dirac evolving wave function, and others a Klein–Gordon evolving wave function.[19] All these theories have therefore deterministic evolution both for the wave function and the particles.[20] They all are explicitly non-local, as they should, and since each particle configuration in the wave function is taken at the same time, they all require a notion of *absolute simultaneity*. So, while the Lorentz invariance of the law can be taken care of by having a suitably evolving wave function, one will also need a global space-time structure to implement non-locality: a preferred foliation of space-time. A possibility is to simply postulate one.[21] Otherwise, the foliation may be suitably defined in terms of the wave function.[22] In both cases (postulated or derived foliation),

[17] See, e.g. Dürr and Lazarovici (2020), Chapters 11 and 12.

[18] Bohm and Hiley (1993), Holland (1993), and Valentini (1997).

[19] For all these theories, see Tumulka (2018) and references therein.

[20] Proposals that include particles creation and annihilation will have a stochastic particle evolution, see Tumulka (2018) and references therein. Nonetheless, one could describe particle creations and annihilations even with a deterministic dynamics (see Oldofredi (2020) and references therein). Galvan (2015) has proposed a Lorentz invariant pilot-wave theory with formally no preferred foliation, which however possesses an absolute synchronization. He argues that this makes his theory more compatible with relativity than the alternatives, while I remain unconvinced, as synchronization is still a global notion (see Drezet (2019) for a similar model).

[21] Dürr *et al.* (1999).

[22] Dürr *et al.* (2014).

the preferred spatiotemporal foliation is not a static object but evolves dynamically according to a Lorentz invariant law, dynamical and Lorentz invariant itself.

Note that such preferred foliation of space-time is not ad hoc: it is in principle undetectable, but there is an explanation of why that is the case, rooted on our lack of access to the particle configurations.[23] Nonetheless, because the preferred foliation is a global spatiotemporal slicing that is not part of the Minkowski spatiotemporal structure, many have looked for alternatives, like spontaneous localization theories, to see whether they can do better.[24]

7.5.2 Relativistic stochastic theories

The first working spontaneous localization theory, also called GRW theory, was originally proposed to build the collapse in the dynamics of the wave function, which now becomes non-linear and indeterministic.[25] In this theory, the wave function evolves according to the Schrödinger equation up to a random time, at which it collapses around a random location with a given accuracy (determined by free parameters, which are seen as new constant of nature). One could turn this into a theory about some spatiotemporal ontology, by suitably postulating one and adding it to the theory. Various proposals have been put forward and respectively called GRWf (for an ontology of spatiotemporal events, or flashes),[26] GRWm (for a matter field ontology),[27] and GRWp (for particles).[28] In these theories the wave function governs the spatiotemporal ontology in a stochastic way. Several relativistic extensions have been suggested, most notably rGRWf and rGRWm.[29] They have in common that they are all Lorentz invariant, non-local and only use relativistic spatiotemporal structure.

7.6 (Nonlocal) quantum relativity?

Within their limitations, all the proposals discussed in the previous section are nonlocal and Lorentz invariant. However, while relativistic pilot-wave theories are linear and deterministic, but they require a preferred spatiotemporal global structure (the foliation), relativistic GRW-type theories are non-linear and stochastic but also foliation free. If requiring exclusively a relativistic spatiotemporal structure is necessary for labelling a theory 'genuinely relativistic', then only GRW-type theory qualifies. Nonetheless, in this section, I argue that, given non-locality, it would be better to proceed with a theory that modifies both theories as minimally as possible, instead of being as faithful as possible to relativity, especially if that implies modifying quantum

[23] Bricmont (2016). Nonetheless, many physicists think that it would be better if one could empirically observe the foliation: 'From our perspective, for as long as we are confined to a state of statistical equilibrium that hides the underlying non-locality from direct view, it seems probable that the argument [non-locality vs. relativistic space-time] will continue to be unresolved' (Valentini 2008).

[24] See Bell (1987) and Tumulka (2006). See also e.g. Ghirardi (2012) and Myrvold (2021), and references therein.

[25] Ghirardi, Rimini, and Weber (1986).

[26] Bell (1987).

[27] Benatti *et al.* (1995).

[28] Allori (2020), and references therein.

[29] Respectively in Tumulka (2006, 2021) and Bedingham *et al.* (2013).

theory in an unnatural way. This leads to identify relativistic pilot-wave theories as the best balance between non-locality and Lorentz invariant spatiotemporal structure.

7.6.1 Combining quantum theory and relativity

To combine quantum theory and relativity we need to combine their essential features. I have argued earlier that quantum theory is a theory of matter moving in space-time and interacting non-locally. So, in short, the spirit of quantum mechanics is the non-locality of interaction. I also have argued earlier that relativity is a theory about space-time and its structure, whose fundamental symmetry is Lorentz invariance. So, in short, the spirit of relativity is Lorentz invariant spatiotemporal structures. Thus, the former is a theory of matter, and the latter is a theory of the arena in which matter interacts. To combine these spirits, we need to build a theory of matter evolving in space-time interacting non-locally whose fundamental symmetry is Lorentz invariance.

7.6.2 Minimal modification of relativity but unnatural modification of quantum theory

Relativistic GRW-type theories provide such a theory by being as faithful as possible to relativity. That is, they are non-local, but they only use relativistic spatiotemporal structure (no global structure). However, what are the costs of keeping only relativistic spatiotemporal structure to accommodate quantum non-locality?

Note that while it would perhaps be simpler or better to have non-local, Lorentz invariant, foliation free theories that are also deterministic, I have argued elsewhere that only stochastic theory can be (nonlocal, Lorentz invariant and) foliation free. I have also argued that this stochasticity leads to the dissolution of the notion of locality.[30] In fact, without continuous worldlines for the spatiotemporal ontology, it is difficult to say in what sense an event has caused another. This threatens the very notion of influence or interaction: how can we make sense of an object influencing another one if we cannot even identify the cause and the effect?[31] In turn, without the notion of influence, it is difficult to make sense of the notion of locality since locality is defined as influence travelling at finite velocity. Relatedly, GRW-type theories display 'supernon-local' behavior. That is, due to the stochasticity of their law, they allow for non-local transfer of matter even in single 'particle' systems. Consider a single 'particle' confined in a box,[32] insert a barrier in the box, splitting the wave function, separate the two half-boxes, and then open one of them. In GRWm, say, given that the system is microscopic (only one 'particle'), the matter field is spread out in both half-boxes until one box is opened, the point at which the wave function instantaneously (thus non-locally) localizes.[33] Nothing like this happens in the pilot-wave theory, where to find non-local effects, one needs at least two particles. This is problematical because

[30] Allori (2022).

[31] For a similar point, see Esfeld and Gisin (2014). This has been questioned by Myrvold (2021).

[32] The scary quotes are needed because, strictly speaking there may be no fundamental particles in these theories, they may have a matter field or a flash ontology. Nonetheless, we can speak about a single 'particle' because the system behaves as if it were composed by a single-particle-like entity.

[33] A similar scenario holds for GRWf and GRWp; for details, see Allori (2022).

locality, or the lack thereof, is a feature of interacting systems of two or more particles. Here instead even single-particle systems display non-locality, and it is unclear how to interpret it. Non-locality is not about influence anymore as there is no interaction within one system. So, this is another example that the stochasticity of the law, responsible of this supernon-locality, undermines the notion of influence upon which the desirability of locality was based. Thus, while these theories only use relativistic spatiotemporal structure, they have to implement non-locality with stochastic laws dissolving the intuitive notion of interaction.

Moreover, these theories are empirically inequivalent to quantum mechanics, predicting for instance that energy increases. Even if one could accommodate this theoretically, the failure of energy conservation may be rightfully considered a drawback of this theory. Even if crucial tests to determine which theory is correct are underway, GRW-type theories have free parameters that are adjusted after every single experiment to avoid falsification. This constant adjustment makes these theories rather ad hoc, if not fine-tuned.

Therefore, to summarize, in my opinion it is useful to explore GRW-type theories in order to probe the consequences of combining quantum non-locality with a minimal relativistic spatiotemporal structure. The result however is that a non-local theory without any additional relativistic spatiotemporal structure leads to the dissolution of the natural notion of non-locality. Moreover, this can only be accomplished by a non-linear and stochastic theory which modifies the beauty of the Schrödinger evolution, with ad hoc parameters, which predicts, among other things, that energy is not conserved. In other words, any quantum (i.e. non-local) theory which tries to use only a minimal relativistic spatiotemporal structure, namely any theory which does not modify relativity (aside from being non-local) ends up unnaturally modifying quantum theory. If this is what 'genuinely relativistic' necessitates (Lorentz invariance and only Minkowski space-time structure), then it seems better to forget about it.

7.6.3 Minimal modification of both theories

The alternative is given by the Lorentz invariant pilot-wave theories, which are deterministic and linear. In these theories, quantum mechanics is naturally completed by the specification of particles positions and relativity is completed by the specification of a preferred spatiotemporal slicing. These theories are not ad hoc, not stochastic, not nonlinear, and they have not so hard-to-swallow predictions. The explicit non-locality of the pilot-wave theory clearly embodies the spirit of quantum theory, and it also clearly suggests how to amend relativity in light of non-locality: add a preferred spatiotemporal foliation. That is, relativistic pilot-wave theories provide the best balance between the quantum and the relativistic spirit, respectively non-locality and Lorentz invariant spatiotemporal structure, by slightly modifying them both: add particles to quantum theory and add a preferred foliation to relativity. All this by remaining deterministic and not ad hoc.

As emphasized above, many people think the preferred foliation is in contrast with the spirit of relativity: a preferred foliation reintroduces the notion of absolute simultaneity that is not part of the Minkowski space-time structure and which is not required by Lorentz invariance. However, since we are trying to put together quantum theory with relativity, Lorentz invariance cannot be the only constraint, as we also need to accommodate non-locality. In other words, relativity was formulated with the idea that interaction travels, and that Lorentz invariance is a constraint to its maximum velocity. Thus, it was natural to suppose that there was no other structure than the one dictated by Lorentz invariance. However, now the situation has changed: we know about quantum non-locality and the natural way to accommodate it is to have an additional structure in Minkowski space-time. This is not going to be a redundant structure, as it is needed to account for non-locality that we know is part of the structure of the world. There are no simpler options than doing that. As we have shown, in GRW-type theories we may drop this preferred foliation at the costs of unnaturally modifying quantum theory in an artificial way. But why would we do that? *Why protect relativity at all costs if we already know that we need to modify it already, due to non-locality?* Why modify only quantum theory? Would not it be best to modify minimally both theories instead? So, to me the choice is clear: relativistic pilot-wave theories provide the best balance between quantum mechanics and relativity, and thus they are the way to go. Relativistic pilot-wave theories provide the minimal modification of both theories to accommodate all the experimental data, including non-locality. They are theories of matter evolving in space-time interacting non-locally, whose fundamental symmetry is Lorentz invariance. These theories add a spatiotemporal ontology to quantum theory, and a preferred spatiotemporal foliation to relativistic space-time. By doing this, they put all the puzzles pieces together: the lack of a spatiotemporal ontology of quantum theory is solved by adding particles which also allow to recognize relativity as a theory of space-time; quantum non-locality is explicitly acknowledged in terms of the wave function and it suggests the missing piece of relativity, namely a preferred foliation. Both quantum theory and relativity are incomplete, and they are both naturally and minimally completed by relativistic pilot-wave theories that therefore seem to me tailor-made, or just what the doctor ordered.[34]

7.6.4 How to understand relativity

If we go in this direction, it does not mean that relativity is irrelevant: laws are Lorentz invariant, and all the relativistic spatiotemporal structures (light cones, Minkowski metric, etc.) will still be there. Both relativity principles still hold. The relativity principle, which is a guide to what a good theory should look like, is still true, as the theories are Lorentz invariant. However, space-time is not as invariant as we thought it would be: there is a preferred frame. But one cannot eliminate what is needed to explain the empirical data: as in the case of classical mechanics, one needs a preferred frame to account for the effects of accelerated motion, so here one needs a preferred frame to deal with non-locality.

The only other implication of having a preferred spatiotemporal foliation, in addition to straightforwardly accounting for non-locality, is that the notion of absolute simultaneity comes back into the picture. But, as discussed above, relativity of simultaneity is a consequence of Lorentz invariance alone, while now we also have the

[34]This is what Detlef Dürr used to say.

non-locality constraint. As a consequence, there seems to be little reason in insisting on requiring that simultaneity is relative. This seems to be the case, especially considering that it is a counter-intuitive feature of relativity which does not seem to contribute to its overall explanatory power, being a consequence, rather than a presupposition, of (unconstrained) Lorentz invariance.

So, in my opinion, the best way of thinking of relativity is as an incomplete theory of space-time structure, which relativistic pilot-wave theories suitably complete. Arguably, Einstein might not have been too surprised or upset to discover that relativity is not the last word. In fact, he originally believed that relativity was a principle theory, and in his view principle theories require a deeper explanation in terms of a constructive theory.[35] Principle theories (like thermodynamics) explain macroscopic phenomena constraining them in terms of principle, while constructive theories (like kinetic theory) explain them in terms of their microscopic components and their dynamics, and provide a deeper understanding of why the principles hold. If relativity is a principle theory, the corresponding constructive theory would be like the one proposed by Lorentz, according to which relativistic effects are explained by actual contraction of matter. Einstein rejected it, not because he thought that it was wrong-headed but rather because he thought we lacked a satisfactory theory of matter.[36] Relativistic pilot-wave theories do not currently provide a constructive explanation of the principles of relativity or their effects. Nonetheless, what we know about matter interaction, which Einstein did not, is that it is non-local. So, something needs to be modified in relativity theory to account for it, regardless of whether it is a principle theory or not.

Perhaps one may think of relativity as a principle theory constraining acceptable theories, rather than phenomena: theories should at least be Lorentz invariant and have a Minkowski spatiotemporal structure, without specifying whether they need to have something else or not. For what it is worth, this seems compatible with what Einstein wrote: 'General laws of nature are co-variant with respect to Lorentz transformations. This is a definite mathematical condition that the theory of relativity demands of a natural law, and in virtue of this, the theory becomes a valuable heuristic aid in the search for general laws of nature'.[37]

7.7 Final considerations

If what is argued here is correct, then the situation is somewhat bitterly ironic. In fact, Einstein's most famous argument for the incompleteness of quantum theory is that if it were complete, it would be non-local; instead, quantum theory is incomplete and non-local. Moreover, relativity turns out to be incomplete too, because it cannot deal with quantum non-locality. Lorentz invariant extensions of the pilot-wave theory, as I have maintained, provide the current best option at completing both theories: the ground quantum theory in space-time and they account for non-locality by adding the needed spatiotemporal structure to relativity.

[35] Einstein (1919).

[36] Pauli and Eddington were also supporters of constructive explanation in relativity (see Brown and Pooley 2004, and references therein).

[37] Einstein (1920).

Naturally, there are many open questions. Quantum non-local interactions are instantaneous. What implication does it have for the notion of interaction? As we have shown, the idea of locality captures the idea that interaction travels: an object takes some time to 'notify its presence' to another object. Since instead we have non-locality, we are back to Newton's problem that lifting my arm instantly influences the motion of Saturn. As we observed, in the classical case we can alleviate the problem because we can use fields to mediate the interaction, which then becomes not instantaneous, and the strength of the interaction decreases with distance so that we can ignore the effects of objects that are far away. In the quantum case instead, as anticipated, we do not have either: no spatiotemporal mediator, and no decrease of strength with distance, even if the interaction of the system with its environment (decoherence) effectively destroys entanglement, giving rise to a seemingly local universe. Could (or should) we still think that interaction travels under these constraints? If it does, it travels at speed greater than the velocity of light, which in turn means that light is not the fastest. If instead we give up the idea that interaction is something that travels, then what are the consequences and how else should we understand interaction? As a side remark, let me note that if we do abandon the notion of traveling interaction, we also lose the need of having fields as mediators, which would make the overall ontology of the theory simpler. Be that as it may, how should we think of entangled particles? One possibility could be to think of them not as two particles interacting non-locally, but as non-separable entities. While the notion of locality has to do with interaction, the notion of separability has to do with the way matter is. That is, if the properties of the whole are completely determined by the properties of its parts, then we talk about separability.[38] But in this context matter is made of particles, so in virtue of what is a pair of entangled particles non-separable? Another possible reaction is to simply maintain that the law governing the behaviour of matter is non-local, and such law does not 'propagate' but rather it is what primitively connects a pair of entangled particles. Presumably, one can think of many more ways to make sense of non-locality, and I do not know which attitude will provide itself to be most fruitful. But, since the word is non-local, everyone, not just those who endorse the pilot-wave theory, needs to start asking this type of question.

References

Albert, D.Z. (1996). Elementary Quantum Metaphysics. In: Cushing, J.T., Fine, A. and Goldstein S. (eds.) *Bohmian Mechanics and Quantum Theory: An Appraisal*, pp. 277–284. Dordrecht: Kluwer.

Allori, V. (2013). Primitive Ontology and the Structure of Fundamental Physical Theories. In: Albert, D.Z. and Ney, A. (eds.) *The Wave Function: Essays in the Metaphysics of Quantum Mechanics*, pp. 58–75. New York: Oxford University Press.

Allori, V. (2020). Spontaneous Localization Theories with a Particle Ontology In: Allori, V., Bassi, A., Dürr, D. and Zanghì, N. (eds.), *Do Wave Functions Jump?*

[38] Howard (1985).

Perspectives on the Work of GianCarlo Ghirardi, pp. 73–93. Cham, Switzerland: Springer.

Allori, V. (2021). Wave-Functionalism. *Synthese* **199**: 12271–12293.

Allori, V. (2022). What is it Like to be a Relativistic GRW Theory? Or: Quantum Mechanics and Relativity, Still in Conflict After All These Years. *Foundations of Physics*, **52**: 79.

Aspect, A., Dalibard, J., and Roger. G. (1982). Experimental Test of Bell's Inequalities Using Time-Varying Analyzers. *Physical Review Letters*, **49**(25): 1804.

Bacciagaluppi, G. and Valentini, A. (2009). *Quantum Theory at the Crossroads: Reconsidering the 1927 Solvay Conference.* Cambridge: Cambridge University Press.

Bedingham, D., Dürr, D.,Ghirardi, G.C., Goldstein, S., Tumulka, R. and Zanghì, N. (2014).Matter Density and Relativistic Models of wave function Collapse. *Journal of Statistical Physics*, **154**(1-2): 623–631.

Bell, J. S. (1964). On the Einstein–Rosen–Podolsky paradox. *Physics* **1**(3): 195–200.

Bell, J. S. (1987). Are there quantum jumps?. In Kilmister, C.W. (ed.) *Schrödinger. Centenary Celebration of a Polymath*, pp. 41–52. Cambridge: Cambridge University Press.

Benatti, F., Ghirardi, G.C. and Grassi, R. (1995). Describing the Macroscopic World: Closing the Circle within the Dynamical Reduction Program. *Foundations of Physics.* **25**: 5–38.

Bohm, D. (1951). *Quantum Theory.* New York: Prentice-Hall.

Bohm, D. (1952). A suggested interpretation of the quantum theory in terms of "hidden" variables. I & II. *Physical Review*, **85**(2): 166–193.

Bohm, D. and Hiley, B.J. (1993). *The Undivided Universe: An Ontological Interpretation of Quantum Theory.* London: Routledge & Kegan Paul.

Bricmont, J. (2016). *Making Sense of Quantum Mechanics.* Berlin: Springer.

Bricmont, J. (2025). The de Broglie–Bohm Theory *is* and *is not* a Hidden Variable Theory. In Oldofredi, A. (ed.) *Guiding Waves in Quantum Mechanics: One Hundred Years of de Broglie–Bohm Pilot-Wave Theory.* Oxford: Oxford University Press.

Brown, H R., and Pooley, O. (2004). Minkowski Space-Time: A Glorious Non-Entity. In Dieks, D. (ed.) *The Ontology of space-time*, pp. 67–89. Amsterdam: Elsevier.

de Broglie, L. (1924) , Recherches sur la théorie des quanta. Ph.D. Thesis, Université de Paris. *Annals de Physique* **10**: 22–128 (1925). English translation, de Broglie, L. (2004). *On the Theory of Quanta*, Foundation of Louis de Broglie.

Drezet, A. (2019). Lorentz-Invariant, Retrocausal, and Deterministic Hidden Variables. *Foundations of Physics*, **49**(10): 1166–1199.

Dürr, D., Goldstein, S. and Zanghì, N. (1992). Quantum Equilibrium and the Origin of Absolute Uncertainty. *Journal of Statistical Physics*, **67**: 843–907.

Dürr, D., Goldstein, S., Münch-Berndl, K. and Zanghì, N. (1999). Hypersurface Bohm–Dirac Models. *Physical Review A*, **60**: 2729–2736.

Dürr, D., Goldstein, S., Norsen, T., Struyve, W. and Zanghì, N. (2014). Can Bohmian Mechanics be Made Relativistic?, *Proceedings of the Royal Society A*, **470**(2162): 20130699.

Dürr, D. and Lazarovici, D. (2020). *Understanding Quantum Mechanics.* Berlin: Springer.

Einstein, A. (1919). *What is the Theory of Relativity?*. Letter to The London Times.

Einstein, A. (1920). *Relativity: The Special and the General Theory.* New York: Henry Holt & Co..

Einstein, A. (1949). Autobiographical Notes. In Schilpp, P.A. (ed.) *Albert Einstein Philosopher-Scientist*, pp. 1–94. New York: Open Court.

Einstein, A., Podolsky, B. and Rosen, N. (1935). Can Quantum-mechanical Description of Physical Reality be Considered Complete? *Physical Review*, **47**(10): 777.

Esfeld, M. and Gisin, N. (2014). The GRW Flash Theory: A relativistic Quantum Ontology of Matter in Space-Time? *Philosophy of Science*, **81**(2): 248–264.

Esfeld, M., Lazarovici, D., Hubert, M. and Dürr, D. (2014). The Ontology of Bohmian Mechanics. *British Journal for the Philosophy of Science*, **65**: 773–796.

Galvan, B. (2015). Relativistic Bohmian Mechanics Without a Preferred Foliation. *Journal of Statistical Physics*, **161**: 1268–1275

Ghirardi, G.C. (2012). Does Quantum non-locality Irremediably Conflict with Special Relativity?*Foundations of Physics*, **40**: 1379–1395.

Ghirardi, G.C., Rimini, A. and Weber, T. (1986). Unified Dynamics for Microscopic and Macroscopic Systems. *Physical Review D*, **34**: 470.

Goldstein, S., Norsen, T., Tausk, D.V. and Zanghì, N. (2011). Bell's Theorem. *Scholarpedia*, **6**(10): 8378.

Holland, P.R. (1993). *The Quantum Theory of Motion. An Account of the de Broglie–Bohm Causal Interpretation of Quantum Mechanics.*. Cambridge: Cambridge University Press.

Howard, D. (1985). Einstein on locality and separability. *Studies in History and Philosophy of Science*, **16**(3): 171–201.

Howard, Don. 1990. "Nicht sein kann was nicht sein darf", Or the Prehistory of EPR, 1909–1935. In A.I. Miller (ed.) *Sixty-Two Years of Uncertainty*, pp. 61–111. New York: Plenum.

Hubert, M. and Romano, D. (2018). The wave-function as a multi-field. *European Journal for Philosophy of Science*, **8**: 521–537.

Maudlin, T. (1994). *Quantum Non-Locality and Relativity.* Oxford: Blackwell.

Maudlin, T. (2019). *Philosophy of Physics: Quantum Theory.* Princeton: Princeton University Press.

Myrvold, W. (2015). What Is a Wave Function? *Synthese*, **192**(10): 3247–3274.

Myrvold, W. (2021). Relativistic constraints on interpretations of quantum mechanics. In Knox, E. and Wilson, A. (eds.) *The Routledge Companion to Philosophy of Physics*, pp. 99–121. New York: Routledge.

Ney, A. (2021). *Finding the World in the Wave Function.* New York: Oxford University Press.

Norsen, T. 2010. The Theory of (Exclusively) Local Beables. *Foundations of Physics*, **40**:1858-1884.

Norsen, T. (2016). Bohmian Conditional Wave Functions (and the Status of the Quantum State). *Journal of Physics: Conference Series*, **701**: 012003.

Oldofredi, A. (2020). Stochasticity and Bell-type Quantum Field Theory. *Synthese*, **197**(2): 731–750.

Przibram, K. (1967). *Letters on Wave Mechanics: Schrödinger, Planck, Einstein, Lorentz.* New York: Philosophical Library.

Suàrez, M. (2015). Bohmian Dispositions. *Synthese*, **192**(10): 3203-3228.

Tumulka, R. (2006). A Relativistic Version of the Ghirardi-Rimini-Weber Model. *Journal of Statistical Physics*, **125**: 821–840.

Tumulka, R. (2018). On Bohmian Mechanics, Particle Creation, and Relativistic Space-Time: Happy 100th Birthday, David Bohm!, *Entropy*, **20**(6): 462.

Tumulka, R. (2021). A Relativistic GRW Flash Process with Interaction. In Allori, V., Bassi, A., Dřr, D., and Zanghì, N. (eds.) *Do Wave Functions Jump? Perspectives of the Work of GianCarlo Ghirardi*, pp. 321–347. Cham, Switzerland: Springer.

Valentini, A. (1997). On Galilean and Lorentz Invariance in Pilot-Wave Dynamics. *Physics Letters A*, **228**(4–5): 215–222.

Valentini, A. (2008). Hidden Variables and the Large-Scale Structure of space-time. In Craig, W.L. and Smith, Q. (eds.) *Einstein, Relativity and Absolute Simultaneity.* Routledge.

Wallace, D. and Timpson, C.G. (2010). Quantum Mechanics on space-time I: space-time State Realism. *British Journal for the Philosophy of Science*, **61**(4): 697–727.

8

Processualism ahead of time: Bohm's understanding of physics

Emilia Margoni

8.1 Introduction

Process philosophy moves from the commonsensical experience of a continuously changing reality to argue that a proper understanding of it requires dismantling the parlance of self-standing, permanent objects, namely substances. Against a well-established paradigm within Western metaphysics that prioritizes static, fixed, individuals, whereby associated dynamic characteristics correspond to 'mere appearance' or are taken as 'ontologically secondary and derivative' (Seibt 2023), process philosophy focuses on dynamicity as the hallmark of a reality conceived of as continuously varying. Therefore, processualism also suggests rethinking the scope of natural language and scientific theories, by shifting the focus from a static to a dynamic conception of reality, whereby occurring, rather than being, gets emphasized: instead of evaluating what a certain entity x is fundamentally, process philosophy addresses how x evolves, and what the effects of such an evolution are.

While substance metaphysics still represents the prevalent metaphysical view, in the past few decades the notion of process has been suggested within a variety of disciplines and research areas.[1] Even though processualism comes in various forms and degrees (see, e.g. Seibt 2023, paragraph 3), there is one critic target that is shared among all versions of process thinking, namely the already emphasized view of concrete, definite substances as the basic entities of physical reality, natural language, and scientific theories. Indeed, processualists argue that individuality should not be necessarily coupled with countability and definiteness: there exist more fundamental entities, namely processes, that are individuals despite not being either countable or definite. In a sort of role shift, it is from them, as will be clarified afterwards, that concrete, definite entities can be abstracted, within specific theoretical contexts, for specific theoretical purposes.

This understanding of processes as more fundamental than concrete, particular entities is generally couched in terms of the relation between objects without temporal parts (such as point-like particles or spatiotemporal events) and processes with temporal parts (such as a radioactive decay or a spatiotemporal region). Based on this view,

[1] To mention just a few noteworthy examples, see, e.g., Rescher (2000), Stengers (2011), Sukumar (2013), and Nicholson and Dupré (2018).

a processual account of reality has it that particular, three-dimensional entities get individuated through an operation of abstraction from non-particular, four-dimensional processes, where the latter are coupled with temporal evolution. It is precisely in this sense that philosophers of biology (Nicholson and Dupré 2018, p. 12) argue that a basic assumption of process thinking is that 'processes are extended in time: they have temporal parts'. The so-called processual turn of biology and philosophy of biology thus invites rethinking the identity of cells, molecules, and organs in terms of the history of concrete, though indefinite, activities, namely processes – where indefiniteness arises from the fact that processes cannot be confined to specific spatiotemporal coordinates. This means that, instead of trying to identify what these entities fundamentally *are*, the primary focus should be to follow them through time, to assess what they *do* in terms of temporal development with respect to other entities appertaining to the same processes.

While the phenomena typically studied in the context of biology (and other research fields) meet a characterization of processes as intimately connected to temporality, and thus allow speaking of a process as a temporally ordered phenomenon, there are other theoretical contexts in which this notion of temporality gets challenged, and thus the so-far sketched characterization of process. It is the case, for instance, of relativistic physics, whose main lesson is generally expressed in terms of a rejection of time as a linear parameter that can be uniquely defined. If that is the case, different events that occur locally along the worldline of different systems are taken to be part of the same process. A major issue becomes what notion, if any, of process can be secured within such theoretical environment. For while there are approaches to fundamental physics that, by arguing that physics should be primarily concerned with processes, precisely question the tenability of the principle of relativity[2], I would like to argue that a broader conceptualization of what a process is, here called *processualism reloaded*, can be secured even within relativistic physics.

The present chapter is meant to capitalize on David Bohm's understanding of quantum mechanics (QM) and the theory of special relativity (STR) to argue that a processual characterization of physics, both in its traditional and reloaded version, is a viable path that is worth exploring. I will try to argue that, against the prevalent account of Bohm's so-called hidden variable approach, his interpretation of QM opens the door to an anti-deterministic and anti-mechanistic understanding of physics that is particularly compliant with processual thinking.[3] I will then focus on how Bohm's reading of STR allows disentangling the notion of process from a unique ordering of events and thereby reformulates it in terms of the way in which entities can be relationally identified. The underlying assumption – which rejects the traditional metaphysics of substances – is that there is no such thing as a permanent object. Rather, only relatively invariant entities can be defined. This conception is of particular importance within relativistic physics, insofar as different observers disagree on which events are simultaneous, and thus which are the fundamental properties (such as length or mass) of a certain object. The parlance of objects should then be rephrased in terms of processes

[2] A remarkable example is Cortês and Smolin (2018).

[3] On a similar vein, see also the contribution by Van Strien (2025) in the present volume.

that are 'organized, ordered, and structured so as to correspond to the characteristics of the material system' under investigation (Bohm 1996 [1965], p. 114).

This chapter is organized as follows. Section 2 provides a sketchy survey of processual thinking in contemporary metaphysics to emphasize its characteristics *vis-à-vis* substance metaphysics. Section 3 introduces three levels of framework dependency of QM with respect to associated ontological commitments to argue that Bohm's hidden variable approach is anti-deterministic and anti-mechanistic and thus agrees with a processual understanding of physics. Section 4 briefly discusses what I called *processualism reloaded*, namely what notion (if any) of process can be secured within a relativistic framework. Section 5 focuses on Bohm's work on STR to argue that it can be framed in processual terms by focusing on three aspects, namely the arbitrary character of reference frames, the parlance of events and especially processes vs objects and their motion, and the notion of relative invariance. I conclude by explaining the advantages of embracing a processual account of Bohm's analysis on foundational aspects of physical inquiry.

8.2 Process philosophy

The starting assumption of all versions of process thinking is that physical reality is dynamic and that this aspect is indispensable to provide a proper understanding of it. Process philosophy moves from the ordinary experience of a continuously changing reality to dismantle the tenability of what Seibt (2023) calls an obsession of Western metaphysics, that is, the focus on self-standing, unchanging substances as the most fundamental entities figuring in the world: despite the commonsensical observation of ever varying states of things, this perspective accords privilege to static, self-identical individuals, whereby their associated dynamic features either correspond to bare facades or are taken as transient, thus ontologically derivative, properties. The underlying premise, which dates to as early as the pre-Socratic Greek philosopher Parmenides, is that being is to be conceived as fundamentally simple, thus internally unvarying and stable.

In her extensive work on substance-based versus process-based theories, Seibt (see, e.g., Seibt 1996, 2002, 2018a) has outlined that the former only admits two types of entities: on the one hand, particles, objects, and spatiotemporal events are portrayed as concrete and individual; on the other hand, attributes and properties are described as abstract and universal. Importantly, it is only concrete, particular entities that can be accorded individuality, where this notion is strictly tied to their countability and determinateness. Based on this viewpoint, hybrid entities, such as masses and activities, that are concrete yet non-countable 'appear as ontological monstrosities – they are neither individuals nor universals, or they are some of both' (Seibt 1996, p. 172). Substance metaphysics boils down to a static view of the world whereby only entities that are (arguably) equipped with independence, countability, and determinateness are identified as the basic building blocks of physical reality, namely substances.

Against such a fixed conceptualization of reality, process philosophy focuses on the coming-into-being of entities that are portrayed as ever-changing and occurring. According to Seibt (2023), it aims at a paradigm shift that can be summarized in three main claims. Firstly, the underlying assumptions of substance metaphysics are dispensable. Secondly, process metaphysics comes equipped to work as well, or even better, than substance metaphysics with respect to some of the traditional philosophical topics identified by the latter (such as the issue of the individuality of entities, the notion of space-time location, or the metaphysical repercussion of specific phenomena within the research fields of biology and chemistry). Thirdly, other relevant philosophical topics (such as the conceptual challenges broad to surface by QM or contemporary accounts of self-organization and cognition) can only be treated within a process metaphysics.

Following this line of though, with her 'General Process Theory', Seibt (see, e.g. Seibt 2009, 2018b) introduces a novel ontological category, namely that of 'general processes' that are described as 'concrete, dynamic entities that are non-particular (i.e. possibly multiply occurrent in space and time), determinable, and 'functionally individuated' (by contextually determined sections of process networks) (Seibt 2023). Put it differently, processes correspond to concrete, yet non-countable and non-particular activities that occur in a 'multiply disconnected spatiotemporal region with fuzzy boundaries' (Seibt 2002, p. 85), thus rejecting the view according to which individuals should already and always be determinate, particular entities. This understanding of processes as concrete, yet indeterminate and non-particular entities suggests a view of an ever-changing reality devoid of clear boundaries, fixed whence and whither.

A context in which such a perspective is becoming increasingly popular is that of biology and philosophy of biology, where the notions of metabolism, multicellular organism, and lifecycles can be fruitfully treated within a metaphysics that outlines interdependence and relational individuation instead of independence and subjecthood. In this sense, the distinction between continuants (entities existing in time without temporal parts, such as animals, particles, and objects) and occurrents (perduring entities with temporal parts, such as the process of growing, drizzling, and rotating) allows taking the latter as 'truth makers for propositions stating the existence of a continuant at a time' (Simons 2018, p. 53). More specifically, Simons argue that occurrents are more fundamental than continuants, insofar as the latter are obtained through a process of abstraction via a 'causal relatedness known [...] as genidentity' (Simons 2018, p. 55), where genidentity corresponds to an equivalence relation between the phases of the continuants and occurrents are processes. A fruitful example would be a cellular compound that gets individuated via the role it plays throughout a more complex physiological process, or an Alpha particle that gets identified as the effect of a radioactive decay.

Two aspects need to be emphasized. Firstly, this view of continuants as precipitates of occurrents implies rejecting the idea of concrete, particular, self-identical entities being fundamental, for their very identification is now conceived as depending on processes. Secondly, based on the specific context at stake, different continuants can be identified. Continuants thus display what Huneman (2021) calls a form of 'weak individuality', whereby a multiplicity of related individuals can be defined through different abstractions following different theoretical purposes. It is one's type of enquiry that is responsible to identify what, in a specific context, serves as an individual, to such an extent that, in the field of biology, 'to be individual is to be somewhere in the space of individuality' (Huneman 2021, p. 46). By assuming that an individual is composed of an assemblage of entities, it is contingent upon the specific theoretical context at stake – which picks out the relevant interactions among such assemblage based on certain reference parameters – to establish what counts as an individual (Huneman 2021, p. 54).

8.3 Bohm's interpretation of quantum mechanics and processualism

In the recent philosophical literature on quantum mechanics, much attention has been devoted to the ontological commitments of specific interpretations – the most startling aspect of QM being generally identified with the measurement problem. According to what was taken as the 'orthodox' interpretation of QM at the time Bohm wrote his two seminal papers (Bohm 1952), the state of a quantum system is completely specified by a wave function that establishes the probabilities of actual results for repeated measurements. A highly non-trivial issue is then to further explore whether this lack of predictivity corresponds to an inherent lack of complete determination of the properties of the systems at the quantum level, or whether this impossibility of establishing precisely definable physical states is to be attributed to the incompleteness of the theory itself. Be this as it may, the orthodox view introduces a shift between the quantum and classical levels, paradigmatically embodied in Bohr's complementarity principle, according to which, in the quantum world, we are confined to complementarity pairs of intrinsically, incompatibly defined concepts, such as particle and wave, position and momentum.

The orthodox view started being questioned as early as the beginning of the 1950s[4] by few dissenters[5] who put forward alternative interpretations, leading to, for example, the de Broglie–Bohm theory, the relative state formulation by Everett, and later on the spontaneous collapse models of GRW type – all these proposals arguably stemming from alternative attempts to recover the gap between the quantum and classical domains.[6] For those who are interested in understanding the relation between QM and its (presumed) ontological commitments, a first level of controversy is the underdetermination of the theory by evidence. As all these formulations are said to be empirically equivalent[7], there is no unequivocal way to favour a certain interpretation *vis-à-vis*

[4]More correctly, the first attempt to offer an alternative interpretation of QM is to be attributed to Louis de Broglie at the 1927 Solvay Congress, as he presented his pilot-wave theory, based on the so-called double solution hypothesis, which only refers to waves and particles. Other well-renowned critical papers are Einstein, Podolski and Rosen (1935), and Schrödinger (1935).

[5]For an interesting historical reconstruction see, e.g. Freire Junior (2015).

[6]According to Egg (2021), one should distinguish between these three advanced interpretations. In the case of the de Broglie–Bohm theory, the quantum formalism is supplemented with additional variables; the relative state formulation goes as far as to adjust our understanding of the world to reconcile it with the unaltered formalism of QM; finally, the spontaneous collapse models present a modification of the dynamics of QM to give an explanation for the disappearance of superpositions.

[7]For a thorough discussion of this topic see, e.g. Cordero (2001), Lyre (2010), and Acuña (2021). Let me note in passing that the modification of the dynamics envisaged by spontaneous collapse models is meant to imply some in principle different predictions *vis-à-vis* other interpretations (for a recent discussion on the topic see, e.g. Bassi and Ulbricht 2014). If that can be empirically tested, the type of associated underdetermination (here dubbed 'first level of framework dependency') may

others. I will call this the *first level of framework dependency* of the interpretation of QM with respect to its ontological commitments.

Authors usually try to sidestep this problem by picking out one interpretation of QM and thus by offering a conditional analysis of alternative possibilities and their consequences. Unfortunately, this strategy does not lead to a univocal, though conditional, determination of the ontological background of a certain interpretation of QM. Two paradigmatic cases are the ontological vs nomological status of the wave function in Bohmian mechanics and the various GRW types of spontaneous collapse models. These two examples show that even if one endorses one interpretation of QM, there is no consensual procedure whereby that interpretation gets univocally mapped to a certain ontology. I call this the *second level of framework dependency* of the interpretation of QM with respect to its ontological commitments. This second problem is arguably more disavantageous than the previous one. For while the latter shows the theory ladenness of the assumptions, entities, and outcomes involved in the various interpretations of QM, the former reveals that even by sticking with a specific interpretative paradigm, there is still room for ontological variability – to the extent that one and the same interpretation can be matched to a multiplicity of ontological alternatives.

These two levels of framework dependency undermine many of the attempts to offer a compelling account of the ontological commitments and consequences of, say, both QM in general and specific versions of it (on a similar vein, see also Wallace 2020). However, an even more serious problem affects these attempts. While the relation between a certain interpretation of QM and its ontological commitments is generally confined to the quantum domain, it is not uncommon to find attempts to generalize the lesson of (various interpretations of) QM to other (more microscopic) domains of physical reality. As a result, present-day metaphysics teems with proposals that move from the assumptions of one or the other interpretation of QM – quantum indeterminism and primitive ontology being two popular examples – and apply these assumptions to different physical regimes. The theoretical strategy whereby one seizes on the lesson of a certain interpretation (even in the case QM could be interpreted in an unequivocal, agreed-upon manner), and tries to extrapolate a view that applies to broader contexts than those in which the theory is meant to apply, should be considered with some suspicion. Existing disagreements on the first and second levels of framework dependency show that the ontological commitments of a certain interpretation of QM can be mapped with diverse ontological alternatives, even more so in case the latter refer to a different domain of physical reality. This is therefore a *third level of framework dependency* of the interpretation of QM with respect to its ontological commitments.

While a more comprehensive account of this type of problematic *modus operandi* would require an in-depth analysis of the realist vs anti-realist account of physical theories, in the present context I set aside such a thorny topic and confine myself to a much more limited objective. As a particularly instructive case for the application of the three levels of framework dependency that I have just singled out, I shortly discuss the way in which their interplay undermines taking Bohm's own interpretation

be eventually circumvented, while there would still be an underdetermination (here termed 'second level of framework dependency') among the various GRW-type theories.

of QM as implying the existence of a fundamental, fixed ontology and rather opens the path to a processual understanding of scientific theories as a way to articulate the dynamism of physical reality.

Bohm's theory (Bohm 1952) is a second-order theory that starts with the Schrödinger equation for a single particle and operates via a decomposition of the wave function, Ψ, in polar form. With this procedure, two coupled equations get defined, respectively corresponding to the quantum Hamilton–Jacoby and the continuity equation, where it is assumed that each particle is both acted upon by a classical and a quantum potential and that the Born probability distribution holds initially. Bohm's theory is equivalent to the orthodox view, provided that one accepts that (Bohm 1952, I, p. 171) the wave function obeys the Schrödinger equation, that a specific value for the initial momentum of the single particle holds, and that we are confined to deal with a statistical ensemble that follows the Born probability distribution.

Now, within the so-called second level of framework dependency described above, a main issue arises as to whether the wave function should be portrayed as a nomological or ontological entity. This issue pivots on the distinction between a certain theory's *primitive ontology*, namely the set of entities to which the theory grants existence, and its associated *dynamical structure*, namely the parameters that are responsible for the evolution of those entities (see, e.g. Allori 2013, Esfeld 2018). A long-standing tradition, which dates to Dürr, Goldstein, and Zanghì 1992a, 1992b, considers the wave function as something law-like, whereas only point-like particles are accorded an ontological status. In its most resolute formulation (see, e.g. Esfeld, Lazarovici, and Lam 2017; Esfeld and Deckert (2017); Esfeld 2020), three-dimensional particles are all that exist, while fields, forces, and potentials are taken as mere nomological entities. While the ontological interpretation of the wave function has been recently defended by Romano (2020), starting by a critical analysis of Dürr, Goldstein, and Zanghì's (1996) analogy between the wave function and the Hamiltonian to support the nomological interpretation, here I would like to stick to Bohm's own point of view.

After the discussion on the decomposition strategy, Bohm comments (Bohm 1952, I, p. 170) there is a strong analogy between the wave function satisfying the Schrödinger equation and the electromagnetic field obeying Maxwell's equations, to such an extent that it seems plausible 'to regard the wave function of an individual electron as a mathematical representation of an objectively real field' (*ibid.*). For what matters in the present context, by embracing an ontological interpretation of the wave function, it follows that, even within the sole domain of quantum physics, the ontological richness of Bohm's theory cannot be exhausted by three-dimensional particles, for there are also classical and quantum fields. In this sense, Bohm's own interpretation of his approach seems incompatible with a primitive ontology that only makes room for point-like particles.

Let us now further explore the third level of framework dependency to make the claim that the theory-restricted ontological commitments of Bohm's theory cannot be convincingly applied to physical domains in which the predictions of QM are not meant to hold. If it is true that ontology is constitutively speculative, it is also true that the conceptual strategy whereby one starts with a theory that is valid at a certain energy scale, then extracts putative ontological indications, and finally extends them to other physical regimes in which the theory is not meant to apply is theoretically hazardous. It particularly happens in the case of Bohm's theory. Firstly, Bohm always considered QM as an incomplete theory (see, e.g. Bohm 1971[1957], pp. 47, 64, 72, 86) – to the extent that the observables of QM, namely those that are obtained with a measuring procedure, do not represent any physical properties belonging to the observed system alone. Secondly, as already mentioned, in the very writings in which Bohm advanced his alternative interpretation of QM, he himself paved the way for a sub-quantum theory in which the entities that figure in QM might eventually be overcome.[8] I now want to focus on this second point.

In the philosophical literature on Bohm, there is a double-standard account of his theoretical effort. On the one hand, Bohm is depicted as the physicist who, against the standard view of QM, intended to recover the gap between the quantum and the classical world by endorsing an alternative, deterministic interpretation of QM. On the other hand, Bohm is portrayed as the thinker who, largely inspired by his durable exchanges with Indian philosopher Jiddu Krishnamurti, speculated about the 'unfathomable' (to adopt Cushing's expression, Cushing 2019, p. 113) notion of the implicate order. In a way, the objective of this final part of the section is to shed new light on the short sightedness of this representation.

To my knowledge, the contribution that best contextualizes Bohm's development of his own causal interpretation is Marij van Strien's (2020), who offers a continuist reading of his work on QM from his 1952 papers through to his last works. In the light of the continuity she identifies in Bohm's overall oeuvre, she argues that Bohm was able to turn the accusation of classicality around by setting himself up to develop a novel interpretation of QM able to move beyond the limits of the orthodox view. While I am strongly sympathetic with Van Strien's reconstruction, her attention is devoted to the conceptual connection between Bohm's earlier and later works. However, as she herself points out, those who are interested in endorsing the causal interpretation tend to keep the (putative) realist, mechanist, and determinist commitments of the 'early Bohm' and to reject his later speculations. Because of this, in the present context, I stick to Bohm's early writings to substantiate the claim that he was never committed

[8] As already emphasized, the equivalence of Bohm's interpretation is ensured by the fact that (i) the wave function obeys the Schrödinger's equation, (ii) the particles' momentum has a particular expression (given by $p = \nabla S$), (iii) the ensemble of particles is distributed according to a probability density that follows Born rule ($p = |\Psi(x)|^2$). However, Bohm was never committed to the idea that these three hypotheses should be considered as absolutely valid – i.e. valid in every physical domain. Quite the contrary, starting from his 1952 papers, he pondered on possible modifications to the mathematical formulation of his theory eventually leading to a new, sub-quantum, physical theory. Indeed, the three hypotheses above summarized can be taken as a restriction to a specific physical domain that is not deep-rooted to the conceptual structure of Bohm's interpretation. On this view, QM is to be regarded as an effective, approximate theory, working only on certain energy (distance) ranges. This is why, after presenting his interpretation of QM, together with a survey of case studies, Bohm advanced the hypothesis that, by properly relaxing one or the other of (i), (ii), (iii), one may extend his interpretation to phenomena associated with the sub-quantum regime. In his 1952 paper, part I, he proposed two possible strategies, respectively corresponding to a modification of point (i) and point (ii). Importantly, this attempt shows that Bohm always struggled hard to overcome the quantum-mechanical perspective to reach other domains of reality in which the very particle/wave duality could be overcome. For a historical reconstruction of Bohm's own reflections on this, see Talbot (2017), pp. 163-166, 174-175, 246). For a recent discussion, see Oldofredi (2023).

to a particular primitive ontology and rather that his theoretical style opened the path to a processual understanding of physics and physical theories.

While discussing what he labels the 'mechanistic philosophy', Bohm (1971 [1957]) argues that this theoretical perspective supports a view of the universe as consisting of a set of ultimately fixed, static 'building blocks'. Historically, this mechanistic philosophy assumed that this role should be accorded to indivisible atoms, conceived of as separately existent mechanical entities whose nature is given once and for all. Even after the discovery of sub-atomic components, such as electrons, protons, and neutrons, the mechanistic philosophy still embodied the prevailing scientific world view according to which there must be basic entity-like units whose (external) interactions give rise to the various levels of physical reality. The only mistake was to postulate the atoms as the ultimate constituents of the world. As Bohm comments on this:

> New developments in modern physics show, however, that this point of view cannot be correct, either. For the further progress of physics has disclosed that even the electrons, neutrons and protons are not immutable, and that, under suitable conditions, they can be transformed into each other and into a whole host of qualitatively different kinds of particles [...]. (Bohm 1971, p. 59)

Naturally, the mechanist point of view can still be endorsed by properly readjusting its commitments based on available evidence, to the extent that there is no final argument whereby the mechanistic perspective can be abandoned. Still, Bohm argues 'the assumption that all qualitative changes are, at bottom, just passive 'shadows' of quantitative changes of some basic set of entities [...] cannot be founded on any conceivable kinds of experimental fact' (Bohm 1971, p. 61). All in all, it looks pretty clear that Bohm was quite dissatisfied with the idea of the universe consisting of ultimate bits of matter – let alone point particles in three-dimensional space – that are believed to be fundamental entities, whose fixed nature is meant to build everything up. By referring to the ultimate experiments in high-energy physics – which revealed the existence of a whole new array of phenomena – Bohm considered the theoretical perspective of elementary particles making up the world as inadequate. Based on his viewpoint, QM is an approximate theory in which point particles in three-dimensional space are to be regarded as fruitful abstractions given with a relatively stable form. While this already undermines the applicability of the primitive ontology framework to Bohm's theory, at least in the radical reading by Esfeld and collaborators, I would like to make a final remark.[9]

In one of the last sections of his 1971 [1957] book, Bohm muses on the idea of there being unvarying, definite properties that can be attributed to a set of entities. Since every theory is to some extent an abstraction, and thus only captures a finite, inevitably limited set of qualities appertaining to the systems under consideration, it follows that 'no given thing can be exactly and in all respects the kind of thing that is defined by any specified conceptual abstraction' (Bohm 1971, p. 154). More importantly, Bohm writes 'carrying the analysis further, we now note that because all of the infinity of factors determining what any given thing is are always changing with time, no such thing can even remain identical with itself as time passes' (Bohm 1971,

[9] For a somewhat different and more 'contextual' account of primitive ontology, see Oldofredi (2022).

p. 157). In conclusion, not only does the notion of point particles apply abstractly, but their very self-identity is also challenged, to such an extent that 'all our concepts are, in a great many ways, abstract representation of matter in the process of becoming' (Bohm 1971, p. 156).

8.4 Process philosophy reloaded

The interplay between the above-defined three levels of framework dependency in the context of QM allows questioning the tenability of those accounts of Bohm's theory suggesting its commitment to the (sole) existence of three-dimensional point particles. Not only does embracing the ontological interpretation of the wave function contradicts such a view even within the context of QM, but also Bohm's more general discussion on the scope and status of physical theories reinforces his anti-mechanistic and anti-deterministic conception of physics broadly construed. And this understanding of physics and physical theories sits particularly well with, and provides further strength to, processual thinking as a way to articulate systems' individuality as contingent upon one's type of inquiry: systems, such as particles, should not be taken as self-standing, static entities. Rather, they are amenable to a process of individuation that is strictly dependent on the specific theoretical context at stake.

Remarkably, the so-far discussed notion of process has to do with the relationship between continuants, namely entities without temporal parts, and occurrents, namely perduring entities with temporal parts. In short, three-dimensional, definite entities are abstracted from four-dimensional, indefinite processes, to such an extent that processualism seems to gesture towards a form of four-dimensionalism, whereby a process corresponds to a four-dimensional space-time region. It seems to be the case, for instance, of biology and philosophy of biology, where the identity of specific compounds, organs, and organisms get traced by their evolution with respect to the surrounding environment: none of them can be identified based on intrinsic properties. Rather, their identity is contingent upon the history of concrete, non-particular processes. Based on this viewpoint, concreteness and non-particularity are typically conceived in terms of evolution, while evolution is intimately associated with becoming. Within such a processual framework, and against substance metaphysics, the identity of entities is taken to depend on more fundamental, non-particular occurrents, namely processes. In this sense, while elaborating on the anti-essentialist stance of processual thinking, Guay and Pradeau (2016) argue that instead of asking what a certain entity x is fundamentally, what is relevant to gain insight on that entity is to follow x through time, namely, to explore its temporal evolution.

However, it should be noted that this type of processualism turns out to be unavailable within certain areas of theoretical physics that question the commonsensical facets of temporality. Indeed, one of the most startling consequences of relativistic physics is precisely the rebuttal of time as an absolute parameter that can be uniquely defined, let alone a container where events display. A major issue arises as to what type, if any, of process can be secured within such a theoretical environment. For if the principle of relativity holds true, different events that occur locally along the worldline of different systems are taken to be part of the same process. Therefore, a question arises as to how to account for the way in which different events can be adopted to make up individual objects, and thus portrayed 'as characterizing the same individual' (Guay and Pradeau 2016, p. 320).

When dealing with such a thorny issue, two main strategies can be adopted. The first, such as the one defended by Cortês and Smolin (2014, 2018) is that, to secure a processual account within fundamental physics, the relativity of simultaneity should be questioned. Accordingly, they argue that a processual account of physics implies a notion of time as fundamental, thus granting a picture of the universe as an ensemble of uniquely identified events that occur and then cease to be – this uniqueness being granted by the existence of an absolute temporal parameter. The second strategy aims to incorporate the main lesson of relativistic physics and tries to rearticulate the notion of process within such a framework by disentangling the notion of process from a unique ordering of events. It is the case, for instance, of Carlo Rovelli's *Evolving Constant Approach* (see, e.g. Rovelli 1991), which promotes the principle of relativity and yet claims that physics is the study of processes. Within such a theoretical context, processualism gets coupled with a particular form of relationism whereby the identity of a system is amenable to a multiplicity of identifications, as no absolute temporal parameter is available.[10]

8.5 Bohm's interpretation of STR and processualism

Moving from the so-far sketched account of processualism within theoretical paradigms that incorporate what is generally considered one of the main lessons of relativistic physics, the scope of the present section is to explore whether Bohm's understanding – in Bohm (1996 [1965]) – of STR can be expressed in such processual terms. As broadly known, STR is premised on two principles. Firstly, the principle of relativity argues that the laws of physics are invariant in every inertial frame of reference. Secondly, the light postulate asserts that the speed of light in vacuum is equal for every observer, independently of the relative motion between the latter and the light source. It qualitatively follows that, whenever two observers move at a different speed, they can agree neither on the time coordinate to be assigned to events nor on the lengths of objects. This is precisely why incorporating the legacy of STR requires resanctioning the notion of process. Keeping this in the background, in the present context I try to show that, based on three argumentative lines traceable in Bohm's text, his interpretation of STR resonates with, and provides further ground to, what I called processualism reloaded. These are (1) the arbitrary character of the frame of reference, (2) the parlance of events and especially processes vs objects and their motion, and (3) the notion of relative invariance. Let us consider them in order:

1. ***The arbitrary character of the frame of reference***: After discussing, in the first chapters, the Michelson–Morley and the Fizeau experiments, as well as the advocated ether hypothesis to provide an explanation for them, Bohm starts focusing on the development of the Lorentz theory and transformations, with a specific focus on the ambiguity of spatiotemporal measurements arising within such a formalism. This so-to-speak ambiguity, which will then prove key for the

[10] I explored this issue in more detail in Margoni (2022).

successive definition of Einstein's theory, comes to the conclusion that spatiotemporal coordinates cannot be taken as absolute and uniquely defined, for they now stand for relationships between events and objects to some type of measuring apparatus, which is fixed by the scientist herself, whereas these spatiotemporal coordinates 'do not actually exist in space but are a purely conceptual invention, an abstraction introduced by us for our own convenience' (Bohm 1996, p. 32). Following this line of thought, the conceptual step performed by Einstein was that of keeping Lorentz transformations while disposing of the ether hypothesis. For while the Lorentz theory already incorporated the light postulate, the notions of "true" distances and 'true' temporal duration with reference to the ether were still, despite their patent unobservability, enigmatically present. By dismantling the conceptual tenability of true spatiotemporal coordinates, thus discarding the ether hypothesis, it follows that even the notion of uniquely defined objects must be dropped. Indeed, the absence of absolute spatiotemporal parameters implies that 'different observers do not agree on which events constitute "the same time", and therefore they do not agree on what are the basic properties of the "objects" (such as length, mass, etc.)' (Bohm 1996, p. 114). *Contra* a tradition that, according to Bohm, should be traced back to Ancient Greek philosophy, and particularly to Aristotle (Bohm 1996, p. 35), which portrays space and time as absolute, relativistic physics shows the inherently arbitrary character of the frame of reference: none should be accorded a special ontological status *vis-à-vis* others. Accordingly, no uniquely defined characterization of an object's properties is viable within relativistic physics.

2. ***Events and processes vs objects and their motion***: The fact that the parlance of entities such as objects and their associated motion, typical of Newtonian physics, should be supplanted by that of events, and especially processes, is emphasized since the very premise of Bohm's text on STR (Bohm 1996, p. IX). As Bohm comments on this, the notion of an object is now described as 'a pattern of events and processes that tends to remain similar to itself over an indefinitely long period of time' (Bohm 1996, p. 114). Based on this perspective, an object that is characterized by varying properties corresponds to a 'pattern that alters and transforms', which begins in some region, 'where the object is "created" ', and ends in some other region, 'where the object is "destroyed" '(*ibid.*). The description of the world as a set of objects moving in three-dimensional space or four-dimensional space-time is thus 'replaced by its analysis in terms of events and processes, organized, ordered, and structured so as to correspond to the characteristics of the material system that is being studied' (*ibid.*). This implies that, within relativistic physics, space and time 'constitute the means by which the characteristics of physical phenomena are to be treated' (*ibid.*). Importantly, while such an identification of space and time as the basic building blocks of relativistic physics may seem to imply a form of processualism that coincides with four-dimensionalism, Bohm clarifies that the Minkowski diagram stands for 'a kind of map of the events in the world, which can correctly give us the order, pattern, and structure of real events, but which is not in itself the world as it actually is' (*ibid.*). Put it differently, both within Newtonian and Einsteinian physics, our ideas of space and time

should be considered as useful conceptual maps that may display 'a structure that is partly true to that of real physical processes and partly false' (Bohm 1996, p. 140). It is precisely in this sense that, within other domains, the notions of space and time – being them absolutely (as in Newtonian physics) or relatively (as in Einsteinian physics) defined – may be supplanted by others, yet-to-be discovered, concepts.

3. ***The notion of relative invariance***: The reference, within relativistic physics, to spatiotemporal events and processes brings to the surface the fact that 'objects, entities, substances, etc., that we actually experience, perceive, or observe have always (thus far) shown themselves to be only relatively invariant in their properties, this relative invariance having often been mistaken for absolute permanence' (p. 86). This attitude to portray entities as permanent substances may mistakenly follow from conservation principles, such as the fact that the total energy of an isolated system is conserved (p. 89). However, comments Bohm, 'Knowing that all these objects can be broken, corroded, melted, burned, or that they are subject to decay, is it not better to refer to them, in effect, as relatively fixed and invariant, rather than as entities with absolutely permanent properties?' (pp. 92–93). Put it otherwise, *contra* substance-based accounts, Bohm suggests embracing a notion of relative invariance, whereby the properties that are attributed to a certain entity are always relative to a specific domain of inquiry: no self-standing, invariant entity has ever been observed. Therefore, adds Bohm, 'Can we not then simply refrain from making assumptions concerning the absolute permanence of what is, in the nature of the case, unknown?' (p. 93). Entities are thus characterized by relatively invariant features, to such an extent that '[i]n the domain in which these entities or structures are relatively invariant, we may refer to their movements and transformations in a way rather similar to that adopted with regard to the objects of everyday experience' (p. 93). But this manoeuver is ultimately conceptual and should thus be confined to a limited domain. To provide an example, Newton's equations are invariant under Galilean transformations, while Maxwell's equations are invariant under Lorentz transformations. Analogously, there may be other levels of physical reality where relativistic principles and notions may have to be overcome, thus proving the only relative nature of invariant quantities figuring in STR.

All in all, Bohm's reading of STR allows disentangling the notion of process from a unique ordering of events and thereby reformulates it in terms of the way in which entities can be relationally identified. The underlying assumption – which rejects the traditional metaphysics of substances – is that there is no such thing as a permanent object. Rather, only relatively invariant entities can be defined. This conception is of particular relevance within relativistic physics, insofar as different observers disagree on which events are simultaneous, and thus which are the fundamental properties (such as length or mass) of a certain object. The parlance of objects should then be rephrased in terms of processes that are ordered and assembled so as to stand for the properties of the specific system under investigation.

8.6 Conclusions

Process philosophy is becoming increasingly popular in a variety of disciplines and research areas. It provides a flexible interpretation of the commonsensical experience of dynamic phenomena aimed at questioning the traditional view of particular, self-standing entities as the most fundamental building blocks of physical reality. Against this snapshot view of the universe, processualists claim concrete, non-particular, processes are prior to concrete, particular objects, and associated states of things. The first intent of this chapter was to capitalize on Bohm's understanding of QM to claim that his own reading of it was never committed to the existence of ultimately fixed bits of matter, let alone three-dimensional point particles. Rather, according to Bohm, QM works only within a limited domain, and should be overcome by a yet-to-be discovered theory (or set of theories) whereby the particle/wave duality may be even superceded. More importantly, Bohm muses on to the idea that every scientific theory is to some extent an abstraction that allows singling out certain entities as basic only for a matter of convenience. While most processual accounts focus on the relation between four-dimensional processes and three-dimensional entities, I then argued that processualism can incorporate the main lesson of relativistic physics and yet retain its commitment to evolution and becoming. From this standpoint, a process stands for the way in which different entities get relationally identified. I then tried to show that Bohm's reading of STR allows disentangling the notion of process from a unique ordering of events and thereby reformulates it in terms of the way in which entities can be relationally identified. The underlying premise – which discards the traditional metaphysics of substances – is that there is no such thing as a permanent object. Rather, only relatively invariant entities can be defined. In sum, Bohm's understanding of physics resonates with, and provides further ground to, process thinking as a way to articulate the relationship between concrete, particular entities, and concrete, yet indefinite, processes.

References

Acuña, P. (2021). Charting the landscape of interpretation, theory rivalry, and underdetermination in quantum mechanics. *Synthese*, **198**: 1711–1740

Allori, V. (2013). Primitive ontology and the structure of fundamental physical theories. In Ney, A. and Albert, D.Z. (eds.) *The Wave Function: Essays on the Metaphysics of Quantum Mechanics*, pp. 58–75. New York: Oxford University Press.

Bassi, A. and Ulbricht, H. (2014). Collapse models: from theoretical foundations to experimental verifications. *Journal of Physics: Conference Series*, **504**: 012023.

Bohm, D. (1952). A suggested interpretation of the quantum theory in terms of "hidden" variables. I & II. *Physical Review*, **85**(2): 166–193.

Bohm, D. (1971 [1957]). *Causality & Chance in Modern Physics*. Philadelphia: University of Pennsylvania Press.

Bohm, D. (1996 [1965]). *The Special Theory of Relativity*. London: Routledge.

Cordero, A. (2001). Realism and underdetermination: Some clues from the practices-up. *Philosophy of Science*, **68**: S301–S312.

Cortês, M. and Smolin, L. (2014). The Universe as a process of unique events. *Physical Review D*, **90**: 084007.

Cortês, M. and Smolin, L. (2018). Reversing the irreversible: From limit cycles to emergent time symmetry. *Physical Review D*, **97**: 026004.

Cushing, J.T. (2019). Bohmian Mechanics and Its Ontological Commitments. In Cordero, A. (ed.) *Philosophers Look at Quantum Mechanics*, pp. 107–118. Springer: Cham.

Dürr, D., Goldstein, S. and Zanghì, N. (1992a). Quantum Equilibrium and the Origin of Absolute Uncertainty. *Journal of Statistical Physics*, **67**(5): 843–907.

Dürr, D., Goldstein, S. and Zanghì, N. (1992b). Quantum Chaos, Classical Randomness, and Bohmian Mechanics. *Journal of Statistical Physics*, **68**(1): 259–270.

Dürr, D., Goldstein, S. and Zanghì, N. (1996). Bohmian mechanics and the meaning of the wave function. In Cohen, R.S., Horne, M. and Stachel, J. (eds.) *Experimental Metaphysics: Quantum Mechanical Studies in Honor of Abner Shimony.* Boston Studies in the philosophy of science, pp. 25–38. Berlin: Springer.

Dürr, D., Goldstein, S. and Zanghì, N. (2013). *Quantum Physics Without Quantum Philosophy.* Berlin: Springer.

Egg, M. (2021). Quantum ontology without speculations. *European Journal for Philosophy of Science*, **11**: 32.

Einstein, A., Podolsky, B. and Rosen, N. (1935). Can Quantum-Mechanical Description of Physical Reality Be Considered Complete?. *Physical Review*, **47**: 777.

Esfeld, M. (2018). Collapse or no collapse? What is the best ontology of quantum mechanics in the primitive ontology framework?. In Gao, S. (ed.) *Collapse of the Wave Function. Models, Ontology, Origin and Implications*, pp. 167–184. Cambridge: Cambridge University Press.

Esfeld, M. (2020). A proposal for a minimalist ontology. *Synthese*, **197**: 1889–1905.

Esfeld, M. and Deckert, D.A. (2017). *A Minimalist Ontology of the Natural World.* New York: Routledge.

Esfeld, M., Lazarovici, D., Lam, V. and Hubert, M. (2017). The physics and metaphysics of primitive stuff. *British Journal for the Philosophy of Science*, **68**: 133–161.

Freire Junior, O. (2015). *The Quantum Dissidents. Rebuilding the Foundations of Quantum Mechanics (1950-1990).* Berlin: Springer.

Huneman, P. (2021). Biological individuals as "weak individuals" and their identity: exploring a radical hypothesis in the metaphysics of science. In Meincke, A.S. and Dupré, J. (eds.) *Biological Identity. Perspectives From Metaphysics and The Philosophy of Biology.* London and New York: Routledge.

Lyre, H. (2010). Why quantum theory is possibly wrong. *Foundations of Physics*, **40**: 1429–1438.

Margoni, E. (2022). Can There be a Process Without Time? Processualism Within Timeless Physics. *Foundations of Physics*, **52**: 48.

Nicholson, D.J. and Dupré, J. (2018). *Everything Flows: Towards a Processual Philosophy of Biology*. Oxford: Oxford University Press.

Oldofredi, A. (2022). Beables, Primitive Ontology and Beyond: How Theories Meet the World. In Allori, V. (ed.) *Quantum Mechanics and Fundamentality: Naturalizing Quantum Theory between Scientific Realism and Ontological Indeterminacy*. Springer Nature.

Oldofredi, A. (2023). Orthodox or dissident? The evolution of Bohm's ontological reflections in the 1950s. *European Physical Journal H*, **48**: 12.

Rescher, D. (2000). *Process Philosophy: A Survey of Basic Issues*. Pittsburgh: University of Pittsburgh Press.

Romano, D. (2021). Multi-field and Bohm's theory. *Synthese*, **198**: 10587–10609.

Rovelli, C. (1991). Quantum evolving constants. Reply to 'Comment on 'Time in Quantum Gravity: An Hypothesis'. *Physical Review D*, **44**: 1339–1341.

Schrödinger, E. (1935). Die gegenwärtige Situation in der Quantenmechanik. *Naturwissenschaften*, **23**: 807–812, 823–828, 844–849.

Seibt, J. (1996). Existence in time: from substance to process. In: Faye, J. (ed.) *Perspectives on Time*. Boston Studies in Philosophy of Science. Dordrecht: Kluwer.

Seibt, J. (2002). Quanta, tropes, and processes: on ontologies for QFT beyond the myth of substance. In Kuhlmann, M., Lyre, H. and Wayne, A. (eds.) *Ontological Aspects of Quantum Field Theory*. Singapore: World Scientific.

Seibt, J. (2009). Forms of emergent interaction in General Process Theory. *Synthese*, **166**(3): 479–512.

Seibt, J. (2018a). "What is a process? Modes of occurrence and forms of dynamicity". In: Stout, R. (ed.) *General Process Theory, Processes, Experiences, and Actions*. Oxford: Oxford University Press.

Seibt, J. (2018b). Ontological Tools for the Process Turn in Biology: Some Basic Notions of General Process Theory. In Dupré, J. and Nicholson, D. (eds.) *Everything Flows: Towards a Processual Philosophy of Biology*, Oxford: Oxford University Press.

Seibt, J. (2023). Process Philosophy. *The Stanford Encyclopedia of Philosophy*.

Stengers, I. (2011). *Thinking with Whitehead*. Cambridge, MA: Harvard University Press.

Sukumar, N. (2013). The atom in a molecule as a mereological construct in chemistry. *Foundations of Chemistry*, **15**: 303–309.

Van Strien, M. (2020). Bohm's theory of quantum mechanics and the notion of classicality. *Studies in History and Philosophy of Modern Physics*, **71**: 72–86.

Van Strien, M. (2024). Why Bohm was never a determinist. In Oldofredi, A. (ed.) *Guiding Waves in Quantum Mechanics: 100 Years of de Broglie–Bohm Pilot-Wave Theory*. Oxford: Oxford University Press.

Wallace, D. (2020). Lessons from realistic physics for the metaphysics of quantum theory. *Synthese*, **197**: 4303–4318.

9

The (un)detectability of trajectories in pilot-wave theory

Johannes Fankhauser

9.1 Introduction

Let's start this chapter with the following question. What does quantum mechanics (QM) state about what is *real*? Unfortunately, about a hundred years after the theory's advent, the answer is still: not much. At best, we have a multitude of proposals in the form of quantum interpretations, some of which try to answer this question while some of which do not. But all come with their problems, and the quantum community is deeply split on which one, if any, to opt for. At worst, up to the present day, the quantum theory continues to be fuelled by confusion about what quantum measurement results supposedly signify.

Despite its tremendous empirical accuracy, one question that the theory does not state is: *What* can be inferred from empirical evidence about the physical system that is measured? In other words, what do quantum measurements signify? And what (if any) information do they convey about the physical system?

The physical world can be defined using two distinct domains. One domain encompasses empirical data, such as observable elements such as measurement outcomes and records in the environment. This domain constitutes what is directly observable – the fundamental phenomenological facts. Solely residing within this *manifest* domain does not necessitate any inherent commitment to what exists beyond the tangible aspects of the physical world. The second domain – the *non-manifest* domain – encompasses elements, objects, and facts that lie beyond direct observation. These contain entities that exist to which a physical theory relates the manifest elements.[1] We may label those elements with the familiar attribute of being 'ontological'. The non-manifest domain, furthermore, contains another category, which is termed the nomological sector, i.e. the laws of physics determining the dynamics of the ontological sector.

Despite the above problems of quantum mechanics, some argue that the key lies in giving up altogether the relationship between the manifest and non-manifest domains. Hence, resorting to operationalism or some sort of instrumentalism (see, for instance, Bogen (2017); instrumentalism in Chakravartty (2017), and references therein). Others

[1] For example, it indicates when a manifest element, e.g. a position measurement record, actually refers to a non-manifest element, e.g. the position of a particle.

attempt to propose a coherent underpinning by appealing to more ontological commitments (see, for instance, Allori 2023). But, unsurprisingly, a half-hearted attempt or wavering between the two will invariably cause trouble as evidenced in modern literature.

As it turned out, QM teaches us that what it is that is 'observed' is not *a priori* entailed by the use of it. As Einstein aptly noted: what is observable is contingent on the theory (cf. Heisenberg 1958).

Appealing to this narrative serves to dispel the proposed experiments discussed in this chapter. As shown later, a naive account of the significance of measurement outcomes in the context of pilot-wave theory serves as a paradigm case for unwarranted claims about the relationship of certain experiments and the physical system. Therefore, insights on what can and cannot be inferred from these experiments have consequences on what is detectable in pilot-wave theory. I examine two proposals: (1) the so-called 'weak values' and 'weak measurements' with regard to the observation and vindication of Bohmian trajectories. The alleged detectability of particle trajectories is discussed in light of weak *velocity* measurements: The temporal evolution of Bohmian particles is governed by a deterministic dynamics. By construction, however, the individual particle trajectories are underdetermined and generically defy detectability in principle. I discuss two proposals in light of which this lore might possibly be called into question: Firstly, the so-called weak measurements. Due to their characteristic weak coupling between the measurement device and the system under study, they permit the experimental probing of quantum systems without essentially disturbing them. It is therefore natural to think that weak measurements of velocity in particular offer the opportunity to actually observe the particle trajectories. If true, such a claim would not only experimentally demonstrate the incompleteness of QM, but it would also provide support of de Broglie–Bohm theory in its standard form, singling it out from an infinitude of empirically equivalent alternative choices for the particle dynamics. (2) The so-called surrealistic trajectories that were invoked in an attempt to falsify and challenge pilot-wave theory: in this case, an experiment was conceived where the Bohmian predictions for the locations of particle differ from the prediction according to standard quantum theory. Where the particle is measured to reside, i.e. its alleged 'actual' position – so the story goes – conflicts with its position derived from the Bohmian laws of motion. Hence, pilot-wave theory is supposed to provide incorrect results about where particles really travel.

The discussion offers two insights: I show that weak velocity measurements help to clarify the nature of particle trajectories but not in a robust verificatory sense: they constitute no new arguments, let alone empirical evidence, in favour of standard de Broglie–Bohm theory. One must not naively identify weak and actual positions. This is revealed by a careful reconstruction of the physical arguments on which the description of weak velocity measurements rests. Pilot-wave theory is entirely dispensable for a coherent treatment and interpretation of weak velocity measurements; they receive a natural interpretation within standard QM as observational manifestations of the gradient of the wave function's phase.

Moreover, a rehearsal of surrealistic trajectories is shown to resemble the same conundrum: what inferences are justified from measurement outcomes to the properties of a particle? The puzzle again resolves by a careful analysis of the significance of measurement results in standard quantum theory and pilot-wave theory.

Thus, the common thread of both instances is the insufficient demarcation of accessible measurement records and what they are supposed to represent ontologically.

This chapter closely follows the author's previously published research, with some passages reproduced. The discussion on weak velocity measurements in Sections 9.2 and 9.3 is based on Fankhauser and Dürr (2021), while the analysis of surrealistic Bohmian trajectories in Section 9.4 draws on Fankhauser (2023).

9.2 Underdetermination in pilot-wave theory

One of the primary interpretative challenges QM has encountered involves comprehending the nature of entanglement: in the so-called EPR paradox, one widely separates the partners of an entangled pair of particles. They can then no longer interact. Hence, we may, according to Einstein, Podolsky, and Rosen, 'without in any way disturbing the system' perform (and expect a well-defined outcome of) a position measurement on one partner and a simultaneous momentum measurement on the other partner (Einstein *et al.* 1935, p. 777). Prima facie, it looks as if thereby we can bypass the uncertainty relations. However, this raises the question of whether QM in its current form is complete: does every element of physical reality have a counterpart in the description of the QM formalism?

As is well-known, Einstein was '[...] firmly convinced that the essentially statistical character of contemporary quantum theory is solely to be ascribed to the fact that this [theory] operates with an incomplete description of physical systems' (Einstein 1949, p. 666).

To date, the most elaborate attempt to thus 'complete' QM dates back to Bohm (1952) – 'Bohmian Mechanics' or, in recognition of de Broglie's earlier proposal, 'de Broglie–Bohm theory'.[2]

But de Broglie–Bohm theory is not free of problems. From its early days on, a principal objection to it[3] targets the unobservability of its particle dynamics. By construction, in de Broglie–Bohm theory, the individual particle trajectories seem to be undetectable *in principle*.[4] Only their statistical averages are observable. They coincide with the standard quantum mechanical predictions. Thereby, standard de Broglie–Bohm theory achieves empirical equivalence with QM.[5]

Empirically, the standard guidance equation thus is not the only option. More precisely, it is not necessary for empirical equivalence with QM. Infinitely many different choices

[2]There exist two *distinct* variants of de Broglie–Bohm theory: the 'quantum potential' school (expounded, e.g. by Bohm and Hiley 1993, or Holland 1993), and the 'first-order formulation', canonised by Dürr, Goldstein, Zanghì, and their collaborators. The present treatment is only concerned with the latter; 'de Broglie–Bohm theory' and 'pilot-wave theory' exclusively refer to this variant.

[3]For subtleties in the early objections to de Broglie–Bohm theory, we refer to Myrvold (2003).

[4]Note, however, that rejecting equilibrium initial configurations in pilot-wave theory would indeed not only permit experimental tests of the theory, but also lead to a violation of signal-locality (cf., for instance, Valentini (1991, 1992). For caveats on the claim that Bohmian non-equilibrium necessarily leads to signalling, see Fankhauser (2023)).

[5]Here, I set aside possible subtleties, see Arageorgis and Earman (2017).

$$\boldsymbol{v}^{\Psi} \mapsto \boldsymbol{v}^{\Psi} + |\Psi|^{-2}\boldsymbol{j} \tag{9.1}$$

are equally possible for otherwise arbitrary vector fields $\boldsymbol{j}$ whose divergence vanishes, $\nabla \cdot \boldsymbol{j} = 0$. They yield coherent alternative dynamics with distinct particle trajectories, while leaving the predictive-statistical content unaltered (Deotto ad Ghirardi 1998).

One need not even restrict oneself to deterministic dynamics (an option expressly countenanced by, e.g. Dürr and Teufel 2009, Chapter 1.2): a stochastic dynamics, with $|\Psi|^{-2}\boldsymbol{j}$, corresponding to a suitable random variable can also be introduced. As a result, the particles would perform random walks, with the r.h.s. of the integral equation

$$\boldsymbol{Q}(t) - \boldsymbol{Q}(t_0) = \int_{t_0}^{t} \boldsymbol{v}^{\Psi} d\tau$$

containing a diffusion term. A proposal of this type is Nelson Stochastics (see, e.g. Goldstein 1987; Bacciagaluppi 2005). In short, de Broglie–Bohm theory's individual particle trajectories are observationally inaccessible by construction.

In consequence, de Broglie–Bohm theory is vastly underdetermined by empirical data: all versions of this theory with guidance equations of the type (9.1) are experimentally indistinguishable. Yet, the worlds described by them clearly differ (as shown in Fig. 9.1).

This issue renders weak value measurements particularly interesting. By (prima facie) allowing measurements of individual particle trajectories, they appear to directly overcome de Broglie–Bohm theory's underdetermination. But would not that contradict the empirical inaccessibility of the trajectories?

9.3 Detecting Bohmian trajectories with weak measurements?

Methods of weak measurement have opened up a flourishing new field of theoretical and experimental developments (see, e.g. Aharanov and Rohrlich 2005; Tamir and Cohen 2013; Svensson 2013; Dressel *et al.* 2014). Broadly speaking, weak measurements generalize strong measurements in that the final states of measured systems need no longer be eigenstates (and are thus a particular case of positive operator-valued measurements (POVM)). Following Aharonov *et al.* (1988), weak measurements are measurement processes (modelled via the von Neumann scheme) in which the interaction between the measurement apparatus ('pointer device') and the particle ('system') is weak: it disturbs the wave function only slightly. As a result, one can combine a weak measurement of one quantity (say, initial momenta) and a subsequent ordinary 'strong' (or projective) measurement (say, positions).

More precisely, after a weak interaction (say, at $t = 0$), the pointer states are not unambiguously correlated with eigenstates of the system under investigation. In contradistinction to strong measurements, the system does not (effectively) 'collapse' onto eigenstates; the particles cannot be (say) located very precisely in a single run of an experiment. This apparent shortcoming is compensated for when combined with a strong measurement shortly *after* the weak interaction: the experimenter is then able not only to ascertain the individual particle's precise location (via the strong measurement), for a sufficiently large ensemble of identically prepared particles with

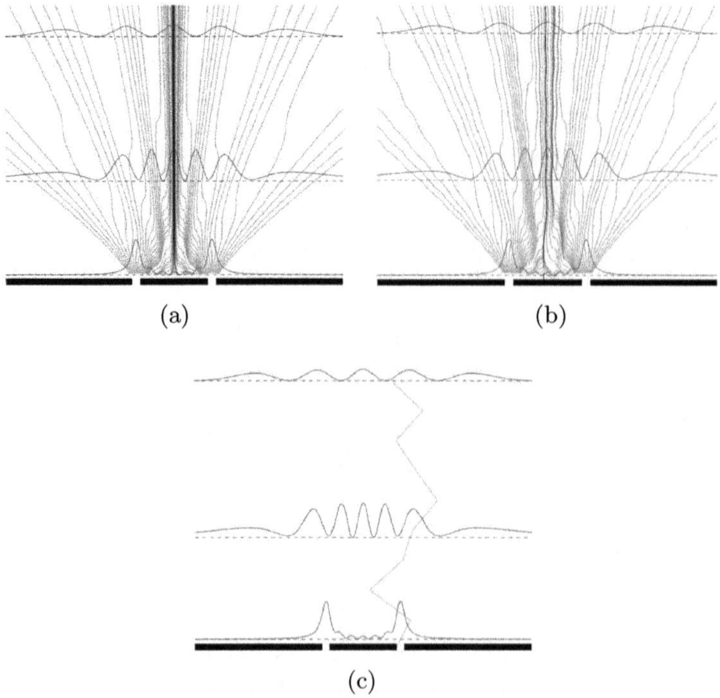

Fig. 9.1: A particle follows different trajectories corresponding to different/non-standard guidance equations. (a) The familiar wiggly deterministic trajectories that lead to the interference pattern in a double-slit experiment determined by the standard guidance equation. (b) Alternative trajectories obtained from adding a divergence-free vector field $\boldsymbol{j} := \frac{1}{x^2+y^2}\begin{pmatrix} -y \\ x \end{pmatrix}$ to the standard Bohmian velocity field. (c) A single stochastic trajectory generated by a random variable sampled according to $|\psi|^2$ – a Bohmian-style theory *without* any particle velocities. For illustration, the probability density $|\psi|^2$ is shown at three different snapshots in time. All choices of the dynamics (i.e. (a), (b), and (c)) are observationally indiscernible: the resulting measurable distributions at the screen at the top of each figure are the same.

initial state ψ_{in} (viz. Gaussian wave packets with a large spread), she can also gain statistical access to the probability amplitude of all subensembles whose final states – the so-called 'post-selected' state – have been detected (in the strong measurement) to be ψ_{fin}:

$$\langle \hat{\mathbf{x}} \rangle_w := \frac{\langle \psi_{fin} | \hat{\mathbf{x}} | \psi_{in} \rangle}{\langle \psi_{fin} | \psi_{in} \rangle}. \tag{9.2}$$

This quantity is called the 'weak position value' (for the position operator $\hat{\mathbf{x}}$). (The concept is straightforwardly applied to other operators, mutatis mutandis.) It can be shown that after many runs, the pointer's average position will have shifted by $\langle \hat{\mathbf{x}} \rangle_w$.[6] Specifically, if we characterize the final/post-selected state via position eigenstates $|\mathbf{x}\rangle$, determined in a strong position measurement and unitary evolution of the initial state, we obtain

$$\langle \hat{\mathbf{x}}(\tau) \rangle_w = \mathrm{Re}\left(\frac{\langle \mathbf{x}|\hat{U}(\tau)\hat{\mathbf{x}}|\psi_{in}\rangle}{\langle \mathbf{x}|\hat{U}(\tau)|\psi_{in}\rangle} \right), \tag{9.3}$$

where $\hat{U}(\tau)$ denotes the unitary time evolution operator during the time interval $[0;\tau]$. Following Wiseman (2007), it is suggestive of construing $\langle \hat{\mathbf{x}}(\tau) \rangle_w$ as the mean displacement of particles whose position was found (in a strong position measurement at $t = \tau$) to be at $\mathbf{x}$. From this displacement, a natural definition of a velocity field ensues:

$$\mathbf{v}(\mathbf{x}, t) = \lim_{\tau \to 0} \frac{1}{\tau}(\mathbf{x} - \langle \hat{\mathbf{x}}(\tau) \rangle_w). \tag{9.4}$$

Note that all three quantities entering this velocity field –τ, $\mathbf{x}$, and $\langle \hat{\mathbf{x}}(\tau) \rangle_w$ – are experimentally accessible. In this sense, the velocity field is 'defined operationally' (Wiseman). In what follows, I refer to the application of this measurement scheme –a strong position measurement in short succession upon a particle's weak interaction with the pointer – for the associated 'operationally defined' velocity field as 'Wiseman's measurement protocol for weak velocity measurements', or simply *'weak velocity measurements'*.

Inspired by Wiseman (2007), eminent Bohmians have advocated such weak measurements as a means of actually observing individual trajectories in standard de Broglie–Bohm theory (e.g. Goldstein, 2017, Section 4). Moreover, they point to already performed experiments (e.g. Kocsis *et al.* 2011; Mahler *et al.* 2016) that appear to corroborate de Broglie–Bohm theory's predictions and claim to show the particle trajectories (see also Oianguren-Asua *et al.* (2025) for another discussion on weak measurements in the theory).

For a better understanding of its salient points, let's now specify such weak velocity measurements in the context of the double-slit experiment.

9.3.1 Weak measurements in the double-slit experiment

Consider the standard double-slit experiment with, say, electrons hitting a screen. It enables the detection of the electrons' positions. This constitutes a strong position measurement. Accordingly, I call this screen the *strong screen.* Let a weak measurement of position be performed between the strong screen and the two slits from which the particles emerge. Let this be called the *weak screen.* The two screens can be moved to measure various distances from the double-slit. Suppose it takes the particles some time $\tau > 0$ to travel from the weak screen to the strong screen.

After weakly interacting, particle and weak screen are entangled. Hence, only the composite wave function – *not* the reduced state of the pointer – evolves unitarily

[6] In general, $\langle \hat{\mathbf{x}} \rangle_w$ is complex, meaning that both the position and the momentum are shifted. Therefore, only the real part of the weak value will be considered in the following discussion.

during time τ. Due to this evolution, the post-selected position x on the strong screen will, in general, differ from the weak value $\langle\hat{x}\rangle_w$, obtained from averaging the conditional distribution of the pointer of the weak screen. The procedure is depicted in Fig. 9.2).

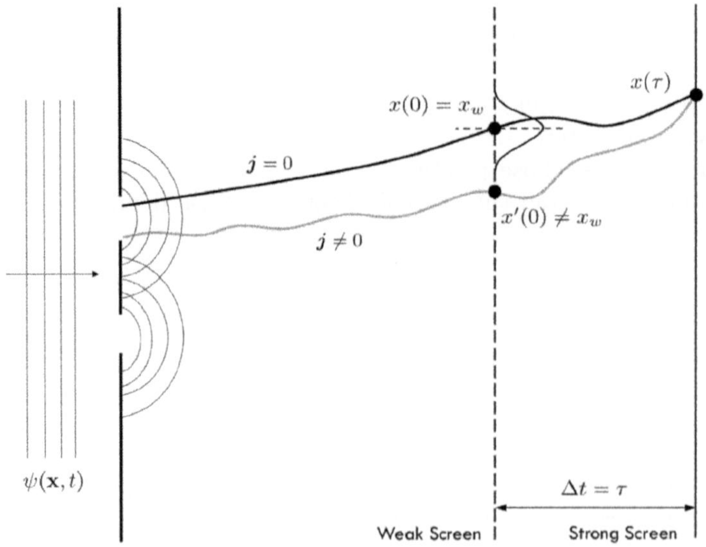

Fig. 9.2: The weak measurement procedure for a given post-selected state $x(\tau) = X_\tau$. The weak value is obtained from the distribution on the weak screen. When the velocity field is that of standard de Broglie–Bohm theory ($\boldsymbol{j} = 0$), the actual position of the particle $x(0)$ matches the weak value x_w. For an alternative guidance equation ($\boldsymbol{j} \neq 0$), it does not: the particle crosses the weak screen at a point $x'(0)$, other than the weak value. This shows that depending on which guidance equation one chooses, the weak value need not yield the actual position of the particle at time 0.

On both screens, the wave function is slightly washed out. It differs from an undisturbed state (i.e. in the absence of the weak screen). To obtain the two position values – the weak and the strong one – strong measurements are now performed both at the weak and the strong screen (i.e. on the pointer variable and the target system). For each position outcome x at the strong screen, let's select a sub-ensemble. We then read out the statistical distribution of the position measurement outcomes at the weak screen for any such sub-ensemble.

We have thus assembled all three observable quantities needed for Wiseman's operationally defined velocity (9.4): the time τ that elapsed between the two measurements, the positions x (obtained as values at the strong screen), and the average value of all positions of the sub-ensemble (obtained as values at the weak screen), associated with (i.e. post-selected for) x. This can now be carried out for different positions of the two screens: move them to various locations along the propagation direction, and repeat the same measurement procedure at each position. With this method, one can eventually map the velocity field for a sufficiently large number of measurements.

Kocsis *et al.* have indeed performed an experiment of a similar kind, using weak measurements of momentum. Their result, whose predictions are depicted in Fig. 9.1(a), qualitatively reproduces the trajectories of standard de Broglie–Bohm theory.[7] Moreover, it can be shown (cf. the Appendix in Fankhauser and Dürr 2021) that weak velocity measurements are measurements of the gradient of the phase of the wave function. Thus, they coincide with the definition of standard Bohmian velocities in the guidance equation

$$\mathbf{v} = \frac{\hbar}{m}\nabla S, \tag{9.5}$$

where S is the gradient of the phase of the wave function, $\psi(\mathbf{x}) = |\psi(\mathbf{x})|e^{iS(\mathbf{x})}$.

Note that only the standard quantum mechanical formalism has been utilized for this. Therefore, we may conclude that – based solely on standard QM – weak velocity measurements permit experimental access to the gradient of the wave function's phase.

9.3.2 Why weak velocity measurements do not measure velocities

Suggestive as these results are, I now show that such measurements could not provide direct experimental evidence displaying the shape of particle trajectories, even if it is assumed that some deterministic particle trajectories exist. They cannot, that is, go anyway to experimentally resolving the underdetermination in putative de Broglie–Bohm theory guidance equations mentioned earlier. I show that a strong assumption is required that would render it question-begging to employ weak velocity measurements in order to infer the particles' *actual* velocities. I argue that this turns out to *presuppose* standard de Broglie–Bohm theory.

When do weak and actual velocities coincide? In the following, x and y denote the position variables of the individual particles to be measured and the measurement apparatus, respectively. For simplicity, consider one dimension only. Let the particles be prepared in the initial state

$$|\psi\rangle = \int dx\ \psi(x)|x\rangle. \tag{9.6}$$

Furthermore, let the pointer device (i.e. the weak screen of the double-slit version of weak measurements in Section 9.3.1) be prepared in the initial state given by a Gaussian with large spread σ, centred around 0:

$$|\varphi\rangle = \int dy\ \varphi(y)|y\rangle = N \int dy e^{-\frac{y^2}{4\sigma^2}}|y\rangle, \tag{9.7}$$

[7] It is worth mentioning that the experiment has been performed for (massless) *photons*. However, standard de Broglie–Bohm theory is a non-relativistic quantum theory for massive particles: it cannot handle photons. The treatment of photons within field-theoretic *extensions* of de Broglie–Bohm theory, capable of dealing with photons (or bosons, more generally), is a delicate matter, which is outside the scope of this chapter. I refer the interested reader to Holland (1993, Chapter 11) and Dürr *et al.* (2013, Chapter 10) (also for further references). Kocsis *et al.*'s experiment hence has no direct bearing on de Broglie–Bohm theory's status. Rather than the trajectories of *individual* photons, Flack and Hiley (2014, 2016) have argued that Kocsis *et al.*'s experiments measure mean momentum flow lines.

where N is a suitable normalization factor. Together, the particle and the pointer form the compound system with the joint initial state

$$|\psi\rangle \otimes |\varphi\rangle = \int dxdy\ \psi(x)\varphi(y)|x\rangle \otimes |y\rangle. \tag{9.8}$$

Now consider the first – the weak – measurement process. It consists of an interaction between the particle and the pointer. Upon completion of this process (say at $t = 0$), the compound system ends up in the entangled state

$$|\Psi(x, y, 0)\rangle = \int dxdy\ \psi(x)\varphi(y - x)|x\rangle \otimes |y\rangle. \tag{9.9}$$

The probability distribution for the pointer variable y, *given* some position X of the particle, is therefore:

$$\rho_X(y) = \frac{|\Psi(X, y, 0)|^2}{|\psi(X)|^2} = |\varphi(y - X)|^2. \tag{9.10}$$

This probability density determines the expectation value

$$\mathbb{E}(y|x = X) = \int dy\ y\rho_X(y) = X. \tag{9.11}$$

That is, the mean value of the pointer distribution, conditional on the particle occupying position X, coincides with that position. This underwrites the following counterfactual:

($\mathbf{C_0}$) *If one were to perform an ordinary (strong) position measurement on the particles immediately after the weak interaction, the expectation value of the weak pointer would yield the actual position of the particle.*

Via $\mathbb{E}(y|x = X)$, the particle position is thus empirically accessible through the statistics of large ensembles of identically prepared particles from which we pick the post-selected outcomes $x = X$.

This thought is further exploited in the final steps of Wiseman's procedure. In the preceding considerations, the strong measurement was performed immediately upon the weak one. Instead, we'll now allow for a small delay. After the particle and the pointer have (weakly) interacted, the total system evolves freely for some small but finite time τ. Its state, then, is

$$|\Psi(x, y, \tau)\rangle = e^{-\frac{i}{\hbar}\tau\hat{H}_0}|\Psi(x, y, 0)\rangle, \tag{9.12}$$

where $\hat{H}_0$ denotes the system's free Hamiltonian.

Eventually, we perform a strong measurement of the particle's position X_τ at $t = \tau$. (The strong coupling between the measurement device and the particle enables precise detection of the latter's actual position.) We thus obtain the expectation value for the pointer variable, conditional on the particle occupying the position X_τ at $t = \tau$:

$$\mathbb{E}(y|x = X_\tau) = \int dy\; y|\Psi(X_\tau, y, \tau)|^2. \tag{9.13}$$

Through the statistics of a sub-ensemble of particles whose strong position measurements yielded X_τ, this expectation value is empirically accessible.

In *analogy* to Equation (9.11), let's define the position:

$$X_0 := \mathbb{E}(y|x = X_\tau). \tag{9.14}$$

Combined with the particle position X_τ, obtained from the strong measurement at $t = \tau$, it thus appears as if we have access to particle positions at two successive moments. Using Equation (9.14), the associated displacement is

$$X_\tau - X_0 = X_\tau - \mathbb{E}(y|x = X_\tau). \tag{9.15}$$

Let's grant one can make it plausible that the particles' trajectories are differentiable. Then, the displacement (Equation (9.15)) gives rise to the velocity field

$$v(X_0) := \lim_{\tau \to 0} \frac{1}{\tau}(X_\tau - \mathbb{E}(y|x = X_\tau)). \tag{9.16}$$

Note that all terms on the r.h.s. of Equation (9.16) are observable. (Hence, presumably, Wiseman's labelling (9.16) as an 'operational definition'.) In conclusion, it seems, via the statistics of an experimental set-up implementing Wiseman's procedure, we can empirically probe this velocity field.

But what does this velocity field signify? It is tempting to identify it with the particles' actual velocities. That is, should this be true, the flow of Equation (9.16) generates the particles' trajectories (assumed to be deterministic and differentiable). Is this identification justified?

By *defining* an X_0 via Equation (9.14), our notation certainly suggests so. Let's assume that this is correct and call this the 'Correspondence Assumption' (COR). That is, suppose that the actual particle position X_τ at $t = \tau$ is connected with the earlier particle position $x(0) = X_0 = \hat{T}_{-\tau}X_\tau$ at $t = 0$, where $\hat{T}_{-\tau}$ denotes the shift operator that evolves backwards particle positions by τ. (In other words, for arbitrary initial positions, $\hat{T}_\tau$ supplies the full trajectory.) Then, according to (COR), the expectation value (9.14) corresponds to the particles' position at $t = 0$. For post-selection of sub-ensembles with $x(\tau) = X_\tau$, (COR) thus takes the form (in the limit of large spread σ)

$$\textbf{(COR)}\; \mathbb{E}(y|x(\tau) = X_\tau) = \hat{T}_{-\tau}X_\tau.^{8} \tag{9.17}$$

In other words, (COR) implies the counterfactual:

($\mathbf{C_t}$) *If one were to perform a strong position measurement at $t = \tau$ (with the weak interaction taking place at $t = 0$), yielding the particles' position at $x(\tau) = X_\tau$, the*

[8] The assumption is in fact what explicitly connects the manifest variables, i.e. the accessible records, to the non-manifest variables of the theory, i.e. the actual Bohmian particle positions. In other words, the expectation value $\mathbb{E}(y|x = X_\tau)$ resulting from the weak measurement procedure, is assumed to *signify* particle configurations at a particular time τ.

weak value would be directly correlated with the particles' earlier position $\hat{T}_{-\tau}X_{\tau}$. That is, upon a strong measurement at $t = \tau$, the expectation value would reveal the particles' true positions:

$$\mathbb{E}(y|x(\tau) = X_{\tau}) = \hat{T}_{-\tau}X_{\tau}. \tag{9.18}$$

On (COR), the weak value thus gives the particle's *actual* position at the weak screen: the expectation value on the l.h.s. is reliably correlated with the particle's earlier positions. But most importantly, this is an *if and only if condition*: If (COR) is satisfied, then we recover the actual position; but if it is not, we do not. As a result, one ought to assume that (COR) is valid for weak position measurements to yield actual particle positions.

Thereby, any set of data compatible with QM appears to corroborate standard de Broglie–Bohm theory: *given* (COR), weak velocity measurements yield standard de Broglie–Bohm theory's velocity field. It thus seems as if standard de Broglie–Bohm theory's empirical underdetermination has been overcome.

Such an apparent possibility of confirming standard de Broglie–Bohm theory would be remarkable. It crucially hinges, however, on the soundness of (COR). As I shall show (COR) is generically false, and there is no plausible argument for why it should hold. This is eventually illustrated with a simple example. Prima facie, (COR) looks like a plausible extrapolation of a strong measurement immediately after the weak interaction (i.e. at $t = 0$). Firstly, (COR) holds in the limit $\tau \to 0^+$. Next, in a deterministic world, it would seem that

$$\mathbb{E}(y|x(\tau) = \hat{T}_{\tau}\kappa) = \mathbb{E}(y|x(0) = \kappa), \tag{9.19}$$

where $\kappa \in \mathbb{R}$ denotes a position.

By appeal to C_0, this would then yield

$$\mathbb{E}(y|x(\tau) = \hat{T}_{\tau}\kappa) = \mathbb{E}(y|x(0) = \kappa) = \kappa, \tag{9.20}$$

as desired.

Yet, this inference is illicit. While it is true that $\{(x(\tau), y) \in \mathbb{R} \times \mathbb{R} : x(\tau) = \hat{T}_{\tau}\kappa\}$ and $\{(x(0), y) \in \mathbb{R} \times \mathbb{R} : x(0) = \kappa\}$ contains the same pointer configurations, this *does not* imply that $\mathbb{P}(y|x(\tau) = \hat{T}_{\tau}\kappa) = \mathbb{P}(y|x(0) = \kappa)$. For this to hold, the conditional probabilities – as defined via post-selection – on both sides must be well defined. That is,

$$\frac{\mathbb{P}(y \wedge x(\tau) = \hat{T}_{\tau}\kappa)}{\mathbb{P}(x(\tau) = \hat{T}_{\tau}\kappa)} \text{ and } \frac{\mathbb{P}(y \wedge x(0) = \kappa)}{\mathbb{P}(x(0) = \kappa)} \tag{9.21}$$

must exist (and coincide).

In classical statistical mechanics, one may take this for granted. However, in a *quantum* context, entanglement complicates the situation: it compromises the ascription of probability measures to certain events. One must heed the time with respect to which the assigned probability measure is defined. This is the case with weak velocity measurements. Recall that Wiseman's measurement protocol only performs the strong measurement at $t = \tau$. This precludes defining the second term in (9.21). That is, no strong measurement is performed – and no attendant 'effective collapse' of the wave function occurs – at an *earlier* time (viz. at $t = 0$). As a result, at the time of the weak interaction ($t = 0$), the wave function of the pointer and particles is entangled. That means, however, that we *cannot* naively assign the event of any particular particle position at $t = 0$ an objective, individual probability measure.

This follows from the fact that $\mathbb{P}(x(0) = \kappa)$ is obtained from the pointer-*cum*-particle system's reduced density matrix (i.e. by partially tracing out the pointer's degrees of freedom). But this transition from the density matrix of a pure state to the reduced density matrix of an 'improper mixture' lacks objective-physical justification (see, e.g. Mittelstaedt 2004, Chapters 3 and 4). Contrast that with the situation of $\frac{\mathbb{P}(y \wedge x(\tau)=X_\tau)}{\mathbb{P}(x(\tau)=X_\tau)}$: this *is* well defined via post-selection. That is, due to the 'effective collapse' (see, e.g. Dürr and Teufel 2009, Chapter 9.2), induced by the strong measurement at $t = \tau$, the event $x(\tau) = X_\tau$ *can* be assigned a well-defined probability measure.

No *independent* reasons have been given so far for believing that (COR) is true. Consequently, counterexamples to (COR) abound – and are perfectly familiar: *any* non-standard variant of de Broglie–Bohm theory of the type of Equation (9.1) (i.e. with non-vanishing, divergence-free vector field $\boldsymbol{j}$). In them, the particle's trajectory generically crosses the weak screen at a point *distinct* from what the weak velocity measurements would make us believe. Fig. 9.2 above illustrates this.

The outcome cannot be considered as representing the actual position of the particle at time t. It is just unknown: it could have traversed *any* location within the support of the Gaussian wave function, centered around the weak value. Still, the operationally defined velocity (obtained from averaging) would not change: as long as the Born rule and the validity of the Schrödinger equation hold, its value remains the same. (In this sense, any guidance equation of the type of Equation (9.1), is compatible with Wiseman's operationally defined velocity.)

In the absence of an independent argument for the correspondence between weakly measured and actual positions (i.e. COR), it remains unclear what – if anything – Wiseman's operational velocity (9.4) signifies non-manifestly, i.e. ontologically.

The assumption C_t, necessary for the correspondence of weak and actual velocities, is, in fact, equivalent – in virtue solely of the quantum mechanical formalism and the supposition of deterministic differentiable particle trajectories – to standard de Broglie–Bohm theory. (Firstly, suppose that C_t is true. Then, the weak velocity measurement yields the actual particle velocities. Wiseman's operationally defined velocity (9.4) uniquely picks out a guidance equation—that of standard de Broglie–Bohm theory.[9]

[9] On a related note, Dürr, Goldstein, and Zanghí identify a characteristic feature of standard pilot-wave theory's velocity field as the 'crucial condition' for why, allegedly, Wiseman's measurement protocol constitutes 'genuine' measurements of Bohmian trajectories (Dürr *et al.* 2013, Chapter 7; see also Dürr *et al.* 2013, Chapter 3, and Goldstein 2017, Section 4 for similar claims). The assumption effectively amounts to the fact that in the standard guidance equation a particle's velocity field only depends on the particle's wave function whenever the particle-*cum*-pointer compound system has the form $\psi(x) \otimes \phi(y - x)$. It uniquely singles out standard de Broglie–Bohm theory. However, DGZ seem to misidentify that 'crucial condition'. Indeed, it plays no obvious role in the attempt to exploit

Conversely, suppose standard de Broglie–Bohm theory to be true. A weak velocity measurement then discloses the actual particle velocities. Thus, C_t holds.)

9.3.3 Non-empirical support for pilot-wave theory?

The main result of Wiseman's paper can be considered as a conditional claim: *if* one adopts his operationally defined velocity (and assumes deterministic, differential particle trajectories), it is uniquely determined as that of standard de Broglie–Bohm theory; on this reading, Wiseman remains neutral – whether it is plausibly satisfied (or not).

Wiseman's envisioned reasoning pertains to a non-empirical motivation for standard pilot-wave theory: Albeit not per se referring to individual particles, the statistically defined operational velocity provides a 'justification for [standard de Broglie–Bohm theory's] law of motion [i.e. the standard guidance equation]' (Wiseman 2007, p. 2). This in turn purports that the theory's standard form 'is preferred over all other on physical grounds' (Wiseman 2007, p. 12). That is, although other velocity fields generate the same (statistically empirically accessible) mean velocity, we ought to believe that the standard velocity field is true – rather than any of its alternatives.

Similarly, Dürr *et al.* turn on the allegedly natural character of his proposal to operationally define velocities via weak values: '(Standard de Broglie–Bohm theory) delivers thus the most natural explanation of the experiments described' (Dürr and Lazarovici 2018, p. 145, my translation).

But as I have illustrated in Section 9.3.2, weak positions do not reveal the actual particle position. What, then, is the proposed relationship between weak velocities and actual particle velocities supposed to be?

The disadvantages of naive realism about weak position values are demonstrated in the so-called Three-Box Paradox (Aharonov and Vaidman 1991; Aharanov and Rohrlich 2005, Chapter 16.5; Maroney 2017). Here, we encounter another illustrative case where the identification of the manifest with the non-manifest is more complex than one would hope.

Imagine a particle and three boxes labelled A,B, and C. Let the particle's initial state be

$$|\psi_{in}\rangle = \frac{1}{\sqrt{3}}(|A\rangle + |B\rangle + |C\rangle), \tag{9.22}$$

where $|A\rangle$ denotes the state in which the particle is in box A, and similarly, $|B\rangle$ and $|C\rangle$. For its final state, on which we post-select, choose

$$|\psi_{fin}\rangle = \frac{1}{\sqrt{3}}(|A\rangle + |B\rangle - |C\rangle). \tag{9.23}$$

Via the definition of weak values, one then obtains the resulting weak values for the projectors onto state $i \in A, B, C$, $\hat{P}_i := |i\rangle\langle i|$:

weak velocity measurements for the standard guidance equation, and nowhere it is invoked explicitly (Fankhauser and Dürr 2021).

$$\begin{aligned}\langle \hat{P}_A \rangle_w &= 1\\ \langle \hat{P}_B \rangle_w &= 1\\ \langle \hat{P}_C \rangle_w &= -1.\end{aligned} \tag{9.24}$$

If one were to believe that weak values invariably reveal the real positions of particles, one would have to conclude that box C contains -1 particles. However, in de Broglie–Bohm theory's ontology (in any of its variants), particles either occupy a position or do not occupy a position.

Consequently, Bohmians must be careful of interpreting weak values as real position values without qualification.

We are thus left with, at best, a considerably weaker position, one close to Bricmont (2016, p. 136): '[Weak velocity measurements via Wiseman's protocol] (are) not meant to 'prove' that the de-Broglie-Bohm theory is correct', because other theories will make the same predictions. Still, the result is nevertheless suggestive because the predictions made here by the de Broglie–Bohm theory are very natural within that theory [...].'

9.4 Declaring Bohmian trajectories as surreal: falsifying Bohmian predictions?

The detection procedure leading to what has (somewhat inappropriately) been coined '*surrealistic*' trajectories presents another particularly illustrative example where a naive conflation of the manifest and non-manifest leads to inconsistencies. (The term 'surrealistic' is misleading as there is nothing in the standard quantum theory that one would refer to as being *real*.) Experimental proposals to measure such trajectories were deemed to rule out pilot-wave theory and present a second attempt to detect, or rather falsify, Bohmian trajectories. We observe that the standard QM lacks a foundation for when a trajectory is supposed to be *measurable*. This, in fact, would block the paradox right from the outset because any claim that some 'path of a particle' was detected in an experiment is unwarranted.

There is no need to repeat the many treatments of this experiment in the literature. Instead, the point here is to present surrealistic trajectories as a case where a consistent underpinning of quantum theory leads to the fact that not all position measurements are genuine measurements of a particle's actual position. Thus, such experiments *cannot* falsify the existence of pilot-wave trajectories.

9.4.1 The surrealistic trajectories experiment

In the surreal trajectory experiment – an instructive variant of which I discuss here[10] – a Mach–Zehnder interferometer is supplemented with a grid of spin systems interacting with the particle. The interaction is designed such that when the particle traversing the interferometer hits, the corresponding spins flip from their ready state $|\uparrow\rangle$ to the state $|\downarrow\rangle$. This allows, supposedly, to track the particle's path. After the experiment

[10] Another simple account is discussed in Barrett (2000), and the original experiment was presented in Englert *et al.* (1992).

is performed, the spin flips are read off, indicating the trajectory the particle took through the experiment (see the set-up in Fig. 9.3).

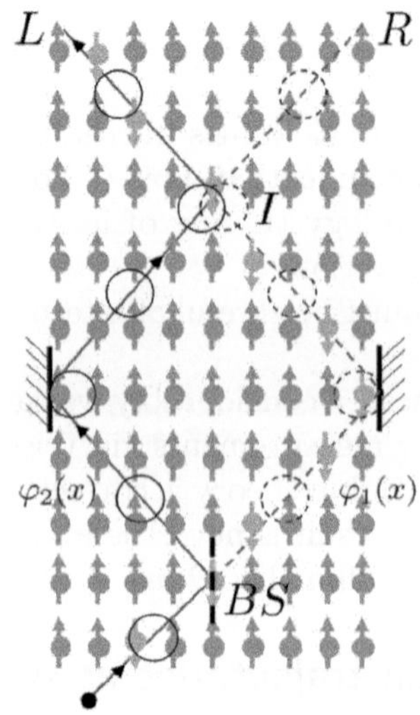

Fig. 9.3: A particle's trajectory is 'detected' by a grid of spins that flip when the particle travels by them. Although both theories produce the same outcomes for the spin flips, as opposed to standard QM, pilot-wave theory predicts the particle to traverse a path that is not indicated by the series of spin flips (the solid black L path to the left).

Assume that before the interaction the experiment consists of n unperturbed spins $|\uparrow\rangle^{\otimes n}$. Without loss of generality, we assign the spins in the path L to the first k slots of the total state and the spins in the alternative path R to the next k slots, respectively. Hence, the state is represented as

$$|\uparrow\rangle^{\otimes n} := |\uparrow_{L_1}\rangle\cdots|\uparrow_{L_k}\rangle|\uparrow_{R_1}\rangle\cdots|\uparrow_{R_k}\rangle|\uparrow\rangle^{\otimes(n-2k)}. \tag{9.25}$$

The quantum description of the experiment looks like so: when the particle travels through the experiment, the total initial state of particle and spin grid

$$\frac{1}{\sqrt{2}}(|\varphi_1(x)\rangle + |\varphi_2(x)\rangle)|\uparrow\rangle^{\otimes n} \tag{9.26}$$

evolves into the final post-interaction state

$$\frac{1}{\sqrt{2}}(|\varphi_1'(x)\rangle|\downarrow_{L_1}\rangle\cdots|\downarrow_{L_k}\rangle|\uparrow\rangle^{\otimes(n-k)} + |\varphi_2'(x)\rangle|\uparrow\rangle^{\otimes k}|\downarrow_{R_1}\rangle\cdots|\downarrow_{R_k}\rangle|\uparrow\rangle^{\otimes(n-2k)}), \tag{9.27}$$

where $|\varphi_1(x)\rangle, |\varphi_2(x)\rangle$ are the two disjoint wave packets of the particle emerging from the beam-splitter BS, and $|\varphi_1'(x)\rangle, |\varphi_2'(x)\rangle$ are the two disjoint wave packets of the particle at the end of the experiment (top region in Fig. 9.3).

The two branches of the wave function are reflected at two mirrors and subsequently overlap in the region I before they travel further on their path. In light of the standard quantum description, it is clear what is going on in terms of how the trace of spin flips arises. Each branch of the particle's wave function causes a series of adjacent spin particles to switch states like a game of dominoes, creating a path a single particle would be expected to take. Since the initial quantum state (Equation (9.26)) is a superposition of those two branches, each track occurs with equal probability. Admittedly, this looks daringly close to the detection of a particle's trajectory. Nothing mysterious so far.

But pilot-wave theory (committed to actual particle positions) tells a different story of where the particle went. To compute the Bohmian prediction, the total wave function of all involved particles is rewritten in position space, and spin degrees of freedom are used in spinor form. The trajectory is then calculated from Bohm's guiding equation.

The total quantum state of an individual spin particle on the grid then reads

$$\Phi(y)\begin{pmatrix}1\\0\end{pmatrix}, \tag{9.28}$$

for spin up, and

$$\Phi(y)\begin{pmatrix}0\\1\end{pmatrix}, \tag{9.29}$$

for spin down, respectively. Here, $\Phi(y)$ is a single spin particle's spatial wave function centred at the position of the particle. Note, importantly, that the wave function of the position degree of freedom is identical for both spin values.

The total final state, thus, is found to be

$$\begin{aligned}\frac{1}{\sqrt{2}}\Bigg(&|\varphi_1'(x)\rangle\Phi_{L_1}(y-y_{L_1})\begin{pmatrix}0\\1\end{pmatrix}\cdots\Phi_{L_k}(y-y_{L_k})\begin{pmatrix}0\\1\end{pmatrix}\\ &\otimes\Phi_{R_1}(y-y_{R_1})\begin{pmatrix}1\\0\end{pmatrix}\cdots\Phi_{R_k}(y-y_{R_k})\begin{pmatrix}1\\0\end{pmatrix}|\uparrow\rangle^{\otimes(n-2k)}\\ &+|\varphi_2'(x)\rangle\Phi_{L_1}(y-y_{L_1})\begin{pmatrix}1\\0\end{pmatrix}\cdots\Phi_{L_k}(y-y_{L_k})\begin{pmatrix}1\\0\end{pmatrix}\\ &\otimes\Phi_{R_1}(y-y_{R_1})\begin{pmatrix}0\\1\end{pmatrix}\cdots\Phi_{R_k}(y-y_{R_k})\begin{pmatrix}0\\1\end{pmatrix}|\uparrow\rangle^{\otimes(n-2k)}\Bigg)\end{aligned} \tag{9.30}$$

with y_{L_i}, y_{R_i} being the locations of the spin particles. Nothing mysterious so far here either. But what happens in the region I according to the Bohmian description?

Throughout the experiment, the spatial degrees of freedom of the spin particles remain untouched since only their spin state flips. In other words, the spin measurement is in this sense incomplete since no record was created in terms of manifest configurations that would displace the spin particles (or further entangle them to spatial degrees of freedom of an apparatus).

Therefore, without a manifest read-out, each particle's spatial wave function is the same irrespective of whether that particle's spin has flipped or not, such that they can be pulled out from the superposition of the final state, i.e.

$$\frac{1}{\sqrt{2}}\prod_{i=1}^{k}\Phi_{L_i}\Phi_{R_i}\left(|\varphi_1'(x)\rangle\begin{pmatrix}0\\1\end{pmatrix}^{\otimes k}|\uparrow\rangle^{\otimes(n-k)}+|\varphi_2'(x)\rangle|\uparrow\rangle^{\otimes k}\begin{pmatrix}0\\1\end{pmatrix}^{\otimes k}|\uparrow\rangle^{\otimes(n-2k)}\right). \tag{9.31}$$

This makes it possible for the two branches of the target particle's wave function to interfere coherently in region I. Plugging the state in I into the Bohmian guiding equation for spinor-valued wave functions reveals that the target particle's velocity has contributions from both branches of its wave function, i.e.

$$mv_x = \hbar\,\mathrm{Im}\,\frac{\Psi^*\nabla_x\Psi}{\Psi^*\Psi} \sim ((\varphi_1^I(x))^*\nabla\varphi_1^I(x) + (\varphi_2^I(x))^*\nabla\varphi_2^I(x)), \tag{9.32}$$

where Ψ is the total wave function of the system. The guiding potential of particle x is independent of all y degrees of freedom. The interaction is designed such that only the spin state evolves when a particle passes by, and the spin particles do not move upon that (otherwise the effect would not occur). If one of the spin particles were to move during the interaction, one of the two contributions would be suppressed. That is, although the two spin states (the left and right branches) of an individual spin particle are orthogonal, i.e. $\langle\Phi(y)|\Phi(y)\rangle\langle\uparrow\,|\,\downarrow\rangle = 0$, they may still overlap in configuration space. Hence, the overlapping spin states in configuration space will affect the target particle's motion as Equation (9.32) suggests.[11]

Equation (9.32) shows that the particle will move straight up when the two branches are made to overlap since the total velocity is the sum of the two contributions coming from the left and right moving wave packet (i.e. horizontal velocity component cancels). When the particle reaches the interference area I, the two wave packets can interfere before they continue to pass through each other and separate again. Moreover, this causes the particle to be reflected from point I and continue travelling on either the experiment's left or right half.

In the last step, a measurement of the final position of the particle is performed, and the spin states are recorded (e.g. with a series of Stern–Gerlach apparatuses). Note that if the spatial wave functions and positions y_i of the spin particles on the grid were to change, they could not be pulled out as an overall factor in Equation (9.32), and would therefore effectively decohere the target particle's wave function into its two branches.

9.4.2 What surrealistic trajectories signify

As a result, the standard quantum framework predicts a chain of spin flips to emerge that *does not* coincide with the particle trajectory as predicted by the Bohmian guiding equation. If we believe that the spin track in the experiment reliably indicates the particle's path, one could elicit a puzzle: the observed path (the flipped spins) is the one on the right, but according to de Broglie–Bohm theory, the particle was never there. That is, the Bohmian trajectory is 'surreal', and therefore the theory cannot be right.

[11] This was highlighted in Maroney (2004) by distinguishing orthogonality of quantum states from 'super-orthogonality', where the latter refers to the feature that two quantum wave functions do not overlap in configurations space.

We first observe that, of course, concerning the state of the spin, whether flipped or unflipped, the hypothetical emergence of a chain of flipped spins is not an actual measurement of anything yet. Although one may claim that some spin state must have changed upon interaction with the particle flying through the experiments, no configurational record exists in the manifest domain indicating a measurement result. This turned out to be crucial. The argument relies on the fact that none of the spin particle's configurations y_i change during the experiment. Otherwise, the guiding equation would not give rise to the described behaviour of the target particle. But this also means that during the experiment, the spin chain must not be observed in terms of manifest configurations. After the interference region I was passed, subsequent measurements may create records thereof.

The existence of the so-called surrealistic trajectories was supposed to be evidence for testable false predictions of pilot-wave theory (see Englert *et al.* 1992). The trajectories predicted, so the story goes, contradict what an actual measurement of the particle trajectory would yield. So, they were used to claim that the Bohmian theory gives wrong predictions for the *actual* position of a particle. If this were true, it would falsify the theory.

However, one firstly recalls that, the predictions of de Broglie–Bohm theory for the measurement outcomes of this thought experiment coincide with the ones of standard QM. Secondly, in the standard theory, no meaning is assigned to the trajectory of a particle so that it remains unclear what such a measurement would even have to do with the dynamics of a particle (cf. Dürr *et al.* 1993; Lazarovici 2020; Bricmont 2016; Dürr *et al.* 2013, Chapter 3). Classical particles do not exist in standard quantum theory, let alone have trajectories. As I alluded, 'surrealistic' trajectories result from an unjustified relationship between the manifest and non-manifest domains. The observed spin track is naively identified with a particle's actual trajectory. Since this claim would go beyond the postulates of standard QM, it is unclear whether that assumption can be consistently made. Furthermore, a theory where a claim about the existence of actual particle trajectories *can* consistently be made is pilot-wave theory. But as the example shows, the trajectory cannot coincide with the one indicated by the chain of spin flips. I therefore conclude that if particle trajectories are assumed to exist in quantum theory, not all 'position' measurements can be interpreted as measurements of a particle's actual position.

There are other analogous instances of 'surreal' quantum observables. For example, as Brown *et al.* (1995) indicate, other properties besides spin, such as mass, charge, and magnetic moment are all inconsistent with what they call the thesis of 'localized particle properties'.

9.5 Conclusions

In the first part of this chapter, weak velocity measurements of Bohmian particles are analyzed. I have explained the status of weak velocity measurements in two ways. On the one hand, they are an interesting application of standard QM in a novel experimental regime (viz., that of weak pointer-system couplings). They allow us to empirically probe the gradient of the system's wave function – irrespective of any particular interpretation of the quantum formalism. On the other hand, however, with respect to the significance of weak velocity measurements, weak velocity measurements shed no light on the status of standard de Broglie–Bohm theory. In particular, on their own, weak velocity measurements do not provide any convincing support – empirical or non-empirical – for standard de Broglie–Bohm theory over any of its alternative versions. The upshot of the analysis by which the puzzle resolves is that one should not naively identify weak positions with actual Bohmian positions. As a result, this highlights the importance of defining the manifest variables of a theory, i.e. the experimentally accessible records, from the non-manifest variables, i.e. ontological variables – the Bohmian positions. The Three-Box Paradox demonstrated the dangers of any naive realism about weak position values.

The discussion continued with the allegedly paradoxical claims commonly made on the so-called 'surrealistic' trajectories in pilot-wave theory, where detectors are purportedly misled. This example served as an another instructive case for inconsistent reasoning on what quantum outcomes are supposed to signify. The analysis showed that a detection of 'surrealistic' trajectories provides no empirical argument against the Bohmian predictions for particle trajectories. Therefore, both in the same vein, neither surrealistic trajectories nor weak velocity measurements qualify as falsification or confirmation in favour of the velocity field, postulated by standard de Broglie–Bohm theory in any robust sense.

While certain advocates of pilot-wave theory were somewhat assertive in claiming that weak measurements substantiated the theory or specific variants of it, similarly, some opponents of the theory have been too assertive. They contend that alternative indirect methods aimed at empirically inferring information about particle trajectories contradict the theory. The fundamental issue lies in the absence of a particle trajectory concept within the standard quantum framework, and the central insight is that quantum measurements have no straightforward bearing on variables in the non-manifest domain.

References

Aharanov, Y. and Rohrlich, D. (2005). *Quantum Paradoxes, Quantum Theory for the Perplexed.* Weinheim: Wiley-VCH Verlag.

Aharonov, Y., Albert, D.Z. and Vaidman, L. (1988). How the result of a measurement of a component of the spin of a spin-1/2 particle can turn out to be 100. *Physical Review Letters*, **60**(14): 1351–1354.

Aharonov, Y. and Vaidman, L. (1991). Complete description of a quantum system at a given time. *Journal of Physics A: Mathematical and General*, **24**(10): 2315–2328.

Allori, V. (2013). Primitive ontology and the structure of fundamental physical theories. In Ney, A. and Albert, D.Z. (eds.) *The Wave Function: Essays on the Metaphysics of Quantum Mechanics*, pp. 58–75. New York: Oxford University Press.

Arageorgis, A. and Earman, J. (2017). Bohmian Mechanics: A Panacea for What Ails Quantum Mechanics, or a Different and Problematic Theory? http://philsci-archive.pitt.edu/12993/.

Bacciagaluppi, G. (2005). A Conceptual Introduction to Nelson's Mechanics. In Buccheri, R., Saniga, M. and Elitzur, A. (eds) *Endophysics, Time, Quantum and the Subjective*, pp. 367–388. Singapore: World Scientific Publishing.

Barrett, J.A. (2000). The persistence of memory: Surreal trajectories in Bohm's theory. *Philosophy of Science*, **67**(4): 680–703.

Bogen, J. (2017). Theory and observation in science. In Zalta, E.N. (ed.) *The Stanford Encyclopedia of Philosophy.*

Bohm, D. (1952). A suggested interpretation of the quantum theory in terms of "hidden" variables. I & II. *Physical Review*, **85**(2): 166–193.

Bohm, D. and Hiley, B. J. (1993). *The Undivided Universe: An Ontological Interpretation of Quantum Theory.* London: Routledge.

Bricmont, J. (2016). *Making Sense of Quantum Mechanics.* Berlin: Springer.

Brown, H., Dewdney, C. and Horton, G. (1995). Bohm particles and their detection in the light of neutron interferometry. *Foundations of Physics*, **25**(2): 329–347.

Chakravartty, A. (2017). Scientific realism. In Zalta, E.N. (ed.) *The Stanford Encyclopedia of Philosophy.*

Deotto, E. and Ghirardi, G. (1998). Bohmian Mechanics revisited. *Foundations of Physics*, **28**(1): 1–30.

Dressel, J., Malik, M., Miatto, F.M., Jordan, A.N. and Boyd, R.W. (2014). Colloquium: Understanding quantum weak values: Basics and applications. *Review of Modern Physics*, **86**: 307–316.

Dürr, D., Fusseder, W., Goldstein, S., Zanghì, N. (1993). Comment on "Surrealistic Bohm Trajectories". *Zeitschrift für Naturforschung A*, **48**(12): 1261–1262.

Dürr, D. and Lazarovici, D. (2018). *Verständliche Quantenmechanik: Drei mögliche Weltbilder der Quantenphysik.* Heidelberg: Springer.

Dürr, D. and Teufel, S. (2009). *Bohmian Mechanics: The Physics and Mathematics of Quantum Theory.* Berlin: Springer.

Dürr, D., Goldstein, S. and Zanghì, N. (2013). *Quantum Physics without Quantum Philosophy.* Berlin: Springer.

Einstein, A. (1949). Reply to criticisms. In Schilpp, P.A. (ed.) *Albert Einstein: Philosopher-Scientist.* Evanston: Library of Living Philosophers.

Einstein, A., Podolsky, B. and Rosen, N. (1935). Can quantum-mechanical description of physical reality be considered complete? *Physical Review*, **47**(10): 777.

Englert, B.-G., Scully, M.O., Süssmann, G. and Walther, H. (1992). Surrealistic Bohm trajectories. *Zeitschrift für Naturforschung A*, **47**(12): 1175–1186.

Fankhauser, J. (2023). *Observability and Predictability and Quantum and Post-Quantum Physics.* PhD thesis, University of Oxford.

Fankhauser, J. and Dürr, P.M. (2021). How (not) to understand weak measurements of velocities. *Studies in History and Philosophy of Science Part A*, **85**: 16–29.

Flack, R. and Hiley, B.J. (2014). Weak measurement and its experimental realisation. In *Journal of Physics Conference Series*, volume **504** of *Journal of Physics Conference Series*.

Flack, R. and Hiley, B.J. (2016). *Weak values of momentum of the electromagnetic field: average momentum flow lines, not photon trajectories.* https://arxiv.org/abs/1611.06510.

Goldstein, S. (1987). Stochastic mechanics and quantum theory. *Journal of Statistical Physics*, **47**(5-6): 645–667.

Goldstein, S. (2017). Bohmian mechanics. In Zalta, E.N. (ed.) *The Stanford Encyclopedia of Philosophy.*

Heisenberg, W. (1958). *Physics and Philosophy: The Revolution in Modern Science.* New York: Harper.

Holland, P.R. (1993). *The Quantum Theory of Motion: an Account of the de Broglie–Bohm Causal Interpretation of Quantum Mechanics.* Cambridge: Cambridge University Press.

Kocsis, S., Braverman, B., Ravets, S., Stevens, M.J., Mirin, R.P., Shalm, L.K and Steinberg, A.M. (2011). Observing the average trajectories of single photons in a two-slit interferometer. *Science*, **332**(6034): 1170–1173.

Lazarovici, D. (2020). Position Measurements and the Empirical Status of Particles in Bohmian Mechanics. *Philosophy of Science*, **87**(3): 409–424.

Mahler, D.H., Rozema, L., Fisher, K., Vermeyden, L., Resch, K.J., Wiseman, H.M. and Steinberg, A. (2016). Experimental nonlocal and surreal Bohmian trajectories. *Science Advances*, **2**(2): e1501466.

Maroney, O. (2017). Measurements, disturbances and the quantum three box paradox. *Studies in History and Philosophy of Science Part B: Studies in History and Philosophy of Modern Physics*, **58**: 41–53.

Maroney, O. (2004). *Information and Entropy in Quantum Theory.* https://arxiv.org/abs/quant-ph/0411172.

Mittelstaedt, P. (2004). *The Interpretation of Quantum Mechanics and the Measurement Process.* Cambridge: Cambridge University Press.

Myrvold, W.C. (2003). On some early objections to Bohm's theory. *International Studies in the Philosophy of Science*, **17**(1): 7–24.

Oianguren-Asua, X., Solé, A., Destefani, C. F., and Oriols, X. (2025). How weak values illuminate the role of "hidden"-variables as predictive tools. In Oldofredi, A. (ed.) *Guiding Waves in Quantum Mechanics: One Hundred Years of de Broglie–Bohm Pilot-Wave Theory* Oxford: Oxford University Press.

Svensson, B.E. (2013). Pedagogical review of quantum measurement theory with an emphasis on weak measurements. *Quanta*, **2**(1): 18–49.

Tamir, B. and Cohen, E. (2013). Introduction to Weak Measurements and Weak Values. *Quanta*, **2**(1): 7–17.

Valentini, A. (1991). Signal-locality, uncertainty, and the subquantum H-theorem. II. *Physics Letters A*, **158**(1): 1–8.

Valentini, A. (1992). *On the pilot-wave theory of classical, quantum and subquantum physics*. PhD thesis.

Wiseman, H.M. (2007). Grounding Bohmian mechanics in weak values and bayesianism. *New Journal of Physics*, **9**(6): 165.

10
The direction of time in Bohmian mechanics

Cristian Lopez and Michael Esfeld

10.1 Introduction

The problem of the direction of time in physics and philosophy has many facets. Some have considered it as evidence of a conflict between a temporally symmetric microphysics and a temporally asymmetric macrophysics (Reichenbach 1956, Price 1996, Callender 1997). Others have rather formulated the problem in terms of whether or not the space-time structure comes equipped with a temporal orientation (Earman 1974, Castagnino and Lombardi 2009). From a metaphysical perspective, the problem could more generally be seen as whether the direction of time is a primitive element in our ontology or not. In this line, there are then two possible views – *primitivism* and *reductionism.* Primitivists believe that the directionality of time is a necessary posit in one's ontology in virtue of its explanatory advantages. It is therefore an irreducible, fundamental feature of the natural world that explains many temporally asymmetric phenomena. Reductionists rather believe that the seeming directionality of time requires an explanation in terms of some non-temporal physical asymmetry to which it can be reduced. In this way, the direction of time is the explanandum, while the non-temporal physical asymmetry is the explanans. Reductionism can be seen as either conservative or eliminativist. According to the former, the direction of time is non-fundamental, but real; according to the latter, the direction of time is unreal. The metaphysical map of the different philosophical attitudes towards the direction of time then looks as follows: primitivism holds that the direction of time is real and fundamental; conservative reductionism maintains that it is real, though reducible to a non-temporal physical basis; eliminativist reductionism claims that it is unreal *tout court.*

In this chapter, we investigate the problem of the direction of time in its metaphysical formulation in Bohmian mechanics (Dürr, Goldstein, and Zanghì 2013; Dürr and Teufel 2009). We show that the theory can accommodate both conservative reductionism and primitivism on the direction of time. Bohmian mechanics can accommodate *conservative reductionism* because it straightforwardly shows how temporal asymmetries (as the counterfactual asymmetry between forward and backward evolutions of the configurations of matter) ultimately reduce to an epistemic asymmetry in our knowledge of the initial particle positions. Bohmian mechanics can also accommodate *primitivism* about the direction of time if Leibnizian relationalism is adopted.[1] According to it, the direction of time is the direction of change of the particle configuration. Change in a Bohmian relational ontology, we argue, is considered as primitively directed in order to distinguish between variation within the particle configuration and change of the particle configuration. In conclusion, Bohmian mechanics does not settle the problem of the direction of time, but as it can accommodates alternative metaphysical views on the direction of time, it can also be employed to provide them physical support.

10.2 Bohmian mechanics and the problem of the direction of time

In its standard formulation, quantum mechanics is just a recipe for predictions (Maudlin 2019). In order for it to be a full-fledged scientific theory, one that yields substantive knowledge about what the world is like, it needs to be properly interpreted and to overcome the famous measurement problem (Albert 1992, Maudlin 1995). This problem basically states that the following three assumptions are mutually incompatible:

1. The quantum state (the wave function) is a complete description of the system.
2. The quantum state always evolves in accord with a linear dynamical equation (such as the Schrödinger equation)
3. Measurements usually have determinate outcomes.

Any so-called interpretation of quantum mechanics is an attempt to overcome the measurement problem by abandoning (or at least modifying) one or more of these assumptions. The so-called additional-variable theories (also known under the misleading label 'hidden-variable theories') are a family of quantum theories that aim to overcome the measurement problem by abandoning assumption (1): the quantum state does not yield a complete description of the system, because there are additional variables; the information about them is not encoded in the quantum state. As Maudlin (1995) showed, the only viable option for such a theory to solve the measurement problem is to admit position as additional variable.

The de Broglie–Bohm–Bell theory shows how to do this. Originally formulated by Louis de Broglie (1928), it was then modified by David Bohm in 1952 as a second-order theory that explains the motion of the particles in terms of the action of a quantum potential generated by the wave function, interpreted as a real field (Bohm 1952). The theory was then simplified and cast as a first-order theory without the commitment to a specifically quantum force or potential by John Bell in the 1960s (see the papers in Bell (1987), in particular Chapters. 4, 7, and 17) and set out in a clear and concise way by Detlef Dürr, Sheldon Goldstein, and Nino Zanghì in the 1990s (see the papers collected in Dürr, Goldstein, and Zanghì 2013). In this theory, called Bohmian mechanics in its contemporary formulation, the motion of the particles is given by a first-order equation that yields the velocity of the particles by means of the

[1] It is worth clarifying that Bohmian mechanics can accommodate relationalism *with respect to time*, which does not mean that Bohmian mechanics can be fully formulated as a relational theory. To the best of our knowledge, it is not clear if Bohmian mechanics can be given a full-fledged relational reformulation.

wave function as output given the particle positions as in input. Hence, the complete state of a quantum system is specified by both the wave function (i.e. the quantum state) and the particle positions:

$$(\psi(t), Q(t)) \tag{10.1}$$

where $\psi(t)$ is the wave function of the system, $\psi(t) = (x_1, ..., x_n, t)$, and $Q(t)$ represents the actual positions of the composing particles, $Q(t) = (Q_1(t), ..., Q_N(t))$. Since particles always have determined positions and they vary in time, particles naturally have trajectories, too. This is how Bohmian mechanics modifies assumption (10.1).

As for assumption (10.2), Bohmian mechanics introduces an additional dynamical equation—the guidance equation. While the Schrödinger equation describes the behaviour of the wave function (ψ) and the wave function always evolves according to the Schrödinger equation, the guidance equation describes the behaviour of the particles (basically, the change of their relative positions in time). It does so by providing the velocity of the particles at any time as a function of the wave function at that time, its spatial derivative, and the position of all particles at that time. According to Bohmian mechanics, the dynamics of quantum systems is then twofold:

$$\Big(- \frac{\hbar^2}{2m} + V(q,t) \Big) \psi(q) = i\hbar \frac{\partial \psi(q)}{\partial t}, \tag{10.2}$$

$$\frac{dQ_k(t)}{dt} = \frac{\hbar}{m_k} Im \frac{\psi^* \partial k \psi}{\psi^* \psi}(Q_1, ..., Q_N). \tag{10.3}$$

To put it simply, Bohmian mechanics is a non-relativistic quantum theory about the motion of particles, whose behaviour is described by a twofold dynamics in terms of the Schrödinger equation and the guidance equation. Since both equations are deterministic, any randomness is epistemic, namely due to our ignorance of the exact particle positions. Finally, Bohmian mechanics is a non-collapse quantum theory, in the sense that quantum systems do not collapse either by measurements or spontaneously. Bohmian mechanics usually introduces a distinction between the "universal wave function" and "effective wave functions" (or "conditional wave functions"). While the former never collapses, the second may be said to collapse as "a pragmatic affair" (Goldstein 2021). In essence, effective wave functions are special wave functions that are decoupled from the universal wave function, which is more fundamental, involving features of the actual configuration of the environment.

It is clear that the primitive ontology of Bohmian mechanics is one of particles moving on trajectories. Less clear is the ontological status of the wave function, $\psi(q,t)$. Some have argued that it is a field in configuration space (Bohm 1952, Valentini 1997) or a multi-field in physical space (Hubert and Romano 2018, Romano 2021). Others have claimed that the wave function is nomological (or quasi-nomological) in the sense that it mainly plays the role of accounting for the motion of the particles

(Dürr, Goldstein, and Zanghì 2013, Chapter 12). While realists about laws of nature would hence consider the wave function as a law-like entity ('governing' the motion of particles), deflationists would consider it as part of the representational tools to describe quantum phenomena (Bell 1987, Chapter 7; Esfeld 2014). Be that as it may, the ontological status of the wave function in Bohmian mechanics is not crucial for the purpose of this chapter, namely the discussion of the direction of time.

In asking whether the direction of time is primitive or derivative in Bohmian mechanics, we wonder whether the direction of time is an irreducible element in its ontology, or whether it is a derivative element in the theory that can be obtained from a directionless basis. The same question arises in standard quantum mechanics as well. If the focus is on the unitary evolution of quantum systems or on interpretations that do not introduce a non-unitary dynamics, the direction of time must be shown to be a derivative element of the theory that is reducible to more basic, physical processes. For instance, in some formulations of the many-world interpretation, the quantum dynamics is exclusively given by the Schrödinger equation, which is time-reversal invariant. The manifest temporal asymmetry then "emerges" from other non-fundamental processes, as the branching structure or decoherence (Bacciagaluppi 2007). In this sense, the direction of time is real, but reduces to a non-fundamental physical process. The fundamental basis therefore lacks a direction of time.

In collapse quantum theories, in turn, the non-unitary, stochastic, and non-linear nature of collapses could well serve as a basis for the direction of time—any temporal asymmetry can be explained in terms of a non-time-reversal invariant dynamics. In spontaneous collapse models, quantum systems undergo spontaneous collapses that are intrinsically irreversible, yielding a direction of time (see Arntzenius 1997, Callender 2000, and North 2011 among others; see Lopez 2022b for some caveats). The same goes for measurement-induced collapses (see Penrose 1989, Arntzenius 1997, Healey 2002, Ellis 2013 among others; see Lopez 2022a for some caveats). Therefore, it could be argued that such theories require to posit a primitive direction of time that provides the right background structure to support non-time-reversal invariant laws (see Horwich 1987).

To come back to the original question, can Bohmian mechanics be interpreted in such a way to accommodate a primitive direction of time? Can it also be alternatively interpreted to accommodate a derivative direction of time? In what follows, we argue that Bohmian mechanics can actually accommodate both philosophical positions. In the next section, we show how Bohmian mechanics can accommodate conservative reductionism on the direction of time in terms of an epistemic asymmetry between initial and final conditions. In Section 10.4, we show how Bohmian mechanics may require a primitive direction of time to account for change in the configuration of matter in a Leibnizian relational framework.

10.3 Bohmian conservative reductionism on the direction of time

Both the Schrödinger equation and the guidance equation are assumed to be time-reversal invariant. Strictly speaking, the guidance equation is time-reversal invariant if the Schrödinger equation is time-reversal invariant. According to conventional wisdom, the Schrödinger equation is time-reversal invariant in its basic expressions (e.g., a free particle model) if the time-reversal transformation is implemented by an anti-unitary operator that transforms the time coordinate, $T : t \longrightarrow -t$, and takes the complex conjugate over the wave function, $T : \psi \longrightarrow \psi^*$ (see Earman 2002, Roberts 2017, Lopez 2021; see Callender 2000, and Lopez 2019 for an alternative view). By modus ponens, it follows that the guidance equation must also be time-reversal invariant. If the Bohmian dynamics is time-reversal invariant, then there are good reasons to suppose that the fundamental evolution of Bohmian particles is temporally directionless. Therefore, any Bohmian explanation of the direction of time should come from non-dynamical elements of the theory.

In standard quantum mechanics plus the collapse postulate, measurement-induced collapses can straightforwardly justify any manifest asymmetry between counterfactual backward and forward evolutions in simple experiments. For instance, suppose a Mach–Zehnder experiment with two sources of electrons (S1 and S2) and two electron detectors after each source (D1 and D2) are at work. When an electron is fired out by S1, the electron detector D1 will emit a flash of light. The same goes for S2 and D2. In the middle of the experimental arrangement, there is a half-silvered mirror. Since D1 and D2 let each electron to pass through, when an electron reaches the half-silvered mirror, it has 0.5 probability of bouncing off and 0.5 probability of passing through it. At the other extreme of the experimental arrangement, there are two electron detectors (F1 and F2), which emit a flash of light when an electron has either been reflected by the half-silvered mirror or when passed through it (see Fig. 10.1).

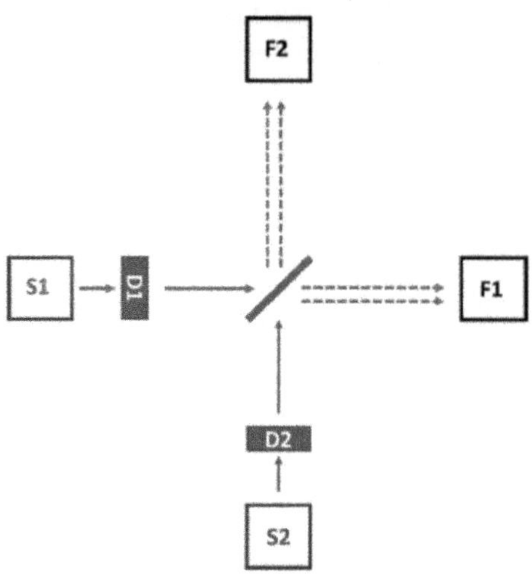

Fig. 10.1: A Mach–Zehnder experiment (adapted from Arntzenius 1997).

This simple experiment allows us to draw some empirically based conclusions about temporal asymmetries. For instance, it is possible to formulate some FORWARD questions and claims such as "what is the probability that F1 detects given that D2 detected?". Questions like this consider certain conditional probabilities of some *later* state given an *earlier* state. Analogously, it is also possible to formulate BACKWARD questions and claims such as "what is the probability that D2 detects given that F1 detected?". Quantum mechanical empirical evidence states that the reply to both questions is asymmetric, in the sense that it will assign different probabilities depending on whether FORWARD or BACKWARD questions are formulated: in the first case the probability is 0.5, while it is 1 in the latter (see Penrose 1989, Arntzenius 1997). This is the FORWARD–BACKWARD asymmetry.

In collapse quantum theories, the temporal asymmetry between FORWARD and BACKWARD can be explained because a quantum system collapses but not de-collapses. That is, in a Mach–Zehnder experiment of this kind, a quantum system undergoes a collapse when measured by F1 (or F2), but it then does not de-collapse when travelling backward (see Penrose 1989; for criticisms see Lopez 2022a). Since Bohmian mechanics is an empirically equivalent theory to standard quantum mechanics plus the collapse postulate (insofar as the Born rule applies), it is to be expected that it makes true the same counterfactuals, recovering the predictions and retrodictions of standard quantum mechanics. That is, it must recover the FORWARD–BACKWARD asymmetry. As mentioned earlier, as Bohmian mechanics dispenses with collapses, such an explanation is not considered. Bohmian mechanics should then be able not only to give the same answers to FORWARD and BACKWARD, but also to explain its asymmetry without relying on collapses.

We submit that the most adequate Bohmian explanation for the FORWARD–BACKWARD asymmetry relies on an asymmetry of our knowledge between initial and final states, in particular our *lack* of knowledge of the precise positions of the particles in their initial states. Therefore, the Bohmian explanation of the FORWARD–BACKWARD asymmetry, which could count as an indication of the temporal asymmetry, ultimately reduces to our epistemic constraint to know about the exact positions of Bohmian particles. Let us consider each case in tandem. As for FORWARD, the physical state of an electron fired by S1 is given by its wave function, $\psi(q_{S1}, t_0)$, *and* the position of the electron in S1. The wave function is non-zero in the region covering the vicinity of the source S1 and can be written as a superposition of "detected by F1" ("$F1$") and "detected by F2" ("$F2$")

$$\psi(q_{S1}, t_0) = \sqrt{\frac{1}{2}}(F1 + F2). \tag{10.4}$$

The exact position of the electron at the source (q_i) completes its initial state. According to Bohmian mechanics, the final state of the system (whether it is detected by F1 or F2) deterministically depends on the exact initial positions. It follows that if the exact position of the system were known, then its final state would also be known. Yet, as a matter of fact, the exact positions are unknown, given the quantum equilibrium distribution. For simplicity, it is possible to imagine that the exact position of the electron could be either on the right side (q_R) or on the left side (q_L) of the region within the which the wave function is non-zero. Since initial positions determine future trajectories, it is possible to claim that if the electron starts off on the left side (q_L), it will be reflected by the half-silvered mirror. If it rather starts off on the right side (q_R), it will get through the half-silvered mirror. This only emphasizes that there is a matter of fact based on the initial position of the electron as to whether it gets reflected or goes through when it reaches the half-silvered mirror. Of course, since this information is inaccessible, the only information is that given by the initial wave function of the electron and the frequency with which the electron begins in either region, which is equal to $\int_q |\psi(q_R, t_0)|^2$ and $\int_q |\psi(q_L, t_0)|^2$, where q_R and (q_L) are the regions within q_{S1} with the initial state of the particle being at t_0. For $\int_q |\psi(q_{S1}, t_0)|^2$, the probability must be 1.

In a single run of the experiment, this is what would happen according to Bohmian mechanics. The electron is fired by S1 at t_0. Next, it is detected by the first electron detector, D1. When it reaches the half-silvered mirror, it can either be reflected or pass through it. Bohmian mechanics correctly predicts that the probability of being reflected or passing through is 0.5 each. When the electron reaches the half-silvered mirror, its wave function splits into two packets corresponding to regions in which it is non-zero: one branch is reflected and the other one passes through. However, only one of these branches (either $F1$or $F2$) will "contain" the actual particle, giving the trajectory that the electron follows. When the final state of the system is known, it is straightforward to infer the initial position of the electron: if F2 detects, then the electron was reflected by the half-silvered mirror, and then it was localized in the region q_L at the source. If the electron had been located in the region q_R, then it would have been detected by F1.

From this simple example, the following temporal sequences can be extracted:

Future-Headed Sequence: $S1\ e(q_R) \rightarrow D1 \rightarrow M \rightarrow F2$ or $S1\ e(q_L) \rightarrow D1 \rightarrow M \rightarrow F1.$ (10.5)

And the usual FORWARD transitions probabilities. For instance,

FORWARD: 'The probability that F2 detects given that D1 detected is 0.5'. (10.6)

Therefore, it is easy to see that Bohmian mechanics can recover standard quantum mechanics predications for FORWARD. But instead of relying on measurement-induced collapses, the story depends on the agent's ignorance about the initial position of the electron. If one knew the initial conditions, say the electron's initial position is $e(q_L)$, then the right sequence would be $S1\ e(q_R) \rightarrow D1 \rightarrow M \rightarrow F2$.

This explanation allows us to see in which way the FORWARD–BACKWARD asymmetry emerges. If the trajectory of the electron through the Mach–Zehnder is exclusively determined by its initial state and the dynamics remains deterministic all along, then the only possible explanation for such an asymmetry is to depend on our knowledge of the initial conditions for FORWARD and BACKWARD. Setting aside any caveats concerning whether time is being reversed adequately, in the FORWARD case the probability that F2 detects given that D1 detects is 0.5, and the probability that D1 detects given that S1 fired out an electron is 1. But in the BACKWARD case, the probability that D1 detects is no longer 1, but 0.5. Furthermore, in FORWARD case the wave function was located in the region of the source S1, $\psi(q_{S1}, t_0)$ and in a superposition of $\sqrt{\frac{1}{2}}(F1 + F2)$. In BACKWARD case, the wave function is now localized in $\psi(q_{F2})$ and in a superposition of $\sqrt{\frac{1}{2}}(D1 + D2)$. More importantly, the electron's position is now within the region of F2, either on the right side (q_R) or on the left side (q_L) of F2. This yields the following temporal sequences for BACKWARD:

Past-Headed Sequence: $F2\ e(q_R) \rightarrow M \rightarrow D1$ or $F2\ e(q_L) \rightarrow M \rightarrow D2.$ (10.7)

BACKWARD: 'The probability that D1 detects given that F2 detected is 0.5'. (10.8)

What is then the source of the asymmetry between FORWARD and BACKWARD conforming with Bohmian mechanics? It is our epistemic constraint to know the exact positions of the particles either in FORWARD or in BACKWARD. To emphasize, if Bohmian mechanics is a quantum theory about particles and their trajectories, and the history of each particle is invariably and completely determined by the twofold Bohmian dynamics and the particles' initial state, then the FORWARD–BACKWARD asymmetry can only emerge from either of them. As mentioned earlier, the dynamics is deterministic and fully time-reversal invariant. So, the only option available is an asymmetry in our knowledge of the initial state. To put it differently, the FORWARD–BACKWARD asymmetry emerges from non-dynamical elements of the theory, and it just reflects our lack of knowledge of the exact positions of the particles in the initial state. This is how Bohmian mechanics can explain many temporal asymmetries as well as recovers the standard quantum mechanical phenomenology. In other words, the FORWARD–BACKWARD asymmetry is part of the quantum-mechanical phenomenology. The subject of disagreement is *how to account for it.* While standard quantum mechanics relies on the collapse postulate (an in-built time-asymmetric process) to explain and predict the asymmetry, Bohmian mechanics relies on our epistemic constraint in exactly knowing the initial state of the system. In this way, it can both explain and predict the FORWARD–BACKWARD asymmetry.

The explanation of the FORWARD–BACKWARD asymmetry in terms of an asymmetry in our knowledge of the initial conditions has also been put forward by Frank Arntzenius (1997) for alternative additional-variable theories. He argues that in the de Broglie–Bohm pilot-wave theory the *constraint* on the initial conditions is what implies the FORWARD–BACKWARD asymmetry. The constraint consists in assuming that the initial wave function is located approximately near to the regions S1 and

D1, and in a superposition of being detected by F1 and F2. In BACKWARD, it is assumed that the initial (final) condition is to be *different*, starting with a wave function approximately near to F2 and in a superposition of D1 and D2. According to him, this assumption explains the usual frequencies that are obtained experimentally. Arntzenius puts it this way:

> This is an assumption that we always make in Bohmian theories, and rarely even notice. Of course, it is a good thing that we do this, for if we did the opposite, we would have an empirically inadequate theory, for it would predict non-invariant forward transitions frequencies and invariant backward one, and that is just wrong (Arntzenius 1997, p. 219).

In the same vein, Callender writes:

> According to some interpretations, such as Bohm's, [FORWARD] is merely derived from the laws and the initial conditions. (...) The guidance equation is TRI [time-reversal invariant] if the Schrödinger equation is. Assuming for the moment the conventional wisdom that the Schrödinger evolution is time symmetric, this implies that the fundamental laws in a Bohmian universe are TRI. The observed asymmetry [i.e., FORWARD–BACKWARD asymmetry] is due to asymmetric initial conditions (Callender 2000, p. 260).

All this implies that in so-called hidden-variable theories whose dynamics is deterministic (Bohmian mechanics among them), there is no reason to suppose that the direction of time is primitive; it is rather derivative. An asymmetry in our knowledge of the initial conditions explains why the quantum mechanical phenomenology looks temporally asymmetric, but it is insufficient to ground a primitive, fundamental direction of time. This does not imply, however, that in Bohmian mechanics the direction of time is unreal, or that it is a timeless theory. The direction of time might still be real, but it is clearly not primitive: it can be explained in terms of non-dynamical aspects of the theory. The fundamental dynamics is indeed undirected.

In this sense we argue that Bohmian mechanics can accommodate conservative reductionism about the direction of time. When the focus is exclusively on the dynamics of the theory, it always produces a space of solutions that can be partitioned in two disjoint sets: forward-in-time and backward-in-time solutions, $W = W^f + W^b$ (where W is the whole space of solutions, W^f contains the forward-in-time solutions, and W^b the backward-in-time ones). Since a time-reversal transformation preserves the space of solutions, it is a structure-preserving mapping that goes from one to the other. At this level, no directedness of time is exhibited in the theory. However, when the initial conditions and our lack of knowledge of them are taken into account (which are essential for the theory to work as intended), the asymmetry immediately appears. Each specific solution relies on initial conditions with precise particle positions. Since our knowledge of these positions is limited, it seems that FORWARD and BACKWARD are asymmetric. To put it differently, in each Bohmian possible world, any agent will account for the same FORWARD–BACKWARD asymmetry because they will have limited knowledge of the exact position of the particles in their world.

10.4 Bohmian primitivism on the direction of time

Primitivism about the direction of time, as mentioned earlier, is the thesis that the direction of time does not call for an explanation because it is a primitive in the ontology. It is a posit required to explain phenomena and features of the theory. For instance, the existence of non-time-reversal invariant laws requires an underlying direction of time as the structural background that can provide support for such laws. One might think that Bohmian mechanics is unable to accommodate a primitive direction of time because its dynamics is fully time-reversal invariant. Nonetheless, we submit that primitivism can be accommodated in the theory under some metaphysical assumptions.

Bohmian mechanics, as discussed before, is a theory of particles and their trajectories. At any temporal instant, the basic ontology of Bohmian mechanics is that of a configuration of Bohmian particles spatially separated from one another. Thus, the ontology is one of matter that is spatially arranged (see Fig. 10.2). Differences in the relative spatial relations between particles give *variation* to the configuration. In Fig. 10.2, for instance, the distances between particles are different, which renders an asymmetry in the configuration of matter: some particles concentrate in some regions, while in others they are farther apart. Spatial relations are world-making relations in the sense that make particles hold together in a configuration that constitutes a possible physical world. Yet, each configuration is changeless, as a snapshot of photography—if two different configurations (that is, two configurations in which the relative distances among particles vary) are shown, it is impossible to say if they are two possible configurations or the same configuration that changed.

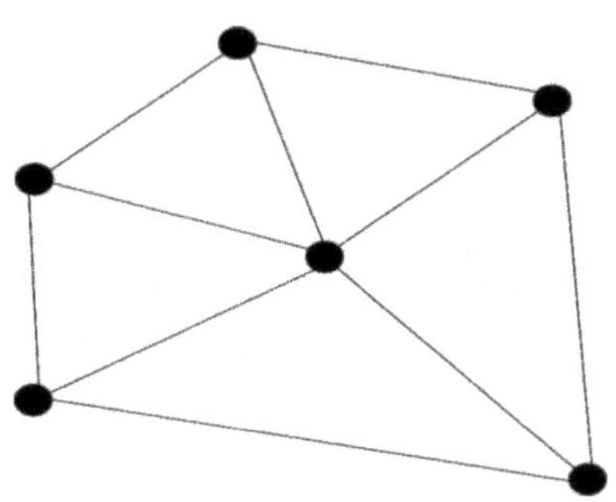

Fig. 10.2: Variation in the configuration of matter.

This figure is nonetheless unsatisfactory for physics. In so far as empirically adequate physical theories are considered, physics is about matter *in motion*. This means that the configuration of matter (i.e., Bohmian particles and their relative spatial distances) not only varies spatially, but also changes in time. Metaphysically, this entails to impose another set of world-making relations—temporal relations. They allow us to distinguish cases in which a configuration of matter changes in time from cases in which two possible configurations of matter are considered. When a configuration of matter changes, the relative distances among Bohmian particles change—they become longer or shorter. So, what is obtained is a series of stacked configurations of matters at different temporal instants with their relative spatial relations becoming longer or shorter (Fig. 10.3).
In Fig. 10.3, the solid lines represent spatial distance relations, while the dotted lines

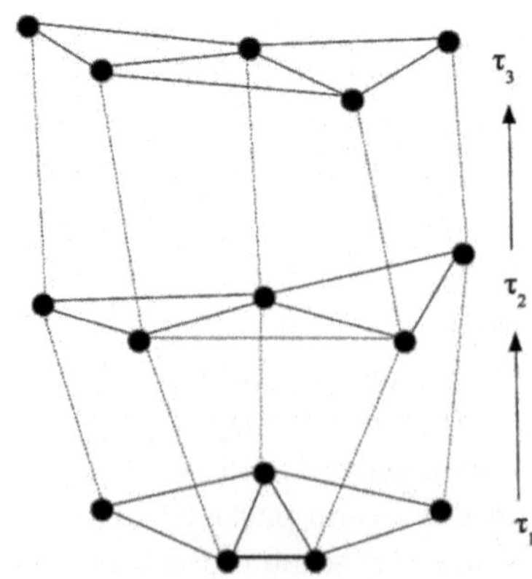

Fig. 10.3: Change in the configuration of matter.

stand for temporal relations. What is metaphysically important is the necessity to distinguish between both relations. One way to do this is by making them to be parasitic on space and time as substances, with time and space existing independently from matter. Spatial and temporal relations between matter supervene on spatial and temporal relations between points and instants, respectively. For instance, in substantivalism, the relative spatial relation between two Bohmian particles supervenes on the absolute spatial relations between the points that the Bohmian particles occupy in substantival space. The same goes for temporal relations and substantival time. In this metaphysical framework, therefore, it is straightforward to distinguish between variation and change since they remit to different substances.

Nevertheless, it can be argued that substantivalism overloads the ontology. Hence, if a comparable distinction between variation and change can be drawn in a more parsimonious framework, then this is to be preferred. In what follows, we argue that Leibnizian relationalism provides the appropriate conceptual tools to accomplish this job. The upshot is that Bohmian mechanics can accommodate a primitive direction of time and implement a distinction between variation of a configuration (space) and change of the configuration (time). Leibnizian relationalism provides the rationale for this.

To begin, the central tenet of Leibnizian relationalism is to reduce space to distance relations among objects that constitute a configuration such as the configuration of matter of the universe (order of coexistence in Leibniz's terms) and to reduce time to the change of these relations (order of succession in Leibniz's terms). Let us consider the main ontological tenets of Leibnizian relationalism. It first endorses a monist ontology, in which there exists only one kind of substances in the physical sphere that stand in spatial and temporal relations. This means mainly two things: firstly, space and time (or space-time) are not a different kind of substances in addition to matter, but they are just an abstraction (or representation) of the spatial and temporal relations that are obtained among the material substances (time and space are *entia rationis* in Leibniz's vocabulary, that is, they are ideal entities that express indeterminate, continuous possibilities); secondly, these are different relations that must also be primitive. It is commonly said that in Leibnizian relationalism spatial and temporal relations among matter are *direct*, in contrast with substantivalism.

While spatial relations deliver variation in the configuration of matter, temporal relations deliver change. It is important to emphasize that spatial relations (i.e. the configuration of matter) are more fundamental than temporal relations (i.e. change). The requirement of temporal relations comes from the necessity to impose change on the configuration of matter. The question of whether the purely spatial ontology changes or not cannot be settled on the basis of the spatial relations alone, but it is to be replied to on the basis of why additional structure is required in the ontology. And the reason to impose change in the ontology is the same reason for why it is necessary to impose any additional structure at all: to explain the phenomena. Then, change enters the ontology as a primitive together with objects standing in distance relations because it is a necessary assumption to explain the behaviour of matter (see also the two axioms proposed and argued for in Esfeld and Deckert (2017), p. 21). Therefore, the Leibnizian basic ontology comprehends matter engaging in spatial and temporal relations.

The claim that change is intrinsically directed is part of Leibnizian relationalism. In analyzing phenomenal change, Leibniz distinguishes between points and instants. In a letter to Louis Bourguet, dated 5 August 1715, he writes:

> I admit however that there is this difference between instants and points—one point of the universe has no advantage of priority over another, while a preceding instant always has the advantage of priority, not merely in time but in nature over the following instants (Leibniz 1989, p. 664; translated from Leibniz 1887, pp. 581-582).

This passage is crucial to understand the primitiveness of the direction of change in Leibnizian relationalism. For Leibniz, an *instant* is a momentary state of the world of successive phenomena (see Anapolitanos 1999, p. 136). The distinctive feature of instants is hence an in-built priority (or directedness), a metaphysical difference that distinguishes them from points – instants stand always in a later/earlier relation; if they did not, they would not be instants, but points in mere variation. They acquire this essence because they are generated by the change of the basic ontology. Instants may then be seen as momentary states of the basic ontology that result from the primitive change of the basic ontology as it unfolds. The Leibnizian understanding of change and temporality has indeed a Lewisian resemblance in this respect: change is nothing but just one single instant following another; this constitutes the unfolding or evolution of the basic ontology. This unfolding is *intrinsically* ordered and directed because it is not made of points, but results in instants (see Arthur 1985, p. 277). For Leibniz, therefore, the notion of undirected change would be meaningless, because being directed is a distinctive feature of what change *is*.

In this metaphysical framework, it is then pertinent to say that since time reduces to change, the direction of time is *nothing* but the direction of change. This suggests a reduction in terms of *identity*. It is not the case that time is unreal or vanishes; nor does it mean that the direction of time is unreal or vanishes. In Leibnizian relationalism, the direction of time is as real as the direction of change because the reduction can be viewed as conservative rather than eliminative: it establishes that time is *identical* to change, so there is no independent matter of fact that distinguishes between the direction of time and the direction of change—they are just the same. It is worth recalling that in Leibniz's metaphysics time (and space) is *entia rationis*, that is, they are ideal entities, which does not mean that they can be eliminated: they are just relations considered in abstraction from determinate relata (see Arthur 1985, p. 285). All this implies that the direction of time is obtained from the explanatory necessity to include change in the basic ontology, to adopt Leibnizian relationalism to reduce time to change, and to the metaphysical distinction between variation and change in terms of the latter being primitively directed. Therefore, the direction of time is primitive, too.

When Bohmian mechanics is complemented with Leibnizian relationalism, a Bohmian primitive direction of time is a reasonable posit. In the primitive Bohmian ontology, all there is are Bohmian particles in a configuration of spatial relations that change in a directed way as a matter of fact. A configuration of particles constitutes an order of coexistence. For Bohmian mechanics to be a physical theory, motion needs to be imposed. It basically means that change in the relative distances between particles is imposed. When change is imposed upon the configuration of Bohmian particles, it changes in one direction. The direction of change is then primitive to Bohmian mechanics when complemented with Leibnizian relationalism. Since the direction of time is nothing but the direction of change, it follows that the direction of time is also primitive. But there is no need for an ontological commitment to substantival time as an independent substance.

To put it differently, spatial relations lack of a direction in the sense that there is no difference between travelling from one Bohmian particle to the other, or the other way around. But, when the configuration of matter changes, it changes from one configuration to a different configuration, that is, from the configuration at τ_1 to the configuration at τ_2, as shown in Fig. 10.3. That is, for it to be change, it must be directed change in the sense of going from one configuration to another, but *not* the other way around. It can be argued that the backward evolution (from τ_2 to τ_1) is possible, too. This may well be so, but both evolutions are different from one another in function of their directionality – it is possible to even distinguish between them due to the fact that their directionality is different; we have two possible evolutions *because* they are different in their directionality.

In this way, Bohmian mechanics can accommodate a primitive direction of time in terms of a primitive direction of change if Leibnizian relationalism is adopted as metaphysical framework. As Leibnizian relationalism shows, when change is imposed as a primitive, temporal relations drive one configuration of Bohmian particles to the next one. Change is then reflected as change in relative spatial relations between Bohmian particles in different configurations. The necessity of distinguishing between variation and change (that is, between spatial and temporal relations) leads to the idea of directed change—while variation is directionless, change must be primitively directed. Otherwise, they would collapse into each other.

As a complementary comment, Leibnizian relationalism applied to Bohmian mechanics is also compatible with presentism and some versions of eternalism in the metaphysics of time. According to Leibnizian relationalism, change is not only primitive, but also eternal. What exists is both the configuration of Bohmian point particles of the universe characterized by the relative distances among them and the change in this configuration, that is the change of the distance relations. The whole change can therefore be taken to simply exist, which goes along with a dynamic view of time. Indeed, as Maudlin highlights, Bohmian mechanics is particularly accommodative to presentism, because its primitive ontology only admits instantaneous positions as primitive, in contrast to Newtonian mechanics that admits both positions and velocities as primitive:

> It is perhaps interesting to note that Bohm's theory is deeply congenial to an ontology which maintains that all which exists is that which exists *now*, i.e., at a point in time classically conceived. Instantaneous velocities can of course be defined, but only as the limit of average velocities over finite periods of time. Those puzzled about the status of velocities in an ontology in which only an instant of time *exists* can happily adopt a Bohmian ontology of particles (with position) and the wave function (Maudlin 2011, p. 113, note 22).

Therefore, Bohmian mechanics and Leibnizian relationalism fit well into both presentist frameworks and dynamic theories of time. Yet, eternalism is not discarded either. Bohmian mechanics and Leibnizian relationalism can adopt the additional thesis that past, present, and future are real, but that there is nonetheless a dynamical present (see, for instance, Broad's moving spotlight theory, Broad 1923). In any case, it can also be claimed that the distinction between eternalism and presentism comes down to an only semantic one; it does not cut any ontological ice (see Dorato 2006). Be that as it may, the combination of persisting configuration of matter given by the relative positions of the particles (order of coexistence) *and* the permanent change of their relative positions leave enough room for alternative metaphysics of time.

It might be argued that our argument for primitivism ultimately fails because Bohmian mechanics is time-reversal invariant. As mentioned earlier, a time-reversal invariant dynamics could be received as an argument against primitivism. Nonetheless, it is possible to deflate the meaning of time-reversal invariance in order to accommodate primitivism (for details, see Lopez and Esfeld 2023). To begin, Bohmian mechanics is built upon the assumption that the dynamics must be time-reversal invariant (see Dürr, Goldstein, and Zanghì 2013, Chapter 2). For instance, in deducing the guidance equation, Bohmian mechanics assumes that it must be time-reversal invariant like the Schrödinger equation. This fact could be symptomatic of a theoretical stipulation to constrain the form of the Bohmian dynamics, but it does not really represent an aspect of reality. It can, for instance, work as a heuristic principle that guides theory construction—it can be seen as a property of the representational tools that physical theories deploy to explain the phenomena.

Mutatis mutandis, this view of time-reversal invariance can be related to similar Bohmian arguments against a realist interpretation of the wave function. For instance, John Bell says:

> One of the apparent non-localities of quantum mechanics is the instantaneous, over all space, 'collapse of the wave function' on 'measurement'. But this does not bother us if we do not

grant beable status to the wave function. We can regard it simply as a convenient but inessential mathematical device for formulating correlations between experimental procedures and experimental results, i.e., between one set of beables and another (Bell 1987, Chapter 7, p. 53).

In the same vein, time-reversal invariance can also be seen as a "convenient but inessential mathematical" feature of physical theories. If this is so, then time-reversal invariance cannot be straightforwardly employed in an argument against or in favour of a direction of time. While the latter is an ontological issue, whether a theory is time-reversal invariant (as Bohmian mechanics) is rather a representational issue about how the dynamics should be written. Both should not be conflated.

10.5 Conclusions

It is frequently assumed that a physical theory can resolve the problem of the direction of time. In general, this assumption has relied either on dynamical considerations (e.g., whether the dynamics is time-reversal invariance) or non-dynamical ones (e.g., an asymmetry in initial conditions). Our aim in this chapter was to analyse the problem of the direction of time in Bohmian mechanics and to show that it can actually accommodate two divergent philosophical positions with respect to the direction of time: conservative reductionism and primitivism. By relying on an epistemic asymmetry between initial conditions in FORWARD and BACKWARD, it was shown that Bohmian mechanics can account for the time asymmetric quantum phenomenology in terms of our lack of knowledge of the exact positions of Bohmian particles in the initial conditions. Bohmian conservative reductionism on the direction of time can thus be an attractive explanation of the direction of time for those that are reticent to primitivist approaches. Bohmian primitivism on the direction of time is not discarded, though. We have shown that Bohmian mechanics metaphysically necessitates a distinction between variation (the order of coexistence of Bohmian particles) and change (the order of succession). Leibnizian relationalism, we have argued, is a suitable framework that can do the job—it imposes that change is intrinsically directed, while variation is not. Since time is identical to change, the direction of time is as primitive as the direction of change is. Hence, those that are also prone to adopt primitivism about the direction of time can employ Bohmian mechanics as physical basis.

References

Albert, D. (1992). *Quantum Mechanics and Experience.* Cambridge, MA: Harvard University Press.

Anapolitanos, D. (1999). *Leibniz: Representations, Continuity and the Spatiotemporal.* Dordrecht: Kluwer.

Arntzenius, F. (1997). Mirrors and the direction of time. *Philosophy of Science*, **64**: 213–222.

Arthur, R. (1985). Leibniz's theory of time. In: Okruhlik, K. and Brown, J.R. (eds.) *The Natural Philosophy of Leibniz*, pp. 263–313. Dordrecht: Reidel.

Bacciagaluppi, G. (2007). Probability, decoherence and the arrow of time. *Studies in History and Philosophy of Modern Physics*, **38**: 439–456.

Bell, J. (1987). *Speakable and Unspeakable in Quantum Mechanics.* Cambridge: Cambridge University Press.

Bohm, D. (1952). A suggested interpretation of the quantum theory in terms of "hidden" variables. I & II. *Physical Review*, **85**(2): 166–193.

Callender, C. (1997). What is "the problem of the direction of time"?. *Philosophy of Science* (Supplement), **64**: S223–S234.

Callender, C. (2000). Is time "handed" in a quantum world?. *Proceedings of the Aristotelian Society*, **100**: 247–269.

Castagnino, M. and Lombardi, O. (2009). The global non-entropic arrow of time: from global geometrical asymmetry to local energy flow. *Synthese*, **169**: 1–25.

de Broglie, L.(1928). La nouvelle dynamique des quanta. In: *Electrons et photons. Rapports et discussions du cinquième Conseil de physique tenu à Bruxelles du 24 au 29 octobre 1927 sous les auspices de l'Institut international de physique Solvay.* Paris: Gauthier-Villars, pp. 105–132. English translation in Bacciagaluppi, G. and Valentini, A. (2009). *Quantum Theory at the Crossroads. Reconsidering the 1927 Solvay Conference*, pp. 341–371. Cambridge: Cambridge University Press.

Dorato, M. (2006). The irrelevance of the presentist/eternalist debate for the ontology of Minkowski spacetime. In Dieks, D. (ed.) *The Ontology of Spacetime*, pp. 93–109. Dordrecht: Elsevier.

Dürr, D., Goldstein, S. and Zanghì, N. (2013). *Quantum Physics without Quantum Philosophy.* Berlin: Springer.

Dürr, D. and Teufel, S. (2009). *Bohmian Mechanics: The Physics and Mathematics of Quantum Theory.* Berlin: Springer.

Earman, J. (1974). An attempt to add a little direction to "The problem of the direction of time". *Philosophy of Science*, **41**: 15–47.

Earman, J. (2002). What time-reversal invariance is and why it matters. *International Studies in the Philosophy of Science*, **16**: 245–264.

Ellis, G.F.R. (2013). The arrow of time and the nature of spacetime. *Studies in History and Philosophy of Modern Physics*, **44**: 242–262.

Esfeld, M. (2014). Quantum Humeanism, or: Physicalism without properties. *The Philosophical Quarterly*, **64**: 453–470.

Esfeld, M. and Deckert, D.-A. with Lazarovici D., Oldofredi A. and Vassallo A. (2017). *A Minimalist Ontology of the Natural World.* New York: Routledge.

Goldstein, S. (2021). Bohmian Mechanics. In Zalta, E. (ed.). *The Stanford Encyclopedia of Philosophy.*

Healey, R. (2002). Can physics coherently deny the reality of time?. In Callender, C. (ed.) *Time, Reality and Experience*, pp. 293–316. Cambridge: Cambridge University Press.

Horwich, P. (1987). On the existence of time, space and space-time. *Noûs*, **12**: 397–419.

Hubert, M. and Romano, D. (2018). The wave function as a multifield. *European Journal for Philosophy of Science*, **8**: 521–537.

Leibniz, G.W. (1887). *Die philosophischen Schriften von G. W. Leibniz. Band 3. Herausgegeben von C. I. Gerhardt.* Berlin: Weidmannsche Verlagsbuchhandlung.

Leibniz, G.W. (1989) [1712]: *Philosophical Papers and Letters.* A selection translated and edited by Leroy Loemker. Dordrecht: Kluwer.

Lopez, C. (2019). Roads to the past: How to go and not to go backward in time in quantum theories. *European Journal for Philosophy of Science*, **9**: 27.

Lopez, C. (2021). The physics and philosophy of time reversal in standard quantum mechanics. *Synthese*, **199**: 14267–14292.

Lopez, C. (2022a). Time's direction and orthodox quantum mechanics: Time asymmetry and measurement. *Journal for General Philosophy of Science*, **53**: 421–440.

Lopez, C. (2022b). Spontaneous collapse theories and temporal primitivism about time's direction. *Foundations of Physics*, **52**: 112.

Lopez, C. and Esfeld, M. (2023). Humean time reversal. *Synthese*, **202**: 31.

Maudlin, T. (1995). Three measurement problems. *Topoi*, **14**: 7–15.

Maudlin, T. (2011). *Quantum Non-Locality and Relativity.* Chichester: Wiley-Blackwell, 1st edition 1994.

Maudlin, T. (2019). *Philosophy of Physics: Quantum Theory.* Princeton: Princeton University Press.

North, J. (2011). Time in thermodynamics. In Callender, C. (ed.) *The Oxford Handbook of Philosophy of Time*, pp. 312–350. Oxford: Oxford University Press.

Penrose, R. (1989). *The Emperor's New Mind.* New York: Oxford University Press.

Price, H. (1996). *Time's Arrow and Archimedes' Point: New Directions for the Physics of Time.* New York: Oxford University Press.

Reichenbach, H. (1956). *The Direction of Time.* Berkeley: University of California Press.

Roberts, B. (2017). Three myths about time reversal invariance. *Philosophy of Science*, **84**: 315–334.

Romano, D. (2021). Multi-field and Bohm's theory. *Synthese*, **198**: 10587–10609.

Valentini, A. (1997). On Galilean and Lorentz invariance in pilot-wave dynamics. *Physics Letters A*, **228**: 215–222.

11 Bohmian mechanics for quantum chemistry

Sebastian Fortin and Olimpia Lombardi

11.1 Introduction

Quantum chemistry is a branch of theoretical chemistry that employs some tools coming from quantum mechanics to describe molecules. Despite the successful development of this area, the deep breakdown between the conceptual basis of standard quantum mechanics (SQM) and some implicit ontological assumptions of quantum chemistry is only rarely acknowledged. The main objective of the present chapter is to argue that Bohmian mechanics (BM) offers an ontological representation much closer to the way that quantum chemists conceive their own field of research than SQM. With this purpose, the chapter is organized in two main sections. In Section 11.2, some aspects of the mentioned breakdown are recalled: the chemical assumption about the components of the atom (Section 11.2.1); the nature of electrons in chemistry and the consequence for the concept of orbital (Section 11.2.2); and the problem of molecular structure (Section 11.2.3). In each case, the features of BM that might overcome the breakdown will be pointed out. Section 11.3 is devoted to comment some technical works that obtain detailed results not available from the standard quantum view: Bohmian trajectories are studied to account for the orbits of the electrons of the hydrogen and helium atoms (Section 11.3.1), to explain the inner mechanism of laser-enhanced ionization of a molecule of hydrogen (Section 11.3.2), and to apply computational methods to describe and analyse the evolution of chemical systems (Section 11.3.3).

11.2 Quantum chemistry and standard quantum mechanics

The relationship between chemistry and physics has been a central topic of discussion in the philosophy of chemistry since its inception. While since the early days of the debate many authors pointed out the difficulties of reducing chemistry to physics (Woolley 1978; Primas 1983, 1998; Amann 1992; Lombardi and Labarca 2005; Labarca and Lombardi 2010), in recent times the idea of reduction reappeared, reinforced by the assumption that quantum chemistry incorporates quantum mechanics to recover and improve the results of traditional structural chemistry. A paradigmatic example of this position is Hinne Hettema with his book *Reducing Chemistry to Physics* (2012). There the author admits that the inter-theoretical links between chemistry and physics require approximations and idealizations that, in many cases, contradict some principles of the supposedly reducing theory (2012, p. 341). Furthermore, Hettema recognizes that the 'ontological discontinuity' between chemistry and physics is 'one of the central problems in the philosophy of chemistry, around which many other problems, such as that of reduction, revolve' (2012, p. 368). Despite these claims, his argument throughout the book is focused on demonstrating how this discontinuity can be resolved, so that the relationship between chemistry and physics becomes a prime example of reduction. His general conclusion is expressed by the last sentence of the book: 'The reduction of chemistry to physics is in this sense indeed a paradigm case for the notion of reduction' (Hettema 2012, p. 410).

Hettema's book is an example of the difficulty in recognizing the deep breach that exists between the conceptual foundations of SQM and the ontological picture that underlies the theoretical work of chemists, even quantum chemists. In the following subsections, we consider various aspects through which the rupture between the two theoretical domains becomes more evident. The precise characterization of SQM is not discussed here, since we refer to features that are unequivocally part of this theory and appear in all textbooks on the subject. The aim of this section is to point out how, from the perspective of BM (de Broglie 1928, Bohm 1952; in its contemporary version, Dürr, Goldstein, and Zanghì 2013; here we will not discuss the different interpretations of BM), the conceptual–ontological discontinuity between the quantum and chemical domains is significantly minimized and, in some cases, even disappears.

The components of the atom

If one asks a quantum chemist what the components of an atom are, they reply that atoms are composed of three types of particles: electrons, protons, and neutrons. Protons and neutrons form the atomic nucleus, and the charged particles interact through a Coulombic potential. In the simplest case of a hydrogen atom, the Hamiltonian operator reads:

$$\hat{H} = \frac{\hat{P}_e^2}{2m_e} + \frac{\hat{P}_p^2}{2m_p} - \frac{e^2}{|\hat{Q}_e - \hat{Q}_p|} \tag{11.1}$$

where e is the electric charge of the electron, m denotes the mass of the particle, $\hat{Q}$ is the position operator, $\hat{P}$ is the momentum operator, and the subscripts e and p refer to the electron and the proton, respectively.

In SQM, a quantum system is represented by a Hilbert space. For instance, electron and proton are two quantum systems represented by $\mathcal{H}_e$ and $\mathcal{H}_p$, respectively, each one with its own Hamiltonian, and the atomic composite system is represented by $\mathcal{H}_A = \mathcal{H}_e \otimes \mathcal{H}_p$. However, a change of variables can be introduced, in terms of the centre of mass coordinates $\hat{Q}_C$ and $\hat{P}_C$, and the relative coordinates $\hat{Q}_R$ and $\hat{P}_R$

$$\hat{Q}_C = \frac{m_e\hat{Q}_e + m_p\hat{Q}_p}{m_e + m_p} \qquad \hat{Q}_R = \hat{Q}_e - \hat{Q}_p, \tag{11.2}$$

$$\hat{P}_C = M\dot{\hat{Q}}_C = \hat{P}_e + \hat{P}_p \qquad \hat{P}_R = \mu\dot{\hat{Q}}_R = \frac{m_p\hat{P}_e - m_e\hat{P}_p}{m_e + m_p}, \tag{11.3}$$

where $M = m_e + m_p$ is the total mass, and $\mu = m_e m_p/(m_e + m_p)$ is the so-called reduced mass. With these new variables, the Hamiltonian of the hydrogen atom (equation (11.1)) can be expressed as $\hat{H} = \hat{K} + \hat{W}$, where $\hat{K}$ is the kinetic energy and $\hat{W}$ is the internal energy:

$$\hat{K} = \frac{\hat{P}_C^2}{2M} \qquad \hat{W} = \frac{\hat{P}_R^2}{2\mu} - \frac{e^2}{|\hat{Q}_R|}. \tag{11.4}$$

Here, it is important to emphasize that, in SQM, this is not an innocuous change of variables as in classical mechanics. In fact, since the kinetic energy $\hat{K}$ only depends on the total momentum, and the internal energy $\hat{W}$ only depends on differences of positions and, eventually, on their derivatives, then $[\hat{K}, \hat{W}] = 0$. Therefore, the total Hamiltonian $\hat{H}$ of the hydrogen atom can be expressed as

$$\hat{H} = \hat{K} + \hat{W} = \hat{H}_K \otimes \hat{I}_W + \hat{I}_K \otimes \hat{H}_W, \tag{11.5}$$

where $\hat{H}_K$ is a kinetic Hamiltonian acting on a Hilbert space $\mathcal{H}_K$, $\hat{H}_W$ is an internal energy Hamiltonian acting on a Hilbert space $\mathcal{H}_W$, and $\hat{I}_K$ and $\hat{I}_W$ are the identity operators of the respective spaces. Furthermore, in a reference frame at rest with respect to the center of mass, $\hat{P}_C = 0$ and, then, $\hat{K} = 0$; as a consequence, the Hamiltonian of the atom becomes $\hat{H} = \hat{W} = \hat{I}_K \otimes \hat{H}_W$.

The above remarks show that, in the context of SQM, there are two equally legitimate ways of decomposing the hydrogen atom: one, as an electron and a proton in interaction, and the other, as a single system characterized by the internal energy $\hat{W}$. The case of the hydrogen atom is an example of the fact that, in general, quantum composite systems can be decomposed into sub-systems in more than one way. Indeed, the so-called 'quantum structure studies' (see, e.g. Dugić and Jeknić-Dugić 2008, Viola and Barnum 2010, Harshman 2012) deal precisely with the different ways in which a quantum system can be decomposed into subsystems, defined by tensor product structures. Furthermore, from the viewpoint of SQM it could be argued that not all decompositions are equivalent, but only those that lead to an invariant description under the symmetry group of the theory are objective. In the example of the hydrogen atom, the single-system picture is the objective representation, since only this description is invariant under the Galilean group (see Ardenghi, Castagnino, and Lombardi 2009). This means that, after the interaction between electron and proton, the hydrogen atom becomes a single system, represented by the Hilbert space $\mathcal{H}_W$, in which electron and proton lose their individuality because they inextricably 'mix' to constitute a new whole entity.

In this regard, BM is completely different from SQM because it assumes a clear 'primitive ontology' (Allori, Goldstein, Tumulka, and Zanghì 2008, Allori 2013) from the very beginning, that is, it posits the kind of basic entities that are the building blocks for anything else in the world. The Bohmian universe is constituted by particles with definite positions at each time. Different types of particles (e.g. protons and electrons) are not distinguished from each other by their 'nature' or 'essence' but only by their behaviour. Nevertheless, in any case, Bohmian particles are individuals that are completely discernible synchronically by their position in the universe configuration and diachronically by their continuous trajectories. This ontological representation erases, in a single attempt, the conceptual difficulties derived from the decomposition of composite systems in SQM: there are not many different ways of decomposing a Bohmian quantum system but only one in terms of the particles that constitute the system under consideration. In the case of atoms, regardless of how electrons and protons interact, they retain their synchronic and diachronic individuality, and the behaviour of atoms is the result of the behaviour of their elemental components. In particular, the hydrogen atom is composed of one electron and one proton, no matter how they interact. This ontological view is more akin to the way that chemists, even in the field of quantum chemistry, conceive of their systems of study, since they do not consider the different tensor product structures corresponding to a composite system: the existence of electrons and protons, which compose atoms, which in turn compose molecules, is a basic assumption of the chemical work.

Orbitals and electrons

In chemistry books, the concept of orbital is usually introduced in mathematical terms, as referring to a one-particle wave function for an electron in an atom or molecule. Nevertheless, the concept of orbital is rapidly substantialized as soon as the talk of 'the shape of an orbital' enters the scene, which is conceived as the shape of a spatial region of high electron density. In turn, the electron density is a measure of the probability of an electron being present at a definite spatial place, and can be computed as the square of the electron's wave function. Despite being a probability, chemists consider that 'the spatial volume occupied by an atom depends on its electronic density' (Grosso *et al.* 2015, p. 1) and that such a region can be visualized (see, e.g. Zuo *et al.* 1999). From this perspective, orbitals supply the basis to explain the geometrical shape of molecules.

In what sense is it stated that a high probability can be visualized? The answer relies on the fact that the electronic density ρ is commonly interpreted as a kind of mean value of the definite positions occupied by the electron in its motion around the nucleus: '[i]f a series of measurements could be made of x without disturbing the motion of the particle, the resulting distribution would be ρ. The latter would then reflect the motion of the particle in the same way in which the density of the image on a long-exposure photograph reflects the motion of a macroscopic object' (Nelson 1990, p. 643). This idea had already appeared in Linus Pauling's famous book originally published in 1939: 'We can accordingly describe the normal hydrogen atom by saying that the electron moves in and out about the nucleus, remaining usually within a distance of about 0.5 , with a speed that is variable but is of the order of magnitude of v_0. Over a period of time long enough to permit many cycles of motion of the electron the atom can be described as consisting of the nucleus surrounded by a spherically symmetrical ball of negative electricity (the electron blurred by a time exposure of its rapid motion)' (Pauling 1960, pp. 14-15). This description is based on conceiving electrons as individual objects, with definite positions and velocities, whose main difference with respect to classical particles is that their behaviour is not governed by classical equations of motion but by an equation that determines their position only in a statistical way. But this view is incompatible with the Heisenberg principle of SQM, according to which quantum 'particles' have no trajectories. Furthermore, the Kochen–Specker theorem (Kochen and Specker 1967) proves that SQM is essentially *contextual*: definite values can be consistently assigned only in a context, determined by the observables that share the same eigenbasis (see Martínez González, Fortin, and Lombardi 2019). This means that the very structure of the theory prevents any assignment of a definite momentum and a definite position to a particle. In other words, according to SQM, a quantum "particle" is not an *individual object* as in the classical world: the fact that it always has some non-definite-valued properties is not a limitation of our knowledge but an ontological fact resulting from the formalism of the theory.

By contrast to SQM, BM is not contextual regarding position, because its primitive ontology fixes a preferred context defined precisely by position. As Michael Esfeld clearly explains: 'Theorems like the one of Kochen and Specker (1967) prove that it is not possible to take the operators or observables of quantum mechanics to have definite values independently of measurement contexts, on pain of violating the predictions of quantum mechanics for measurement outcome statistics. Nonetheless, these theorems leave the possibility open to admit one additional parameter of the quantum objects that has a definite value, without the precise information about that value figuring in the wave function. The natural choice for the additional parameter then is position, since all measurement outcomes consist in macroscopically recorded positions of discrete objects' (Esfeld 2019, p. 234). Since Bohmian particles always have definite position and velocity and follow definite trajectories, the interpretation of the electronic density as the mean value of the definite positions occupied by the electrons in the atom or the molecule turns out to be natural.

Nevertheless, the naive 'long-exposure photograph' analogy to interpret the electronic density must be revised. As Bohm already pointed out in his first papers on his proposal (Bohm 1952, 1953), in order to obtain the electron density of, say, a hydrogen atom in the ground state, it is necessary to make many measurements of the electron position in one of the following two situations. Either the measurement is repeated on the same atom: in this case each interaction perturbs the position of the electron which, for the new measurement, will be in a new initial condition, different from the original one. Or a measurement is made on an ensemble of atoms in the same quantum state but whose electrons are initially in different positions. In both cases, the distribution of positions yields the same electron density as in the SQM case, which is not surprising since SQM and BM are empirically equivalent. However, according to the Bohmian ontological picture the electronic density of an atom or a molecule can be meaningfully interpreted as the mean value of the electrons' definite positions for different initial conditions, in a way that is not discordant with the way quantum chemists conceive their ontology. In turn, the quantum probability involved in the concept of orbital has the same status as the probability in classical statistical mechanics: it measures ignorance about the precise configuration of the particles or the particular trajectories.

The concept of molecular structure

The concept of molecular structure is certainly the central concept in molecular chemistry. As Hans Primas says: 'The alpha and omega of molecular chemistry is the doctrine that molecules exist as individual objects and that every molecule has a shape, characterized by its molecular frame' (Primas 1994, p. 216). Or in the words of Guy Woolley, molecular structure is 'the central dogma of molecular science' (Woolley 1978, p. 1074). According to Robin Hendry, 'molecular structure is so central to chemical explanation that to explain molecular structure is pretty much to explain the whole of chemistry' (Hendry 2010, p. 183). In fact, many properties of chemical substances are explained in terms of the structure of their molecules. The *problem of molecular structure*, which stands at the heart of the discussions about the relationship between physics and chemistry, is whether molecular structure can be explained exclusively by means of the concepts and the laws of quantum mechanics.

In chemistry, molecular structure can be characterized in different ways. In the context of molecular chemistry, the term 'molecular structure' refers to the spatial arrangement of the atoms in the molecule and the chemical bonds that hold them together. From a more detailed perspective, a molecule is a system in physical space, composed of nuclei placed in definite positions and surrounded by inert inner-shell electrons, and bonds explained in terms of interacting 'bonding' electrons; the structure of this system is defined by these elements in their spatial arrangement. From a more quantum viewpoint, a molecule is a system in which the nuclei, placed in definite positions, create a potential in which quantum electrons move; the structure of this system is defined by these elements in their spatial arrangement. What is important to highlight is that, in all cases, the concept implies a strong spatial connotation related, at least, to the positions of the nuclei. And it is at this point that the disagreement with SQM begins.

The Born–Oppenheimer (B-O) approximation plays a central role in the discussions on the problem of molecular structure because it makes possible to replace the full Coulombic Hamiltonian of the molecule by an effective Hamiltonian that separates the behaviour of the nuclei from that of the electrons. The first step of the B-O strategy is the *clamped nuclei approximation*: electrons move on the Coulombic potential generated by nuclei at rest, 'clamped' at definite positions. But from the viewpoint of SQM, this is an assumption that contradicts one of the principles of the theory. As Hasok Chang clearly explains: 'In this "clamping-down" approximation, the atomic nuclei are treated essentially as classical particles; [...] this picture is non-quantum in a very fundamental way as the simultaneous assignment of fixed positions and fixed momenta (namely, zero) to them violates the Heisenberg uncertainty principle' (Chang 2015, p. 198; for a detailed analysis of the assumptions underlying the B-O approximation, see Lombardi and Castagnino 2010). In other words, B-O models are based on a counterfactual approximation, since it cannot be replaced by a factual one that is compatible with the theoretical framework in which the models are developed.

It is quite clear that, in the context of the problem of molecular structure, the B-O 'simply assume the facts about molecular structure that ought to be explained' (Hendry 2010, p. 186). Moreover, from the perspective of SQM, a quantum system is identified by its Hamiltonian; thus, changing the Hamiltonian amounts to changing the identity of the system itself. For chemists, by contrast, given a molecular system with its full Coulombic Hamiltonian, the effective Hamiltonian resulting from the B-O strategy is merely an approximate description of the same original system.

In recent times, some works have dealt with the description of molecules from a 'beyond the B-O approximation' approach. However, this move does not cancel the difficulties, since new issues arise in terms of the full Hamiltonian. Under the influence of Guy Woolley and Brian Sutcliffe (1977), Robin Hendry (2010) characterizes what he calls '*the symmetry problem*' in the following terms. Since the interactions embodied in the full Hamiltonian of the molecule are Coulombic, the solutions of the Schrödinger equation (eigenstates of the Hamiltonian) are always rotationally symmetric. This means that, according to SQM, no directional property can be assigned to an isolated molecule in an eigenstate of the Hamiltonian; for example, in such a state, the expectation value of the electric dipole moment of the molecule is zero. However, several chemical properties are explained by the rotational asymmetry of the molecule (see Sutcliffe and Woolley 2012); for example, the hydrogen chloride molecule has an asymmetrical charge distribution that explains its acidic behaviour and boiling point. Another example is the ammonia molecule, whose electric dipole moment is explained by chemists in terms of its tetrahedral structure – with the negative nitrogen in one of the vertices and the positive hydrogens in the other three – a geometrical structure completely alien to SQM.

Rotational symmetry is not the only symmetry of the full Hamiltonian from the viewpoint of SQM. Since the Coulombic interaction only depends on the distance between the interacting particles, the Hamiltonian is also symmetric under spatial reflection. As a consequence, its eigenstates, in particular the ground state, have parity symmetry. Moreover, this feature is preserved during the time evolution of the system because parity is a constant of motion of the molecular system. However, chemists perfectly know the existence and the properties of optical isomers (or enantiomers), which can be distinguished by their ability to rotate plane-polarized light by equal amounts but in opposite directions. Optical isomers are chiral molecules: the structure of each member of a pair of enantiomers is the mirror image of the structure of the other. But precisely due to this parity asymmetry, the ground state of the chiral molecules cannot be an eigenstate of the parity-symmetric full Hamiltonian. This leads to the so-called Hund's paradox (Hund 1927): why do certain chiral molecules display an optical activity that is stable in time, associated to a well-defined chiral state that cannot be an eigenstate of the molecule's Hamiltonian?

Some authors consider that the problem of molecular structure is nothing more than a version of the quantum measurement problem (Franklin and Seifert 2024). From this perspective, decoherence has been appealed to as the link between the domain of SQM and that of classical molecular chemistry (Trost and Hornberger 2009, Scerri 2011, 2013). However, many authors agree in that decoherence does not solve the measurement problem, but only selects the preferred basis, given by the physical quantities perceived as determinate (see, e.g. Schlosshauer 2004). As a consequence, decoherence does not provide a way out the problem of molecular structure (Fortin, Lombardi, and Martínez González 2016, Fortin and Lombardi 2021).

It is true that, if a solution to the measurement problem is provided in the context of SQM, it is possible to give an answer to the symmetry problem, for example, by explaining why a definite value of electric dipole moment or the chirality of a molecule is observed (Fortin, Lombardi, and Martínez González 2018). But this only accounts for the definite value of a quantity, and is still far from solving the problem of molecular structure, which requires explaining, at least, the positions of all the nuclei in the molecule. In other words, there are many different ways to being, e.g. a chiral molecule, there are many different geometrical configurations of the nuclei that realize, say, a left-handed molecule. This means that solving the measurement problem involved in Hund's paradox amounts to explaining the chiral nature of the molecule, but it is still far from solving the problem of the molecular structure as a whole (Fortin and Lombardi 2021).

All these problems, faced by chemists when trying to use SQM for the description of molecules, have essentially a common origin. Given a molecular system, its full Hamiltonian is expressed as:

$$\hat{H} = \sum_{g}^{A} \frac{\hat{P}_g^2}{2m_g} + e^2 \sum_{g<h}^{A} \frac{Z_g Z_h}{\hat{R}_{gh}} + \sum_{i}^{N} \left(\frac{\hat{P}_i^2}{2m_e} - e^2 \sum_{g}^{A} \frac{Z_g}{\hat{R}_{ig}} \right) + e^2 \sum_{i<j}^{N} \frac{1}{\hat{R}_{ij}} \tag{11.6}$$

where e and m_e are the charge and the mass of the N electrons, respectively; $\hat{P}_i$ is the momentum operator of the electron i, with $i, j = 1, 2, ..., N$; the m_g are the masses of the A nuclei; $\hat{P}_g$ is the momentum operator of the nucleus g; Z_g is the atomic number (the number of protons) of the nucleus g; $\hat{R}_{ij}$ is the 'distance' between the electron i and the electron j; $\hat{R}_{gh}$ is the 'distance' between the nucleus g and the nucleus h, and $\hat{R}_{ig}$ is the 'distance' between the electron i and the nucleus g. The crux of the matter here is that, in SQM, the so-called 'distances' are not such: although the differences $\hat{R}_{\alpha\beta} = |\hat{R}_\alpha - \hat{R}_\beta|$ are usually called 'distance between particles', they are actually quantum observables, which in general have no definite values. In other words, the full Hamiltonian is 'blind' to the particles' definite positions and, as a consequence, cannot account for the asymmetries required by chemistry.

This "blindness" to the particles' positions is a very generic form of symmetry, which has the consequence that, in the case of any form of isomerism (not only optical, but also structural), all isomers corresponding to a given chemical formula have the same full Hamiltonian. Such a symmetry is inherited by the ground state $|\Psi_0\rangle$ of the molecule. Alexander Franklin and Vanessa Seifert (2024) propose to conceive the symmetric ground state $|\Psi_0\rangle$ as a superposition of the asymmetric $|\phi_i\rangle$ corresponding to the different isomers related to the same formula:

$$|\Psi_0\rangle = \sum_i c_i|\phi_i\rangle. \tag{11.7}$$

However, it is a mathematical fact that only by superposing all possible asymmetries with the same weight can a symmetry be reconstructed. Thus, in order to recover the symmetry of $|\Psi_0\rangle$ from the superposition of the asymmetric $|\phi_i\rangle$, it is necessary to include *all the mathematically possible isomeric structures* corresponding to a given formula in the superposition, that is, all the mathematically possible nuclear configurations, *with the same probability*. In other words, in equation (11.7), the $|\phi_i\rangle$ must represent all the possible relative positions of the nuclei – besides all the possible electronic configurations – and all the coefficients c_i must be equal. This theoretical fact should be observationally manifested by the empirical fact that all the possible nuclear configurations are obtained – at least, approximately – with the same frequency in effective experiments. The problem is that this is not the case: given the number and the nature of the elements that compose a molecule, in general only few nuclear configurations (when their rotations and translations are assimilated) are experimentally measured, that is, only few isomers are considered effectively 'real'. For example, there are only three structural isomers of the hydrocarbon C_3H_4, propadiene, propyne, and cyclopropane, and not the potentially infinite compounds with different possible structures. Of course, there are many chemical reasons for the existence of certain isomers and not of others. But, from the viewpoint of SQM the question is: Why are certain isomeric structures obtained in the laboratory and not all those structures necessary to reconstruct the symmetry of the ground state? (for a full argument, see Fortin and Lombardi 2021).

The several obstacles to account for molecular structure in the framework of SQM have been pointed out for several decades now. For example, Woolley considers that the impossibility of determining the geometry of a molecule by means of SQM is a proof of the fact that molecular structure is only a 'powerful and illuminating metaphor' (Woolley 1982, p. 4; see also Sutcliffe and Woolley 2022). In his many works on the matter, Hendry also concludes that there is an in-principle hindrance to account for molecular structure exclusively in SQM terms and, as a consequence, the case of molecular structure is an example of ontological emergence (Hendry 1998, 2004, 2008, 2010).

However, when the obstacles faced by chemists in accounting for molecular structure are understood at their source, it is not difficult to observe that BM can provide a theoretical framework to overcome them in a conceptually clear and general way. Indeed, if the origin of the problem lies in the impossibility of defining the spatial configuration of quantum particles from SQM, the Bohmian perspective erases the problem at a stroke since particles always have determinate positions and velocities, and therefore trajectories. In this way, the asymmetries of the molecules can be naturally explained in terms of the geometrical arrangement of their elements, in particular, the positions of the nuclei that define the potential 'surface' on which the electrons move.

Of course, this does not mean that BM alone can explain the 'correct' molecular structure as defined by chemists: like SQM, BM cannot account for the geometrical arrangement of the nuclei in the molecule, and this should not be surprising since the two theories are empirically equivalent. Nevertheless, from the Bohmian ontological perspective, the common strategy of quantum chemists to compute the energy of the molecule turns out to be justified. In fact, they usually consider many different configurations of the nuclei, that is, many different potentials, and retain the configuration that minimize energy as the description of the molecular structure in the ground state. This strategy of clamping the nuclei from the very beginning is in tension to SQM, but it makes perfect sense in the context of BM since particles always have definite positions.

From a more general perspective, the origin of the problem of molecular structure lies in the deep ontological rupture between the domain of molecular chemistry and the peculiarities of SQM. In contrast, the primitive ontology of BM is much closer to the classical world. As Esfeld states, 'What is the minimal deviation from classical mechanics that is necessary in order to obtain quantum mechanics? BM shows that the physical ontology can remain the same – point particles moving on continuous trajectories – and that the status of probabilities can remain unchanged. What has to change is the dynamics, that is, a wave function parameter has to be introduced with the wave function binding the evolution of the particle positions together independently of their distance in physical space' (Esfeld 2019, p. 239).

11.3 Bohmian trajectories in quantum chemistry

When chemists describe chemical reactions, the essential information for them is how electrons behave in an atom and when a molecule is formed. Such behaviour is what explains the elementary reactions in the so-called reaction mechanisms, that is, the step by step sequence of elementary reactions by which overall chemical change occurs. SQM does not supply chemists with that information as it only provides transition probabilities. On the contrary, Bohmian trajectories may allow them not only to rationalize known processes but also to understand phenomena that have no explanation in the framework of SQM.

In this conceptual framework, it is not surprising that, with the revival of the interest in BM since the 1980s, the theory has been used for the description of atoms and molecules. For example, in the description of the 1s state of the electron in a hydrogen atom, the obtained result is that the electron remains at rest. Although its exact position cannot be computed because the initial condition is unknown, the distance between the electron and the proton can be calculated and the result is that such a distance is equal to the Bohr radius (see Holland 1993, p. 148). In turn, when the σ bonding orbital of the hydrogen molecule is described with BM, calculations show that the electrons are at rest in the region close to the middle point of the line joining the two nuclei (see Holland 1993, p. 319). Surprisingly, BM offers a picture of chemical bond similar to the Lewis bond but in the present case on the basis of a fully articulated physical theory and detailed computations (see Fortin, Lombardi, and Martínez González 2017).

Despite these interesting results and the fact that SQM and BM are empirically equivalent, it must be admitted that the image of electrons at rest in atoms and molecules looks strange to the eyes of chemists, accustomed to think of electrons somehow 'orbiting' around nuclei. As described in this section, more recent applications of BM provide results that are much more familiar, and even more useful, for chemists.

The account of orbitals' shape

In the Bohmian formalism, the motion of a quantum particle is determined by its wave function ψ

$$\psi(x,t) = R(x,t)e^{iS(x,t)/\hbar}, \tag{11.8}$$

where R is the amplitude and S is the phase of the wave function and both are real-valued. Then, the momentum is computed as

$$\mathbf{p} = \nabla S. \tag{11.9}$$

In the hydrogen atom, equation (11.9) predicts that, since $\nabla S = 0$, electrons in an energy eigenstate are at rest. However, already Peter Holland (1999) pointed out that the above equation is valid only for spin-0 particles. By contrast, for particles with spin, the condition of Lorenz covariance of the Schrödinger law of motion implies that the momentum of a particle of spin s is given by

$$\mathbf{p} = \nabla S + \nabla\rho \times s. \tag{11.10}$$

where $\rho = \psi^*\psi$. On the basis of this proposal, Caroline Colijn and Edward Vrscay (2003a) computed the trajectories of the electron of the hydrogen atom in different energy eigenstates. They show that the momentum of the electron is perpendicular to the direction of the steepest increase of ρ and, then, its trajectories lie in the level surfaces of ρ. On the other hand, the momentum of the electron is also perpendicular to the direction of the spin, say, z. Therefore, the shape of the Bohmian trajectories of the electron can be obtained from the intersection between the level surfaces of ρ and the planes of constant z. This analysis allows the authors to conclude that the Bohmian trajectories are circular orbits about the direction of the spin in the cases 1s and 2s and periodic orbits for the 2p orbitals. Moreover, by numerically integrating the trajectories corresponding to the wave function ψ_{2p_x} for several initial conditions in the xz plane, the authors also show a good qualitative agreement between these trajectories and the shapes of the $2p_x$ orbitals as usually depicted in textbooks. In a second part of their paper, they also describe the trajectories of the electron when the wave function is a linear combination of 1s and $2p_0$ eigenstates evolving in time under the Hamiltonian of the hydrogen atom. In chemical terms, they offer an internal detail of what an electron does within an orbital resulting from an sp hybridization.

Following the strategy of taking into account the spin of the electron, Colijn and Vrscay (2003b) obtain the trajectory of the electron of the hydrogen atom as it undergoes the transition from state 1s to state $2p_0$ as a consequence of the action of an oscillating electron field. More recently, and explicitly as a continuation of the just mentioned works, Timko and Vrscay (2009) obtain the Bohmian trajectories for the two electrons in a helium atom.

Regardless of their formal and computational details, these works show that BM provides chemists with an image that matches their way of conceptualizing their object of study. BM describes electrons that not only orbit around nuclei, but move in a definite way as they make transitions to higher energy states. Those processes, which could only be thought of in statistical terms in SQM, acquire here a spatio-temporal characterization that facilitates chemists' conceptual understanding.

Ionization of molecules

The formation and breaking of chemical bonds is influenced by the motion of electrons and nuclei on an attosecond timescale. In particular, the ionization of molecules is a central process in chemistry as it can be the first step of a chemical reaction. The advent of sub-femtosecond laser pulses during the past decades has made possible to study the electron motion in chemical reactions, and the ionization process has been enhanced by driving the electron dynamics by intense femtosecond laser pulses. The central interest of this type of phenomena lies in the fact that the laser field can allow the control of chemical reactions.

Some recent studies show that the ionization probability of a hydrogen molecule depends on the internuclear distance and reaches a strong peak at a certain critical distance longer than its equilibrium value (Sawada, Sato, and Ishikawa 2014). In this context, ionization is traditionally conceived as a result of a tunnelling effect: the electron crosses the barrier created by the binding potential of the ionic core and the electric potential of the laser field. According to this description, the probability of the electron escaping from the molecule is expected to be greatest at the peak of the electric field. However, recent calculations show that molecules exhibit not only one but multiple bursts of ionization within a half-cycle of the laser field. In turn, such multiple ionization bursts are induced by an ultrafast localization of the electron density at one of the atoms (Takemoto and Becker 2010). Schematically, the process could be represented as follows:

$$\mathrm{H}_2 \rightarrow \mathrm{H}^+\mathrm{H}^- \rightarrow \mathrm{H}_2^+ + e^-. \tag{11.11}$$

Firstly, the electron density undergoes an abrupt localization in one of the atoms, and then the electron escapes and ionization occurs.

These theoretical results were obtained by SQM, by calculating the dynamic behaviour of the electron density. However, as noted above, knowing how electrons behave in chemical reactions is essential for chemists, and those computations do not supply that knowledge. It is precisely to understand the detailed dynamics of electrons in these processes that Norio Takemoto and Andreas Becker (2011), following previous works (e.g. Lai, Cai, and Zhan 2009a, 2009b, Botheron and Pons 2010), apply BM: the calculation of Bohmian trajectories allows them to visualize the electron transfer between the two nuclei. In the authors' words: 'However, results of numerical simulations for the flow of the electron probability density often do not reveal many details and cannot provide much further insights into the internal electron dynamics in the molecular ion. We therefore use the concept of Bohmian trajectories to provide a complementary picture of the dynamics. By analyzing the motion of the trajectories, we furthermore clarify that it is the phase difference of the local wave functions at the two protons that is the origin of the force, which is sometimes driving the electron in the direction opposite to the strong electric field of the laser light' (Takemoto and Becker 2011, p. 1).

Takemoto and Becker assume fixed nuclei at a given distance, and obtain the Bohmian trajectories of the electron for different initial conditions. The result they obtain agrees with the SQM calculations: regions of high electron density correspond to high densities of Bohmian trajectories. Furthermore, by computing the proportion of trajectories in which the electron leaves the molecule, they find the ionization probability. In this situation, they analyse the behaviour of the electrons and verify that the electrons do indeed concentrate around one of the nuclei before ionization. These results are also in agreement with those obtained by SQM.

However, the Bohmian approach makes it possible to visualize the trajectories, showing that, in many cases, the electron transfers back and forth between the two protons at an ultrafast timescale shorter than a half-cycle of the laser field. Furthermore, calculations in Bohmian terms make it possible to predict at which precise instants the trajectories change direction. From the chemical point of view, this description in terms of the precise spatio-temporal behaviour of electrons provides a more complete understanding of the phenomena underlying chemical reactions, and thus provides theoretical elements for their control, which are completely alien to the purely statistical approach of SQM.

Computational methods

A frequent objection to BM is that, by describing precise trajectories, it leads to very complicated computations that can be avoided from the SQM viewpoint. In other words, since BM makes exactly the same experimental predictions as SQM, the criticism continues, the increased complexity is not justified. This subsection challenges this objection by showing that, in certain cases, thinking in terms of Bohmian trajectories offers a pragmatic advantage in obtaining computational results.

Let us consider the wave function in polar form of equation (11.8). When $\psi(x,t)$ is substituted into the time-dependent Schrödinger equation,

$$i\hbar\frac{\partial\Psi}{\partial t} = \left(-\frac{\hbar^2}{2m}\nabla^2 + U\right)\Psi \tag{11.12}$$

it leads to a system of two coupled real equations:

$$\frac{\partial R^2}{\partial t} + \nabla\cdot\left(R^2\frac{\nabla S}{m}\right) = 0 \tag{11.13}$$

$$\frac{\partial S}{\partial t} + \frac{\nabla S^2}{2m} + U - \frac{\hbar^2}{2m}\frac{\nabla^2 R}{R} = 0. \tag{11.14}$$

Bohm noted that, in the classical limit of $\hbar \to 0$ equation (11.13) reduces to the Hamilton–Jacobi equation. Then, equation (11.8) can be conceived as a quantum Hamilton-Jacobi equation that governs the motion of quantum particles through a guidance equation:

$$\mathbf{v_B} = \dot{\mathbf{x}} = \frac{\nabla S}{m}. \tag{11.15}$$

Now, an equivalent description can be obtained by assuming, as formerly proposed by Erwin Madelung (1926), that the evolution of the quantum system is the diffusion of a quantum fluid with density ρ and current density $\mathbf{J}$. These two magnitudes are related by the continuity equation:

$$\frac{\partial \rho}{\partial t} + \nabla \cdot \mathbf{J} = 0 \tag{11.16}$$

where

$$\rho = R^2 \qquad \mathbf{J} = R^2 \frac{\nabla S}{m}. \tag{11.17}$$

Since the hydrodynamic velocity is $\mathbf{v_B} = \mathbf{J}/\rho$, it turns out to be equal to the Bohmian velocity:

$$\mathbf{v_H} = \frac{\mathbf{J}}{\rho} = \frac{\nabla S}{m} = \mathbf{v_B}. \tag{11.18}$$

The time integration of equation (11.17) leads to trajectories that describe the flow of the probability in configuration space, analogous to those found in classical hydrodynamics. On the basis of this analogy, this hydrodynamic version of BM is known as Madelung quantum hydrodynamics (see Wyatt 2005).

The idea of analysing chemical systems by means of a quantum hydrodynamic language goes back to the late 1960s and early 1970s. However, it underwent a remarkable boost by the 1990s (see, e.g. Dey and Deb 1995, 1997, 1998, 1999) on the basis of conceiving the approach as a resource that allows appealing to new computational propagation techniques in which the key element is precisely the Bohmian trajectory (for a recent survey, see Sanz and Miret-Artés 2014; for a compact overview, see Sanz (2019)). In fact, regardless of the metaphysical connotations of Bohmian trajectories, it cannot be denied that they are precisely defined from a formal viewpoint. Given the possibility of decoupling the phase and the amplitude of the wave function, the Schrödinger equation can be reformulated as a differential equation that can be solved by the method of characteristics: Bohmian trajectories are just the curves resulting from such a method, which is commonly used in mathematics as a tool to solve finite-order partial differential equations. This means that criticisms based on the computational complexity of BM are unfair: in many cases, Bohmian trajectories facilitate calculations by allowing the application of well-known mathematical tools. As Ángel Sanz, specialist in applications of BM, clearly states: 'A question that naturally arises is why Bohmian mechanics now receives more attention than about half a century ago, when it was hardly considered a worthless hidden-variable theory almost relegated to oblivion [...]. Perhaps a suitable answer can be given taking on a pragmatic view on what Bohmian mechanics is and what Bohmian mechanics is not in simple terms, that is, at a very basic (that is, pragmatic) level' (Sanz 2019, p. 259).

11.4 Final remarks

At present it is quite clear that, despite few opinions to the contrary, BM is not an interpretation of quantum mechanics: BM and SQM are different theories as they provide deeply different descriptions of the quantum domain. In the present chapter, the aim has been to argue that the ontological picture underlying the work of quantum chemists fits much better with BM than with SQM. To this end, we have pointed out the most relevant ontological pitfalls that SQM faces in describing the domain that chemistry in general, including quantum chemistry, deals with. In turn, we have commented some works that rely on Bohmian trajectories in order to provide spatio-temporal descriptions that allow chemists to visualize certain relevant processes and to give detailed explanations unavailable from SQM.

Given the usefulness of BM for chemistry, the question arises about the relationship between the chemical and physical fields. Of course, those who consider BM as the theory that provides the most conceptually reasonable and comprehensible description of the quantum world will argue in favour of a reductionist perspective: the domain of structural chemistry will be successfully reduced to a purely quantum description if the underlying quantum theory is BM. However, another perspective is possible. The advocates of the various pluralist viewpoints that populate the current philosophy of science (for some of the many examples, see Cartwright 1999, Torretti 2000, Hacking 2002, Lombardi and Labarca 2005, 2006, Chang 2012, Lombardi 2023) may consider that each scientific discipline constitutes its own ontology through its theoretical and experimental praxis. From this perspective, while quantum physicists trained in SQM can continue to work as they are accustomed to and seek the solution to the conceptual problems of the theory in one of its various interpretations, quantum chemists are enabled to conceive of the ontology underlying their field of study in terms of BM. In any case, whether from a reductionist or a pluralist position, what is difficult to deny is that chemists would feel more comfortable conceptually if the quantum mechanics they use in their work were BM instead of SQM.

Acknowledgements

The authors would like to remember and acknowledge Professor Brian Sutcliffe, who recently passed away: he was an inspiration for our work for many years. This chapter was partially supported by grants ID: 2002019020097BA from Universidad de Buenos Aires, PIP 11220200100483CO from CONICET, PICT-2020-SERIA-00782 and PICT-2018-04519 from FONCyT, and "Relaciones interteóricas entre la mecánica cuántica y otros dominios teóricos" from Universidad Austral.

References

Allori, V. (2013). Primitive ontology and the structure of fundamental physical theories. In Ney, A. and Albert, D.Z. (eds.) *The Wave Function: Essays on the Metaphysics of Quantum Mechanics*, pp. 58–75. New York: Oxford University Press.

Allori, V., Goldstein, S., Tumulka, R. and Zanghì, N. (2008). On the common structure of Bohmian mechanics and the Ghirardi-Rimini-Weber theory. *The British Journal for the Philosophy of Science*, **59**: 353–389.

Amann, A. (1992). Must a molecule have a shape?. *South African Journal of Chemistry*, **45**: 29–38.

Ardenghi, J.S., Castagnino, M., and Lombardi, O. (2009). Quantum mechanics: modal interpretation and Galilean transformations. *Foundations of Physics*, **39**: 1023–1045.

Bohm, D. (1952). A suggested interpretation of the quantum theory in terms of "hidden" variables. I & II. *Physical Review*, **85**(2): 166–193.

Bohm, D. (1953). Proof that probability density approaches $|\psi|^2$ in causal interpretation of the quantum theory. *Physical Review*, **89**: 458–466.

Botheron, P. and Pons, B. (2010). Self-consistent Bohmian description of strong field-driven electron dynamics. *Physical Review A*, **82**: 021404(R).

Cartwright, N. (1999). *The Dappled World: A Study of the Boundaries of Science.* Cambridge: Cambridge University Press.

Chang, H. (2012). *Is Water H_2O? Evidence, Pluralism and Realism*, Boston Studies in the Philosophy of Science. Dordrecht: Springer.

Chang, H. (2015). Reductionism and the relation between chemistry and physics. In Arabatzis, T., Renn, J. and Simoes, A. (eds.) *Relocating the History of Science: Essays in Honor of Kostas Gavroglu*, pp. 193–210. New York: Springer.

Colijn, C. and Vrscay, E.R. (2003a). Spin-dependent Bohm trajectories associated with an electronic transition in hydrogen. *Journal of Physics A*, **36**: 4689–4702.

Colijn, C. and Vrscay, E.R. (2003b). Spin-dependent Bohm trajectories for Pauli and Dirac eigenstates of hydrogen. *Foundations of Physics Letters*, **16**: 303–323.

de Broglie, L. (1928). "La nouvelle dynamique des quanta". In *Electrons et photons. Rapports et discussions du cinquième Conseil de Physique tenu à Bruxelles du 24 au 29 octobre 1927 sous les auspices de l'Institut International de Physique Solvay,* pp. 105–132 Paris: Gauthier-Villars. English translation in Bacciagaluppi G. and Valentini, A. (eds.) *Quantum Theory at the Crossroads. Reconsidering the 1927 Solvay Conference*, pp. 341–371. Cambridge: Cambridge University Press.

Dey, B. Kr and Deb, B.M. (1995). Time-dependent quantum fluid-dynamics of the photoionization of the He atom under an intense laser field. *International Journal of Quantum Chemistry*, **56**: 707–732.

Dey, B. Kr and Deb, B.M. (1997). A theoretical study of the high-order harmonics of a 200 nm laser from H^{-2} and HeH^{+}. *Chemical Physics Letters*, **276**: 157–163.

Dey, B. Kr and Deb, B.M. (1998). Stripped ion-Helium atom collision dynamics within a time-dependent quantum fluid density functional theory. *International Journal of Quantum Chemistry*, **67**: 251–271.

Dey, B. Kr and Deb, B.M. (1999). Direct ab initio calculation of ground-state electronic energies and densities for atoms and molecules through a time-dependent single hydrodynamical equation. *Journal of Chemical Physics*, **110**: 6229–6239.

Dugić, M. and Juknić-Dugić, J. (2008). What is system: The information-theoretic arguments. *International Journal of Theoretical Physics*, **47**: 805–813.

Dürr, D., Goldstein, S. and Zanghì, N. (2013). *Quantum Physics without Quantum Philosophy*. Berlin: Springer.

Esfeld, M. (2019). Individuality and the account of nonlocality: The case for the particle ontology in quantum physics. In Lombardi, O., Fortin, S., Lopez, C. and Holik, F. (eds.) *Quantum Worlds. Perspectives on the Ontology of Quantum Mechanics*, pp. 222–244. Cambridge: Cambridge University Press.

Fortin, S. and Lombardi, O. (2021). Is the problem of molecular structure just the quantum measurement problem?, *Foundations of Chemistry*, **23**: 379–395.

Fortin, S., Lombardi, O. and Martínez González, J.C. (2016). Isomerism and decoherence. *Foundations of Chemistry*, **18**: 225–240.

Fortin, S., Lombardi, O. and Martínez González, J.C. (2017). The relationship between chemistry and physics from the perspective of Bohmian mechanics. *Foundations of Chemistry*, **19**: 43–59.

Fortin, S., Lombardi, O. and Martínez González, J.C. (2018). A new application of the modal-Hamiltonian interpretation of quantum mechanics: The problem of optical isomerism. *Studies in History and Philosophy of Modern Physics*, **62**: 123–135.

Franklin, A. and Seifert. V. (2024). The problem of molecular structure just is the measurement problem. *The British Journal for the Philosophy of Science*, **75**: 31–59.

Grosso, B., Cooper, V., Pine, P., Hashibon, A., Yaish, Y. and Adler, J. (2015). Visualization of electronic density. *Computer Physics Communications*, **195**: 1–13.

Hacking, I. (2002). *Historical Ontology*. Cambridge MA: Harvard University Press.

Harshman, N.L. (2012). Observables and entanglement in the two-body system. *AIP Conference Proceedings*, **1508**: 386–390.

Hendry, R.F. (1998). Models and approximations in quantum chemistry. In Shanks, N. (ed.) *Idealization in Contemporary Physics*, pp. 123–142. Amsterdam-Atlanta: Rodopi.

Hendry, R.F. (2004). The physicists, the chemists, and the pragmatics of explanation. *Philosophy of Science*, **71**: 1048–59.

Hendry, R.F. (2008). Two conceptions of the chemical bond. *Philosophy of Science*, **75**: 909–920.

Hendry, R.F. (2010). Ontological reduction and molecular structure. *Studies in History and Philosophy of Modern Physics*, **41**: 183–191.

Hettema, H. (2012). *Reducing Chemistry to Physics. Limits, Models, Consequences.* Groningen: Rijksuniversiteit Groningen.

Holland, P. (1993). *Quantum Theory of Motion.* Cambridge: Cambridge University Press.

Holland, P. (1999). Uniqueness of paths in quantum mechanics. *Physical Review A*, **60**: 4326–4330.

Hund. F. (1927). Zur Deutung der Molekelspektren. III. *Zeitschrift für Physik*, **43**: 805–826.

Kochen, S. and Specker, E. (1967). The problem of hidden variables in quantum mechanics. *Journal of Mathematics and Mechanics*, **17**: 59–87.

Labarca, M. and Lombardi, O. (2010). Why orbitals do not exist?. *Foundations of Chemistry*, **12**: 149–157.

Lai, X., Cai, Q. and Zhan, M. (2009a) Above-threshold ionization photoelectron spectrum from quantum trajectory. *The European Physical Journal D*, **53**: 393–396.

Lai, X., Cai, Q. and Zhan, M. (2009b) From a quantum to a classical description of intense laser–atom physics with Bohmian trajectories. *New Journal of Physics*, **11**: 113035.

Lombardi, O. (2023). A Kantian-rooted pluralist realism for science. In Soto, C. (ed.) *Current Debates in Philosophy of Science: In Honor of Roberto Torretti*, pp. 77–95. Dordrecht: Springer-Synthese Library.

Lombardi, O. and Castagnino, M. (2010). Matters are not so clear on the physical side. *Foundations of Chemistry*, **12**: 159–166.

Lombardi, O. and Labarca, M. (2005). The ontological autonomy of the chemical world. *Foundations of Chemistry*, **7**: 125–148.

Lombardi, O. and Labarca, M. (2006). The ontological autonomy of the chemical world: a response to Needham. *Foundations of Chemistry*, **8**: 81–92.

Madelung, E. (1926). Quantentheorie in hydrodynamischer Form. *Zeitschrift für Physik*, **40**: 322–326.

Martínez González, J.C., Fortin, S. and Lombardi, O. (2019). Why molecular structure cannot be strictly reduced to quantum mechanics. *Foundations of Chemistry*, **21**: 31–45.

Nelson, P. (1990). How do electrons get across nodes? *Journal of Chemical Education*, **67**: 643–647.

Pauling, L. (1960). *The Nature of the Chemical Bond.* Ithaca: Cornell University Press; (1st edition 1939).

Primas, H. (1983). *Chemistry, Quantum Mechanics and Reductionism.* Berlin: Springer.

Primas, H. (1994). Hierarchic quantum descriptions and their associated ontologies. Iin Laurikainen, K.V., Montonen, C. and Sunnarborg, K. (eds.) *Symposium on the Foundations of Modern Physics 1994*, pp. 201–220. Gif-sur-Yvette: Editions Frontières.

Primas, H. (1998). Emergence in exact natural sciences. *Acta Polytechnica Scandinavica*, **91**: 83–98.

Sanz, A.S. (2019). Bohmian pathways into chemistry: A brief overview. In Oriols, X. and Mompart, J. (eds.) *Applied Bohmian Mechanics. From Nanoscale Systems to Cosmology*, pp. 257–330. Singapore: Jenny Stanford Publishing.

Sanz, A.S. and Miret-Artés, S. (2014). *A Trajectory Description of Quantum Processes. II. Applications. A Bohmian Perspective.* Berlin-Heidelberg: Springer-Verlag.

Sawada, R., Sato, T. and Ishikawa, K.L. (2014). Analysis of strong-field enhanced ionization of molecules using Bohmian trajectories. *Physical Review A*, **90**: 023404.

Scerri, E.R. (2011). Editorial 37. *Foundations of Chemistry*, **13**: 1–7.

Scerri, E.R. (2013). Philosophy of chemistry: where has it been and where is it going. In Llored, J.P. (ed.), *The Philosophy of Chemistry: Practices, Methodologies, and Concepts.* Newcastle: Cambridge Scholars Publishing.

Schlosshauer, M. (2004). Decoherence, the measurement problem, and interpretations of quantum mechanics. *Reviews of Modern Physics*, **76**: 1267–1305.

Sutcliffe, B.T. and Woolley, R.G. (2012). Atoms and molecules in classical chemistry and quantum mechanics. In Hendry, R.F. and Woody, A. (eds.) *Handbook of Philosophy of Science. Vol. 6, Philosophy of Chemistry*, pp. 387–426. Oxford: Elsevier.

Sutcliffe, B.T. and Woolley, R.G. (2022). Is chemistry really founded in quantum mechanics?. In Lombardi, O., Martínez González, J.C. and Fortin, S. (eds.) *Philosophical Perspectives in Quantum Chemistry*, pp. 173–202. Berlin-Heidelberg: Springer.

Takemoto, N. and Becker, A. (2010). Multiple ionization bursts in laser-driven hydrogen molecular ion. *Physical Review Letters*, **105**: 203004.

Takemoto, N. and Becker, A. (2011). Visualization and interpretation of attosecond electron dynamics in laser-driven hydrogen molecular ion using Bohmian trajectories. *Journal of Chemical Physics*, **134**: 074309.

Timko, J.A. and Vrscay, E.R. (2009). Spin-dependent Bohmian electronic trajectories for helium. *Foundations of Physics*, **39**: 1055–1071.

Torretti, R. (2000). Scientific realism and scientific practice. In Agazzi, E. and Pauri, M.(eds.) *The Reality of the Unobservable: Observability, Unobservability and their Impact on the Issue of Scientific Realism*, pp. 113–122. Boston Studies in the Philosophy of Science. Dordrecht: Kluwer.

Trost, J. and Hornberger, K. (2009). Hund's paradox and the collisional stabilization of chiral molecules. *Physical Review Letters*, **103**: 023202.

Viola, L. and Barnum, H. (2010). Entanglement and subsystems, entanglement beyond subsystems, and all that. In Bokulich, A. and Jaeger, G. (eds.) *Philosophy of Quantum Information and Entanglement*, pp. 16–43. Cambridge: Cambridge University Press.

Woolley, R.G. (1978). Must a molecule have a shape? *Journal of the American Chemical Society*, **100**: 1073–1078.

Woolley, R.G. (1982). Natural optical activity and the molecular hypothesis. *Structure and Bonding*, **52**: 1–35.

Woolley, R.G. and Sutcliffe, B.T. (1977). Molecular structure and the Born-Oppenheimer approximation. *Chemical Physics Letters*, **45**: 393–398.

Wyatt, R.E. (2005). *Quantum Dynamics with Trajectories: Introduction to Quantum Hydrodynamics*. New York: Springer.

Zuo, J. Kim, M., O'Keefe, M. and Spence, J. (1999). Direct observation of d-orbital holes and Cu-Cu bonding in Cu_2O. *Nature*, **401**: 49–52.

Part IV

Non-Locality, Extensions of the de Broglie–Bohm Theory, and Applications

12

Local causality in the works of Einstein, Bohm, and Bell

Aurélien Drezet

12.1 Introduction and motivations

David Bohm is a central figure in the history of quantum mechanics (QM). As it is well known, in 1952 he proposed a deterministic hidden-variable theory that is able to complete standard QM, and to reproduce its statistical predictions and results (Bohm 1952). Moreover, the theory he developed was actually a rediscovery of the old pilot-wave theory introduced by Louis de Broglie in 1927 at the fifth Solvay conference. Seventy years later retrospectively, the most original contribution of Bohm compared to what de Broglie did is perhaps his analysis of the Einstein–Podolsky–Rosen (EPR) paradox concerning non-locality and completeness of QM.[1] Indeed, the goal of the EPR article (Einstein *et al.* 1935) was to show that if we assume the principle of Einstein locality, QM must be incomplete. More precisely, EPR shows that either QM is incomplete or quantum mechanics is non-local, i.e. it violates Einstein's locality principle. In this context, Bohm (1952) showed that his own hidden-variable theory able to complete QM is explicitly non-local. While this does not contradict the EPR results seen as a theorem, Bohm approach was certainly the option 'which Einstein would have liked least'.[2] Of course, this was just the beginning of the story: in 1964, John Bell, based on EPR work, discovered his famous theorem (Bell 2004, chapter 2) firmly establishing that QM (irrespective of being complete or incomplete) must be non-local.

The previous summary reflects the position advocated by some physicists and philosophers, including Goldstein *et al.* (2023), Maudlin (2014), Bricmont (2016), Norsen (2006, 2011), Albert and Galchen (2009), and of course Bell (2004), but this is not the majority view. The majority view claims that Bell's theorem is concerned with 'local realism', i.e. that it forces us to either abandon realism or locality. Since most physicists would disapprove relaxing locality, they actually consider the theorem as a strong indication that one must abandon realism. Most confusions concerning EPR, Bell's theorem, and non-locality arise from wavy and vague arguments associated with

[1] In this context, we point out that de Broglie rejected non-locality and criticized Bell's theorem (cf. de Broglie 1974).

[2] For a discussion of the status of special relativity and non-locality in relativistic dBB theories, see the contribution of Valia Allori in this volume.

the definitions of Einstein locality and realism. In order to celebrate the work of Bohm on non-locality, the present author thinks it could be useful to summarize once more the EPR–Bohm–Bell connection.

12.2 The story

Everything started with Einstein's disappointment about the way QM was built and axiomatized as a complete indeterministic theory. According to Einstein, the fact that quantum mechanics is a statistical theory should be explained by a deeper dynamical approach (probably deterministic) that is able to complete the statistical predictions by a classical-like type of mechanical and realistic explanations. One should not forget that Einstein was a pioneer of classical statistical mechanics with his famous 1905 interpretation of Brownian motion as resulting of an underlying mechanical process. He also developed general relativity which is a fundamentally deterministic theory of gravitation and space-time. Hence, Einstein could certainly not believe that God plays dice at the microscopic scale. Moreover, for him realism was a central prerequisite (more important even than determinism). In 1953, in a *Festschrift* book to honour de Broglie achievements Einstein wrote: '*I am not blushing to put the concept of "real state of a system" at the center of my meditation*' (Einstein 1953, p. 7).

In 1927, during the Solvay conference, he debated with Niels Bohr on the possibility to beat or defeat the Heisenberg principle using correlations between a microscopic quantum particle (involved in a double-slit interference experiment) and some macroscopic *a priori* classical devices entangled with the particle. Bohr recollection of these discussions shows that Einstein at that time underestimated the coherence of QM and did not realize that the entanglement between particles and macroscopic systems in his 'which-path' experiment only makes sense if the macroscopic apparatus are also analysed within QM (cf. Bohr 1949).[3] In 1930, at the sixth Solvay conference, Einstein and Bohr continued this debate with the famous 'photon-box' *Gedankenexperiment* where Einstein attempted to circumvent the Heisenberg principle (in the time–energy domain) by using correlations between a particle in a box and a device measuring its time and energy (a detailed analysis of the paradox is provided in Nikolić (2012)). Once more, Bohr showed to Einstein that it is not allowed to contradict QM with this kind of approach since QM must be used for describing every parts of the whole indivisible system in agreement with the complementarity principle. Moreover, these two encounters with Bohr convinced Einstein that QM is mathematically and physically self-consistent, and that it is unlikely to try defeating Heisenberg's principle directly. However, he did not stop the fight: in 1935 with his two collaborators Podolsky and Rosen he proposed the famous paradox involving entanglement between two remote particles (cf. Einstein *et al.* 1935). The core of the EPR article is the completeness of QM. By a complete physical theory, they meant that 'every element of the physical reality must have a counterpart in the physical theory'. Questioning the completeness of QM is therefore suggesting that some effects or correlations having an empirical

[3] Retrospectively, one can say that the most astonishing fact for Einstein was perhaps the existence of Heisenberg's shift cut or split allowing one to describe any measuring apparatus as quantum or classical in order to remove paradoxes involving the uncertainty principle. Heisenberg also applied this approach in his unpublished response to EPR; see Crull and Bacciagaluppi (2011).

nature are not predicted or explained by the theory. Of course, since QM is supposed to describe everything this would be impossible. Yet, EPR showed that by associating with QM a very natural and intuitive feature of the classical world (i.e. that nobody would like to abandon), we obtain a contradiction and therefore conclude that QM cannot be complete and at the same time satisfy this natural property. This natural and intuitive property that EPR added is of course Einstein locality: a fundamental feature of causality in relativistic space-time.

The whole EPR deduction is based on the position and momentum observables for two entangled particles, but it is common – and we will follow this strategy here – to use instead the example proposed by Bohm in 1951 of two spin-$\frac{1}{2}$ particles 1 and 2 entangled in the singlet state:

$$|\psi\rangle = \frac{1}{\sqrt{2}}\Big(|+1_{\hat{\mathbf{z}}}\rangle_1|-1_{\hat{\mathbf{z}}}\rangle_2 - |-1_{\hat{\mathbf{z}}}\rangle_1|+1_{\hat{\mathbf{z}}}\rangle_2\Big) \tag{12.1}$$

where $|+1_{\hat{\mathbf{z}}}\rangle_j, |-1_{\hat{\mathbf{z}}}\rangle_j$ are the two eigenstates of the $\sigma_z^{(j)}$ Pauli matrices for the jth particle. As a consequence, we have $(\sigma_z^{(1)} + \sigma_z^{(2)})|\psi\rangle = 0$ that expresses the perfect anticorrelation between the two z-spin components. Importantly, by symmetry we have also $(\sigma_x^{(1)} + \sigma_x^{(2)})|\psi\rangle = 0 = (\sigma_y^{(1)} + \sigma_y^{(2)})|\psi\rangle$ expressing the perfect spin anticorrelation in every direction x, y, etc.. Due to this perfect anticorrelation between the particles, experimentalists Alice and Bob recording respectively the projected spin of particle 1 and 2 along a common axis $\hat{\mathbf{n}}$ will naturally obtain the joint probability

$$P(\alpha, -\alpha/\hat{\mathbf{n}}_1, \hat{\mathbf{n}}_2 = \hat{\mathbf{n}}_1, |\psi\rangle) = \frac{1}{2}, \; P(\alpha, \alpha/\hat{\mathbf{n}}_1, \hat{\mathbf{n}}_2 = \hat{\mathbf{n}}_1, |\psi\rangle)) = 0 \tag{12.2}$$

where $\alpha = \pm 1$, and the probabilities are conditioned on the common analysis direction $\hat{\mathbf{n}}_1 = \hat{\mathbf{n}}_2 = \hat{\mathbf{n}}$ of the two independent Stern–Gerlach splitters used by Alice and Bob and by the singlet wave function $|\psi\rangle$. These conditions explicit the perfect anticorrelation at the core of EPR work.

Now comes the crux: EPR introduce the Einstein-locality assumption based on natural features of the classical world picture concerning correlations and relativistic causality. The main idea is that a local operation made by Alice on particle 1 at space-time point x_1 should not influence what is happening to the second particle recorded by Bob at space-time point x_2 if the two events are space-like separated (so that no signal could propagate between these two points). As Einstein wrote in 1949:

> But on one supposition we should, in my opinion, absolutely hold fast: the real factual situation of the system S_2 is independent of what is done with the system S_1 which is spatially separated from the former (Einstein 1949, p. 85).

Assuming this natural relativistic hypothesis, the EPR deduction is easy to understand. From Equation 12.2 we have perfect anticorrelation and therefore if Alice is recording the spin-1/2 component of particle 1 along the direction $\hat{\mathbf{n}}_1 = \hat{\mathbf{z}}$ with the result $+1$ (respectively -1), we know that Bob must necessarily record in his lab the result -1 (respectively) $+1$) for the same settings $\hat{\mathbf{n}}_2 = \hat{\mathbf{z}}$. Moreover, nothing obliges

Bob to measure the spin projection of his particle along the same direction and he could for instance decide to record the spin along the x-direction: $\hat{\mathbf{n}}_2 = \hat{\mathbf{x}}$. Now Bob or Alice could take their decision at the last moment, and since the locality principle is assumed to occur no influence that could modify their results are allowed to propagate. As a direct consequence, we can make a counterfactual reasoning: if Bob records along the x-direction and obtains for example $+1$ and Alice obtained for example the result -1 along the z-direction, then we certainly know that Alice should have obtained the result -1 along the x-direction and Bob the result $+1$ along the z-direction, even though these experiments have not actually been done. Crucially, the legitimacy of the counterfactual reasoning of EPR is mandated by Einstein locality and is not an independent hypothesis. This justifies the famous introduction of 'elements of reality' by EPR:

> If, without in any way disturbing a system, we can predict with certainty (i.e., with probability equal to unity) the value of a physical quantity, then there exists an element of physical reality corresponding to this physical quantity (Einstein *et al.* 1935, p. 777).

Of course, EPR understood perfectly well that counterfactual reasoning is in general forbidden for a single particle using the usual approach to QM. This is because for a single particle one could invoke Heisenberg's principle to impose a strong form of complementarity and contextuality: it is impossible to record in one single experiment the spin projection for the x- and z-direction because σ_x and σ_z do not commute. Therefore, one must choose between one experiment or the other, and if one is doing sequential experiments it is known that dispersion will occur in agreement with Heisenberg's principle. To speak about determinations and counterfactual hidden properties that could be observed but actually are not is considered as useless in standard QM. To cite Asher Peres: 'Unperformed experiments have no results' (Peres 1978). However, employing Einstein locality EPR found a clean way to go around Heisenberg's principle limitations. Assuming locality we can know counterfactually the spin components of the two particles along orthogonal directions x and z even though we only actually measured the spin of particle 1 (respectively 2) along the z-direction (respectively x-direction). Similarly, with two measurements and assuming locality, we actually know four spin projection values along directions x–z for the two particles, and this is impossible if we assume that QM is complete. In other words, assuming that QM is local (QM-L) and complete (QM-C), we conclude that QM is incomplete (QM-IC).

This is a wonderful logical contradiction that can be formally written: 'QM-L AND QM-C = False' or equivalently '$\neg$ QM-L OR $\neg$ QM-C = True', i.e., 'QM-NL OR QM-IC = True' with $\neg$ QM-L is the negation of QM-L, i.e. QM is non-local and similarly $\neg$ QM-C is actually QM-IC. EPR theorem is unavoidable: it implies that if QM-L is true, then we must have QM-IC true; and contra positively if QM-C is true we must have QM-NL true:

$$\text{QM-L} \Longrightarrow \text{QM-IC, and } \text{QM-C} \Longrightarrow \text{QM-NL}. \tag{12.3}$$

Importantly, EPR leads to three alternatives:

$$
\begin{aligned}
&\text{(i)} && \text{QM-L AND QM-IC} \\
&\text{(ii)} && \text{QM-NL AND QM-IC} \\
&\text{(iii)} && \text{QM-NL AND QM-C}
\end{aligned} \tag{12.4}
$$

where (i) was favoured by Einstein, (ii) by Bohm, and (iii) by Bohr and followers.[4] The beauty and logic of the EPR deduction/theorem is often under-appreciated and the fact that counterfactuality and determinism are actually derived and not inferred by EPR is still nowadays misunderstood.

12.3 Bell's beables

An important comment must be made concerning EPR theorem. Indeed, while this result is rigorous, EPR did not present it in a very formal way. The notions of locality and determinism were used in a very intuitive sense as shown, for instance, in the previous quote of Einstein concerning locality. Therefore, one would have to be more precise on the definition of locality. This step was taken by Bell (as discussed below). Furthermore, the path of EPR to incompleteness implied determinism, meaning that we actually have QM-L $\Longrightarrow$ QM-D, i.e. if QM is local, then it is deterministic, and from that QM-D $\Longrightarrow$ QM-IC. But this derived notion of determinism is not very clearly discussed by EPR. For example, considering the actual z-component of the spin-1/2 particle measured by Alice, we see that the 'element of reality' $v(\sigma_z^{(1)})$ associated with particle 1 implies through locality the existence of a counterfactual element of reality $v(\sigma_z^{(2)}) = -v(\sigma_z^{(1)})$ for the second particle even if Bob actually measured $v(\sigma_x^{(2)})$. Importantly, EPR do not require that the observed values $v(\sigma_z^{(2)})$ pre-exist before the measurement. For example, in the dBB theory the spins are observables that are predetermined by the contextual dynamics. Therefore, in a deterministic and local theory, we only assume that something before the measurement predetermined the elements of reality observed (or not) by Alice and Bob. Writing λ as this initial condition/predetermination, we see that actually EPR determinism presupposed only that we have two functions $v(\sigma_n^{(1)}) := A(\lambda, \hat{\mathbf{n}}, |\psi\rangle)$, $v(\sigma_n^{(2)}) := B(\lambda, \hat{\mathbf{n}}, |\psi\rangle) = -A(\lambda, \hat{\mathbf{n}}, |\psi\rangle)$, with $\sigma_n^{(i)} = \mathbf{n} \cdot \boldsymbol{\sigma}^{(i)}$ being the spin operator of particle $i = 1$ or 2 along the analysis direction $\mathbf{n}$.

In order to write the EPR reasoning in a formal way and discuss locality for deriving determinism, we must use Bell notations for the probabilities of beables. Bell introduced the concept of 'beable' (Bell 2004, chapters 5, 7, 16, and 24) to characterize every physical (i.e. actual or real) properties belonging to the system that must influence the measured correlations. Crucially, here the beables must include the 'classical' Stern–Gerlach devices with (in general different) directions $\hat{\mathbf{n}}_j$ (i.e. associated with external fields and potentials acting on the two parts of the quantum system under study), the possible hidden supplementary variables λ (e.g. the particle positions in the dBB pilot-wave theory), and the wave function $|\psi\rangle$. It is unfortunate that most

[4] A different way to explain this is that either QM is incomplete (i.e. regrouping options (i) and (ii) of Equation. 12.4) *or* (and this or is exclusive) QM is complete and non-local, i.e. option (iii) of (12.4).

of Bell's commentators fail to realize that $|\psi\rangle$ actually belongs to the fundamental beables listed by Bell (as an exception see Oldofredi (2022)). Now it is always possible to write the quantum joint probabilities $P(\alpha, \beta/\hat{\mathbf{n}}_1, \hat{\mathbf{n}}_2, |\psi\rangle) = |\langle\alpha_{\hat{\mathbf{n}}_1}, \beta_{\hat{\mathbf{n}}_2}|\psi\rangle|^2$ as:

$$P(\alpha, \beta/\hat{\mathbf{n}}_1, \hat{\mathbf{n}}_2, |\psi\rangle) = \int_\Lambda P(\alpha, \beta/\lambda, \hat{\mathbf{n}}_1, \hat{\mathbf{n}}_2, |\psi\rangle)\rho(\lambda, \hat{\mathbf{n}}_1, \hat{\mathbf{n}}_2, |\psi\rangle)d\lambda \tag{12.5}$$

where $\alpha, \beta = \pm 1$ and the integral spans over the hidden-variable space Λ.[5] At each run of the quantum experiment, an actual value of λ is selected. Bell used this notation in 1964 and 1971 (Bell 2004, chapters 2 and 4) to derive his theorem but actually we could also generalize it as

$$P(\alpha, \beta/\hat{\mathbf{n}}_1, \hat{\mathbf{n}}_2, |\psi\rangle) = \int_\Omega P(\alpha, \beta/\omega, \hat{\mathbf{n}}_1, \hat{\mathbf{n}}_2, |\psi\rangle)\rho(\omega, \hat{\mathbf{n}}_1, \hat{\mathbf{n}}_2, |\psi\rangle)d\omega \tag{12.6}$$

which is formally the same as the original formula Equation 12.5 used by Bell in his 1964 paper. We introduced the beable $\omega := (\lambda, \theta)$, where θ is a new beable describing the quantum state and Ω is the full 'ontic' space. Again a value of θ is actualized at each run of the experiment. Of course, since the probability description contains already $|\psi\rangle$ as a condition, we expect θ to be somehow redundant. For instance, according to Beltrametti and Bugajski for a complete theory (i.e. without λ), we can always write:

$$P(\alpha, \beta/\hat{\mathbf{n}}_1, \hat{\mathbf{n}}_2, |\psi\rangle) = \int_\Theta P(\alpha, \beta/\hat{\mathbf{n}}_1, \hat{\mathbf{n}}_2, |\theta\rangle)\delta(|\theta\rangle - |\psi\rangle)d|\theta\rangle \tag{12.7}$$

where now the beable $|\theta\rangle$ belongs to the Hilbert space of the two-spin system equivalent to Θ (Beltrametti and Bugajski 1995; Drezet 2012). We have $P(\alpha, \beta/\hat{\mathbf{n}}_1, \hat{\mathbf{n}}_2, |\theta\rangle) = |\langle\alpha_{\hat{\mathbf{n}}_1}, \beta_{\hat{\mathbf{n}}_2}|\theta\rangle|^2$ and $\rho(\hat{\mathbf{n}}_1, \hat{\mathbf{n}}_2, |\psi\rangle, |\theta\rangle) = \delta(|\theta\rangle - |\psi\rangle)$.[6] This representation corresponds to what Harrigan and Spekkens (2010) called an ontic description of the quantum state where the Dirac distribution associates a strongly localized density of probability with $|\psi\rangle$. It is interesting to note that this is not the only ontic representation of the EPR state. For example, inspired by Wigner phase distribution, Scully (1983) developed an angular representation of the $|\psi\rangle$ state also involving a Dirac distribution. Scully specifically considered the joint probability $P(\alpha, \beta/\hat{\mathbf{n}}_1, \hat{\mathbf{n}}_2, |\psi\rangle) := P(\alpha, \beta/\varphi_1, \varphi_2, |\psi\rangle)$, where $\hat{\mathbf{n}}_1, \hat{\mathbf{n}}_2$ are two *a priori* different unit vectors contained in the x–z plane (the particles move along the two opposed $\pm y$ directions and $\varphi_j := \widehat{\hat{\mathbf{z}}, \hat{\mathbf{n}}_j}$ are the angles between the vector $\hat{\mathbf{n}}_j$ characterizing the spin analysers and the vertical common axis $\hat{\mathbf{z}}$). He found

[5]We have $\int_\Lambda \rho(\lambda, \hat{\mathbf{n}}_1, \hat{\mathbf{n}}_2, |\psi\rangle)d\lambda = 1$ and $\sum_{\alpha,\beta} P(\alpha, \beta/\lambda, \hat{\mathbf{n}}_1, \hat{\mathbf{n}}_2, |\psi\rangle) = 1$.

[6]More rigorously writing $|\theta\rangle = \sum_{\alpha,\beta} \theta_{\alpha,\beta}|\alpha_{\hat{\mathbf{n}}_1}, \beta_{\hat{\mathbf{n}}_2}\rangle$ with $\theta_{\alpha,\beta} = \theta'_{\alpha,\beta} + i\theta''_{\alpha,\beta} := \langle\alpha_{\hat{\mathbf{n}}_1}, \beta_{\hat{\mathbf{n}}_2}|\theta\rangle \in \mathbb{C}$, a generally complex valued amplitude (with $\alpha, \beta = \pm 1$), we have $\delta(|\theta\rangle - |\psi\rangle) := \prod_{\alpha,\beta} \delta(\theta'_{\alpha,\beta} - \psi'_{\alpha,\beta})\delta(\theta''_{\alpha,\beta} - \psi''_{\alpha,\beta})$ with $\psi_{\alpha,\beta} := \langle\alpha_{\hat{\mathbf{n}}_1}, \beta_{\hat{\mathbf{n}}_2}|\psi\rangle$, and $d|\theta\rangle := \prod_{\alpha,\beta} d\theta'_{\alpha,\beta}d\theta''_{\alpha,\beta}$. We also have $P(\alpha, \beta/\hat{\mathbf{n}}_1, \hat{\mathbf{n}}_2, |\theta\rangle) = |\theta_{\alpha,\beta}|^2$.

$$P(\alpha,\beta/\varphi_1,\varphi_2,|\Psi\rangle)=\iint P(\alpha/\theta_1,\varphi_1,|\Psi\rangle)P(\beta/\theta_2,\varphi_2,|\Psi\rangle)\rho(\theta_1,\theta_2/\varphi_1,\varphi_2,|\Psi\rangle)d\theta_1 d\theta_2 \tag{12.8}$$

with $P(\alpha/\theta_i,\varphi_i,|\psi\rangle) = \frac{1+\alpha\cos(\varphi_i-\theta_i)}{2}$ and $\rho(\theta_1,\theta_2/\varphi_1,\varphi_2,|\Psi_{\text{EPR}}\rangle) = \frac{1}{2}\delta(\theta_2-\theta_1-\pi)[\delta(\theta_1-\varphi_1)+\delta(\theta_1-\varphi_1-\pi)]$. This leads to $P(\alpha,\beta/\varphi_1,\varphi_2,|\psi\rangle) = \frac{1-\alpha\beta\cos(\varphi_1-\varphi_2)}{4}$ which is the quantum prediction for the singlet state.

In later works, Argaman and Di Lorenzo rediscovered the model with a more symmetric probability density $\rho(\theta_1,\theta_2/\varphi_1,\varphi_2,|\Psi_{EPR}\rangle) = \frac{1}{4}\delta(\theta_2-\theta_1-\pi)[\delta(\theta_1-\varphi_1)+\delta(\theta_1-\varphi_1-\pi)+\delta(\theta_2-\varphi_2)+\delta(\theta_2-\varphi_2-\pi)]$ (cf. Argaman 2010; Di Lorenzo 2012). We will come back to the interesting properties of these models concerning statistical independence and local-causality. Moreover, for the moment the crucial issue is that these various θ variables have not to be interpreted as hidden or supplementary variables λ (even if it was the interpretations made by the authors of these models). Instead, these models provide ontic representations of a complete quantum theory. In other words, the examples provided by Beltrametti and Bugajski, or Scully and others explicitly show that Equation 12.6 always makes sense even for approaches where QM is supposed to be complete. This is the reason why Equation 12.6 can be seen as the natural generalization of Bell's formalism.

In his later writings of 1976 and 1991, Bell emphasized the importance of the generality of the beable concepts and made an explicit use of Equation 12.6 for general beables ω (Bell 2004, chapters 7 and 24). In his famous article *Bertlmann's socks and the nature of reality* Bell wrote

> It is notable that in this argument [i.e. Bell and EPR theorems] nothing is said about the locality, or even localizability, of the variable λ [our ω]. These variables could well include, for example, **quantum mechanical state vectors**, which have no particular localization in ordinary space-time. It is assumed only that the outputs A and B, and the particular inputs a and b [i.e., $\hat{\mathbf{n}}_1,\hat{\mathbf{n}}_2$], are well localized (Bell 2004, pp. 153-154).

In my opinion, beside the issue of locality, this important point concerning ω notations answers some authors who unfortunately still continue to believe that hidden variables are a prerequisite of both Bell and EPR deductions. For example, in their 2014 QBist manifesto Fuchs, Mermin and Schack wrote:

> The parameter λ [again our ω] is undefined. It does not appear in the quantum theory. Nor has anybody ever suggested what in the experience of an agent λ [ω] might correspond to. In QBism this puts it outside the scope of physical science (Fuchs *et al.* 2014).

Clearly, the counterexamples of Beltrametti and Bugajski, or Scully illustrate the error. Furthermore, this proves that Bell's formalism used for discussing the non-locality of QM is actually independent of statements about realism or hidden variables. Therefore, this implies that repeated claims opposing locality and realism are based on a complete misunderstanding of EPR and Bell's work.

Moreover, once we accept Equation 12.6, we can go back to the EPR–Bell reasoning. This allows us to provide a rigorous definition of Einstein-locality or as Bell says 'local-causality'. Considering the elementary probability $dP(\alpha,\beta,\lambda,\theta/\hat{\mathbf{n}}_1,\hat{\mathbf{n}}_2,|\psi\rangle)$, we have always

$$dP(\alpha,\beta,\lambda,\theta/\hat{\mathbf{n}}_1,\hat{\mathbf{n}}_2,|\psi\rangle) = P(\alpha,\beta/\omega,\hat{\mathbf{n}}_1,\hat{\mathbf{n}}_2,|\psi\rangle)\rho(\omega,\hat{\mathbf{n}}_1,\hat{\mathbf{n}}_2,|\psi\rangle)d\omega$$
$$= P(\alpha/\beta,\omega,\hat{\mathbf{n}}_1,\hat{\mathbf{n}}_2,|\psi\rangle)P(\beta/\omega,\hat{\mathbf{n}}_1,\hat{\mathbf{n}}_2,|\psi\rangle)\rho(\omega,\hat{\mathbf{n}}_1,\hat{\mathbf{n}}_2,|\psi\rangle)d\omega. \quad (12.9)$$

Now, Bell showed through several important papers that the good definition of local causality actually implies three mathematically precise conditions:

$$P(\alpha/\beta,\omega,\hat{\mathbf{n}}_1,\hat{\mathbf{n}}_2,|\psi\rangle) = P_1(\alpha/\omega,\hat{\mathbf{n}}_1,|\psi\rangle) \quad (12.10)$$
$$P(\beta/\omega,\hat{\mathbf{n}}_1,\hat{\mathbf{n}}_2,|\psi\rangle) = P_2(\beta/\omega,\hat{\mathbf{n}}_2,|\psi\rangle) \quad (12.11)$$
$$\rho(\omega,\hat{\mathbf{n}}_1,\hat{\mathbf{n}}_2,|\psi\rangle) = \rho(\omega,|\psi\rangle). \quad (12.12)$$

These conditions have been the subject of intense debates concerning locality and causality. The last (Equation 12.12) is often named 'setting' or 'measurement independence' in the literature (and sometimes 'freedom of choice condition') meaning that the distribution of beable ω prepared initially at the source is naturally expected in any 'good' causal theory to be independent of the directions $\hat{\mathbf{n}}_1, \hat{\mathbf{n}}_2$ that could be selected by macroscopic devices located in the remote past along the backward cones, e.g. by photons coming from far-away quasars existing billions of light years away of Alice and Bob. If we relax this condition, we would get some fatalistic or superdeterministic theories (including for instance retrocausal models). Accepting Bell's condition, Equation 12.12 implies a rejection of these fatalistic possibilities or loopholes (this natural assumption is accepted even in the non-local model of de Broglie and Bohm). The two other conditions involve some hypotheses about outcome and parameter/setting independence. In particular, assuming again no superdeterminism and assuming that the space-time points x_1, x_2 of the two recording events by Alice and Bob are space-like separated, we obtain with Equations 12.10 and 12.11 the more precise formulation of Einstein-locality quoted before forbidding spooky and parasitic faster-than-light communications.[7]

The following step for re-deriving the EPR theorem is now straightforward: by assuming Einstein/Bell locality, i.e. Equations 12.10-12.12, we deduce for every ω such that $\rho(\omega,|\psi\rangle) \neq 0$

$$P_1(\alpha/\omega,\hat{\mathbf{n}},|\psi\rangle)P_2(\alpha/\omega,\hat{\mathbf{n}},|\psi\rangle) = 0 \quad (12.13)$$
$$P_1(\alpha/\omega,\hat{\mathbf{n}},|\psi\rangle)P_2(-\alpha/\omega,\hat{\mathbf{n}},|\psi\rangle) \geq 0 \quad (12.14)$$

These conditions look harmless, but now suppose that for a given actualized ω the state $|+1_{\hat{\mathbf{n}}}\rangle_1|-1_{\hat{\mathbf{n}}}\rangle_2$ was recorded as an outcome. From Equation 12.14 it means that $P_1(+1/\omega,\hat{\mathbf{n}},|\psi\rangle) \neq 0$, $P_2(-1/\omega,\hat{\mathbf{n}},|\psi\rangle) \neq 0$, and thus, from Equation 12.13, we must have $P_1(-1/\omega,\hat{\mathbf{n}},|\psi\rangle) = P_2(+1/\omega,\hat{\mathbf{n}},|\psi\rangle) = 0$, and from probability conservation $P_1(+1/\omega,\hat{\mathbf{n}},|\psi\rangle) = P_2(-1/\omega,\hat{\mathbf{n}},|\psi\rangle) = 1$. A similar situation would occur if for the given ω the result $|-1_{\hat{\mathbf{n}}}\rangle_1|+1_{\hat{\mathbf{n}}}\rangle_2$ is obtained. In other words, we derive a deterministic theory: the EPR state requires probabilities

[7]That was clearly the great achievement of Aspect's group in the 1980s to have experimentally developed such a configuration, closing the communication loophole.

$$P_1(\alpha/\omega, \hat{\mathbf{n}}_1, |\psi\rangle) = \delta_{\alpha, A(\lambda, \hat{\mathbf{n}}_1, |\psi\rangle)} = \frac{1 + \alpha \cdot A(\lambda, \hat{\mathbf{n}}_1, |\psi\rangle)}{2} \tag{12.15}$$

and similarly for $P_2(\beta/\omega, \hat{\mathbf{n}}_2, |\psi\rangle)$ with values zero or one and beables[8] $A(\lambda, \hat{\mathbf{n}}, |\psi\rangle) = -B(\lambda, \hat{\mathbf{n}}, |\psi\rangle) = \pm 1$. From this derivation of determinism, all the previous EPR deductions follow and in particular the logical results expressed by Equations (12.3) and (12.4). Most importantly, the definition of QM-L and its negation QM-NL are transparent: this concludes our re-derivation of EPR theorem.

We stress that in the case QM-C using beables ω is *a priori* not mandatory to derive the EPR contradiction (cf. Norsen 2006, 2011). Indeed, accepting locality is equivalent to assume the constraint $P(\alpha, \beta/\hat{\mathbf{n}}_1, \hat{\mathbf{n}}_2, |\psi\rangle) = P(\alpha/\hat{\mathbf{n}}_1|\psi\rangle)P(\beta/\hat{\mathbf{n}}_2|\psi\rangle)$, which is obviously wrong (Norsen 2006; Zukowski and Brukner 2014). Moreover, using ω leads to a more general and complete proof.

12.4 The local realism rhetoric

Before concluding, there is still one issue that we must discuss: the local realism controversy debated in several places. The local realism rhetoric originates from a very serious definition of 'objective local theory' or 'local realistic theory' by Clauser, Horne, and Shimony in the 1970s as an honest substitute to the pejorative term 'local hidden variables' (Laudisa 2023). Moreover, the idea that one can actually decouple realism from locality in the EPR and Bell's theorems is very strange and results from a misunderstanding. Indeed, from the three alternatives listed in (12.4) only (iii), i.e. 'QM-C AND QM-NL' refers to QM being complete. With Bell's theorem rejecting alternative (i), we are left with (ii) 'QM-IC AND QM-NL', and (iii). But, note that the meaning of non-locality changes from theory to theory. An advocate of complementarity will, following Bohr, more probably speak about quantum wholeness, indivisibility of phenomena, or non-separability in order to stress the difference with the action-at-a-distance semantic of Newtonian classical mechanics or hidden variables à la dBB. Therefore, with such rephrasing the two surviving alternatives of Bell's theorem read:

$$\begin{aligned} &\text{(ii)} \;\; \text{QM-NL AND QM-IC} \\ &\text{(iii)}' \;\; \text{QM-NS} \;\; \sim \;\; \text{QM-C} \end{aligned} \tag{12.16}$$

where QS-NS is an abbreviation for quantum non-separability or anything similar, and the '$\sim$' symbol is introduced to emphasize the vague equivalence or link between completeness and Bohr's non-separability/indivisibility. With this sloppy definition, we have indeed to choose between non-local hidden variables (ii), i.e. a particular form of realism, and between a version of QM (iii)' considered as being complete and where

[8] Generally from Equations 12.10-12.12 we can define local conditional mean value $\bar{A}(\lambda, \hat{\mathbf{n}}_1, |\psi\rangle) = \sum_{\alpha=\pm 1} \alpha P_1(\alpha/\omega, \hat{\mathbf{n}}_1, |\psi\rangle)$, and $\bar{B}(\lambda, \hat{\mathbf{n}}_2, |\psi\rangle) = \sum_{\beta=\pm 1} \beta P_2(\beta/\omega, \hat{\mathbf{n}}_2, |\psi\rangle)$ with $\sum_{\alpha=\pm 1} P_1(\alpha/\omega, \hat{\mathbf{n}}_1, |\psi\rangle) = \sum_{\beta=\pm 1} P_2(\beta/\omega, \hat{\mathbf{n}}_2, |\psi\rangle) = 1$ and thus $|\bar{A}(\lambda, \hat{\mathbf{n}}_1, |\psi\rangle)| \leq 1$, $|\bar{B}(\lambda, \hat{\mathbf{n}}_2, |\psi\rangle)| \leq 1$. Moreover, in the particular case of a deterministic theory, we have $P_1(\alpha/\omega, \hat{\mathbf{n}}_1, |\psi\rangle) = 1$ or 0 and therefore $\bar{A}(\lambda, \hat{\mathbf{n}}_1, |\psi\rangle) := A(\lambda, \hat{\mathbf{n}}_1, |\psi\rangle) = \pm 1$ (and similarly for Bob side). Therefore, we deduce Equation 12.15.

the notion of Einstein locality is analysed as too dogmatic and naive. The advocates of this rhetoric will eventually rephrase (iii)' as just quantum non realism or antirealism (in relation with some form of positivism, instrumentalism, and/or operationalism) in order to contrast their view with the 'naive' classical realism defended by Einstein or even Bohm. Therefore, we end up with either assuming a realist non-local quantum world or a quantum non-realist approach where 'non separability' is the rule.

However, in order to understand how locality is (falsely) proposed as an alternative to realism, it is crucial to see that the advocates of sloppy language often confuse the precise definition of local-causality (i.e. Equations 12.10-12.12, proposed by Bell after Einstein) with also the precise notion of local commutativity used in quantum-field and quantum measurement theory (for other related issues see also the Appendix). Referring to this, Bell was very clear that these two notions should not be confused (Bell 2004, chapters 7 and 24). Local commutativity $[\mathcal{O}_A, \mathcal{O}_B] = 0$ of two local Hermitian operators $\mathcal{O}_A$ and $\mathcal{O}_B$ defined in two space-like separated spatial regions can be used to justify some averaged statistical independence of local measurements made by Alice and Bob. More precisely, consider for example the case where Alice is considering the evolution of the mean value $\langle \mathcal{O}_A(t) \rangle$ between times t and $t + \delta t$ when Bob disturbs locally his settings. Assuming in the interaction picture that $\mathcal{O}_A(t+\delta t) = \mathcal{O}_A(t)$ and that the quantum state evolves as $|\Psi(t+\delta t)\rangle = e^{-i\delta t \mathcal{O}_B(t)}|\Psi(t)\rangle$, we obtain if $[\mathcal{O}_A, \mathcal{O}_B] = 0$

$$\langle \mathcal{O}_A(t+\delta t) \rangle = \langle \Psi(t)| e^{+i\delta t \mathcal{O}_B(t)} \mathcal{O}_A(t) e^{-i\delta t \mathcal{O}_B(t)} |\Psi(t)\rangle = \langle \mathcal{O}_A(t) \rangle. \qquad (12.17)$$

This condition shows that a local measurement made by Bob on his side cannot affect statistical observables of Alice. This is used to demonstrate the non-signaling theorem which is a key feature of relativistic QM: $\sum_\beta P(\alpha, \beta/\hat{\mathbf{n}}_1, \hat{\mathbf{n}}_2, |\psi\rangle) = P(\alpha/\hat{\mathbf{n}}_1, \hat{\mathbf{n}}_2, |\psi\rangle) = P(\alpha/\hat{\mathbf{n}}_1, |\psi\rangle)$. Local-causality (i.e. Equations 12.10-12.12) implies non-signaling, but the opposite is not always true. Still it is possible for a quantum non-realist to use locality in the weak sense of non-signaling and local commutativity and at the same time to consider quantum theory as requiring an indivisibility or non-separability of phenomena à la Bohr, i.e. what EPR and Bell define as (iii) 'QM-C AND QM-NL'. Clearly, it is the lack of precision in the language that allows something to be local and non-local at once. It is therefore not surprising that the false alternative between relinquishing locality or realism occurs if we use locality with such a wavy definition in the EPR and Bell theorems.

12.5 Conclusion: from EPR to Bohm and Bell

Going back to the EPR three alternatives (i), (ii), and (iii) of (4) we see that abandoning locality implies to relax at least one of the three conditions (12.10)-(12.12). Moreover, it is remarkable that from the three possible alternatives of (12.4), only physical examples of case (ii) and (iii) are available.

First consider the case (iii) where QM is complete and non-local: we have already two representations given by Equation (12.7) and (12.8). In the representation of Beltrametti and Bugajski of Equation (12.7) we see that Equation (12.12) is preserved, but we relinquish Equations (12.10), (12.11). Using the representation of Scully *et al.* (i.e., Equation (12.8)), we now see that Equations (12.10), (12.11) are preserved, but actually it is Equation (12.12) which is abandoned. This explains why this representation can be said to imply superdeterminism or retrocausality. Moreover, here the conditional probabilities $P(\alpha/\theta_i, \varphi_i, |\psi\rangle) = \frac{1+\alpha\cos(\varphi_i-\theta_i)}{2}$ can only take value 0 or 1 because of the specific form of $\rho(\theta_1, \theta_2/\varphi_1, \varphi_2, |\Psi_{EPR}\rangle)$. The theory is thus 'effectively' deterministic even though $P(\alpha/\theta_i, \varphi_i, |\psi\rangle)$ is in general different from 0 or 1. This was expected because the derivation leading to Equation 12.15 does not require Equation (12.12) to hold true, but only Equations (12.10), (12.11). Furthermore, we observe that once we add to Equation (12.15) the statistical independence given by Equation (12.12) we obtain the EPR contradiction leading to the rejection of 'QM-C AND QM-L' theories.

Now concerning case (ii): there is no known example of quantum theory involving hidden variables λ that is at the same time local in Einstein–Bell sense. For instance, as it is well known the deterministic pilot wave theory of de Broglie and Bohm preserves the statistical independence of Equation (12.12) but relax Equations (12.10), (12.11) in order to allow for action-at-a-distance with instantaneous forces. As Bohm wrote:

> Thus the "quantum-mechanical" force may be said to transmit uncontrollable disturbances instantaneously from one particle to another through the medium of the ψ-field (Bohm 1952, p. 186).

Bohm subsequently remarked that in his hidden-variable theory there is clearly a strong difference with the case of a single particle where the uncertainty principle explains why disturbance can locally preclude measurements of non-commuting observables within one single experimental protocol. For entangled particles in the EPR case, assuming hidden variables one should perhaps always expect a kind of non-local action-at-a distance affecting the two systems. This would somehow save the logics of the Heisenberg principle but would be in tension with special relativity.

It was moreover Bell and not Bohm who answered the question: is it possible to find an example of a quantum theory incomplete and local (i.e. case (i) of Equation (12.8))? As we all know, now the answer he provided was 'no'. It is not my aim here to review the derivation of Bell's theorem which only requires the validity of the three local causality conditions Equations (12.10)-(12.12). We point out that Bell's theorem is easily generalizable for stochastic hidden-variable theories. This makes sense if the singlet state is interacting with non-efficient detectors involving losses. In this regime the EPR derivation leading to determinism is not generally valid, but Bell's theorem and the contradiction with locality is preserved. The open and important question is of course which condition must be relaxed in order to develop a more complete theory involving hidden variables.

This problem will not be discussed here, but just to mention that the choice followed by Bohm involves action-at-a-distance and a preferred space-time foliation conflicting with the goal of special and general relativity which was to remove or prohibit such a preferred foliation from physics (see Drezet (2019) for more on that issue and on how we can make sense of special foliations in the dBB theory). Moreover, in Bohm's approach, the mediation of such supra-luminal forces requires a kind of 'subquantum Aether' with remarkable properties. Indeed, Bohm's quantum potential acting between particles has several peculiar properties: (i) quantum potentials do not decay with the distance (i.e. unlike gravitational Newtonian forces), (ii) quantum potentials are highly selective (e.g. entanglement and non-locality can concern two remote particles 1 and 2 but in turn completely ignore other 'spectator' particles located near particles 1 and 2). Therefore, the quantum potential does not apparently spread in space unlike gravitational forces. (iii) Furthermore, the quantum potential can correlate/entangle particles whatever the obstacles and barriers located between two particles. The quantum force is thus one of the most penetrative things of the Universe (Elitzur 1990). (iv) At the same time, we never directly detected the presence of such a quantum potential (e.g. all attempts to directly detect 'empty waves' of such a quantum potential have for now failed). Therefore, the quantum potential seems not to carry energy by itself, i.e. it is not a field existing independently of entangled particles and is apparently unseparable of the particles involved in the entangled wave function. Clearly, even in a Newtonian Universe, with instantaneous action-at-a-distance, such properties seem extraordinary or miraculous. Therefore, a sub-quantum Aether would be highly non-classical. Moreover, even if one could understand and model the physical properties of such a sub-quantum Aether (attempts have been made for instance by Bohm and Vigier), this would probably be seen as a form of regression to pre-relativistic physics (because of a preferred foliation), and for many, including the present author, this would be too hard to swallow. Of course, this does not invalidate the dBB theory. Following Bell, a number of prominent researchers have advocated a minimalistic position, i.e. accepting the results of 'Bohmian' mechanics in the configuration space and avoiding physical discussions about the physical nature or origin of the quantum potential and its remarkable non-local properties. With this minimalistic approach, we can recover all the statistical predictions of standard quantum mechanics including of course the experimentally observed Bell's inequality violations. However, if according to de Broglie, Bohm or Einstein the goal of physics is not only to predict, but also to explain, then the minimalistic dBB theory is not a complete theory.

For this reason, the present author progressively advocated an alternative 'superdeterministic' approach where the action-at-a-distance implied by the pilot-wave of de Broglie and Bohm is preserved as an effective description (cf. Drezet 2023). At a lower 'sub-quantum level', the theory is fully local in the sense that no signal can propagate faster than light but an underlying superdeterministic and time-symmetric fundamental field driving and synchronizing the particles allows us to relax Equations (12.10)-(12.12) without contradicting the spirit of Einstein relativity. More precisely, in this new theory inspired by de Broglie 'double-solution' work (de Broglie 1925) particles are singularities of a classical field moving in the 4D space-time. The trajectories of the singularities obey the dBB pilot-wave dynamics, and the classical field is the half sum of retarded and advanced waves emitted by the singularities. The time-symmetry of the field allows us to derive non-locality of the pilot-wave from an underlying superdeterministic and local theory.

12.6 Appendix: Fine's theorem

A recurrent but misleading objection against the EPR–Bell deduction of non-locality is based on a theorem attributed to Fine (1982) (already anticipated by G. Lochak in collaboration with de Broglie, cf. Lochak 1976). The claim (already debunked in Drezet (2009)) is that assuming Bell's factorizability for probability generally implies the existence of joint probabilities

$$P(A_1, A_2, B_1, B_2) = \int P(A_1/\omega)P(A_2/\omega)P(B_1/\omega)P(B_2/\omega)\rho(\omega)d\omega \quad (12.18)$$

where A_1 and A_2 (respectively B_1 and B_2) are two incompatible observables (i.e., spin components along the x- and z-directions) for particle 1 (respectively particle 2). Joint probabilities used in Bell's derivation like $P(A_i, B_j) = \int P(A_i/\omega)P(B_j/\omega)\rho(\omega)d\omega$ are just marginals obtained from $P(A_1, A_2, B_1, B_2)$. Moreover, we can similarly obtain $P(A_1, A_2) = \int P(A_1/\omega)P(A_2/\omega)\rho(\omega)d\omega$, and since QM prohibits to define probability for incompatible/complementary observables for a same particle we have a clear contradiction.

The claim of Lochak-Fine that has been repeatedly used by many authors (e.g. Zukowski and Brukner 2014; Kupczynski 2023, and references therein), is thus that Bell's reasoning implicitly takes for granted that such non-contextual and non-physical joint probabilities like $P(A_1, A_2, B_1, B_2)$ do exist. As claimed in Fine (1982), 'the existence of deterministic [Local] hidden variables violates the quantum mechanical condition that joint probability distributions are well defined only for commuting observables'. Again the goal is to show that EPR–Bell local causality is a realist assumption which is dropped in QM (see for instance Zukowski and Brukner 2014). However, this is misleading: nowhere in Bell's derivation using local causality it is mentioned that joint probability like $P(A_1, A_2, B_1, B_2)$ or $P(A_1, A_2)$ should exist (or if they exist they must not be interpreted as physical probabilities associated with measurements, Drezet 2009). Nowhere it is assumed that the non-contextual factorizability used in Equation (12.18) should be accepted. A formula like Equation (12.18) can certainly be obtained for some non-contextual hidden-variable models, but these models cannot agree with QM (see Drezet 2009).

References

Albert, D.Z. and Galchen, R. (2009). Was Einstein Wrong?: A quantum threat to special relativity. *Scientific American*, **300**(3).

Argaman, N. (2010). Bell's theorem and the causal arrow of time. *American Journal of Physics*, **78**: 1007–1013.

Bell, J.S. (2004). *Speakable and Unspeakable in Quantum Mechanics*, 2nd edn. Cambridge: Cambridge University Press.

Beltrametti, E.G. and Bugajski, S. (1995). A classical extension of quantum mechanics. *Journal of Physics A: Mathematical and General*, **8**: 3329–3343.

Bohm, D. (1952). A suggested interpretation of the quantum theory in terms of "hidden" variables. I & II. *Physical Review*, **85**(2): 166–193.

Bohr, N. (1949). Discussions with Einstein on epistemological problems in atomic physics. In Schilpp, P.A. (ed.) *Albert Einstein Philosopher Scientist.* Open Court; 3rd edn 1988.

Bricmont, J. (2016). What did Bell really prove?. In Bell, M. and Gao, S. (eds.) *Quantum Non-Locality and Reality (50 Years of Bell's Theorem)*, pp. 49–78. Cambridge: Cambridge University Press.

Crull, E. and Bacciagaluppi, G. (2011). Translation of: W. Heisenberg, "Ist eine deterministische Ergänzung der Quantenmechanik möglich?". *PhilSci. Archive.*

de Broglie, L. (1926). *Ondes et Mouvements.* Paris: Gauthier-Villars.

de Broglie, L. (1974). Sur la réfutation du théorème de Bell. *Comptes rendus*, Série B, **278**: 721–722.

Di Lorenzo, A. (2012). A simple model for the spin singlet: Mathematical equivalence of non-locality, slave will and conspiracy. *Journal of Physics A: Mathematical and Theoretical*, **45**: 265302.

Drezet, A. (2009). Concerning an old (but still quite alive) rebuttal of the theorem of John Bell. arXiv:0909.4200.

Drezet, A. (2012). Should the wave-function be a part of the quantum ontological state?. *Progress in Physics*, **4**: 14–17.

Drezet, A. (2019). Lorentz-invariant, retrocausal, and deterministic hidden variables. *Foundations of Physics*, **49**: 1166–1199.

Drezet, A. (2023). A time-symmetric soliton dynamics à la de Broglie. *Foundations of Physics*, **53**: 72.

Einstein, A., and Podolsky, B., and Rosen, N. (1935). Can quantum-mechanical description of physical reality be considered complete?. *Physical Review*, **47**: 777.

Einstein, A. (1949). Autobiographical notes. In Schilpp, P.A. (ed.) *Albert Einstein Philosopher Scientist.* Open Court; 3rd edn 1998.

Einstein, A. (1953). Remarques préliminaires sur les concepts fondamentaux. In *Louis de Broglie physicien et penseur.* Paris: Albin Michel.

Elitzur, A. (1990). On some neglected thermodynamic peculiarities of quantum non-locality. *Foundations Physics Letters*, **3**: 525–541.

Fine, A. (1982). Hidden variables, joint probability, and the Bell inequalities. *Physical Review Letters*, **48**: 291–295.

Fuchs, C.A., Mermin, N.D. and Schack, R. (2014). An introduction to QBism with an application to the locality of quantum mechanics. *American Journal of Physics*, **82**: 749–754.

Goldstein, S., Norsen, T., Tausk, D.V. and Zanghì, N. (2023). Bell's theorem. *Scholarpedia.*

Harrigan, N. and Spekkens, R.W. (2010). Einstein, incompleteness, and the epistemic view of quantum states, *Foundations of Physics*, **40**: 125–157.

Kupczynski, M. (2023). Contextuality or non-locality: What would John Bell choose today?. *Entropy*, **25**: 280.

Laudisa, F. (2023). How and when did locality become 'local realism'? A historical and critical analysis (1963–1978). *Studies in History and Philosophy of Science*, **97**: 44–57.

Lochak, G. (1976). Has Bell's inequality a general meaning for hidden-variable theories?. *Foundations of Physics*, **6**: 173–184.

Maudlin, T. (2014). What Bell did. *Journal of Physics A: Mathematical and Theoretical*, **47**: 424010.

Nikolić, H. (2012). EPR before EPR: A 1930 Einstein–Bohr thought experiment revisited. *European Journal of Physics*, **33**: 1089–1097.

Norsen, T. (2006). Bell locality and the non-local character of nature. *Foundations of Physics Letters*, **19**: 633–655.

Norsen, T. (2011). J.S. Bell's concept of local causality. *American Journal of Physics*, **79**: 1261–1275.

Oldofredi, A. (2022). Beables, primitive ontology and beyond: How theories meet the world. In Allori, V. (ed.) *Quantum Mechanics and Fundamentality. Naturalizing Quantum Theory between Scientific Realism and Ontological Indeterminacy.* Cham: Synthese Library, Springer.

Peres, A. (1978). Unperformed experiments have no results. *American Journal of Physics*, **46**: 745–747.

Scully, M.O. (1983). How to make quantum mechanics looks like a hidden-variable theory and vice versa. *Physical Review D*, **28**: 2477–2484.

Zukowski, M. and Brukner, C. (2014). Quantum non-locality—it ain't necessarily so... *Journal of Physics A: Mathematical and Theoretical*, **47**: 424009.

13

A vision for a Bohm-style theory of quantum electrodynamics

Roderich Tumulka

13.1 Introduction

Despite many successes of quantum electrodynamics (QED), we do not presently have a good understanding of this field of physics. QED has all of the foundational problems that standard non-relativistic quantum mechanics has, and further ones in addition. I discuss some of these problems and some options for what a Bohm-style theory of QED, with an ontology in space and time, could look like. I also point out why the proposal made by Bohm himself in 1952 for QED is not quite convincing. Finally, I outline the kind of Bohm-type theory of QED that I would consider convincing, and report about recent progress toward this kind of theory.

In this chapter, I would like to outline why quantum electrodynamics in its usual formulation is not very satisfactory, and what a satisfactory theory could look like. It is one of the goals of physics to find the fundamental laws of nature; I tend to take seriously the possibility that the correct laws could have a Bohmian character, meaning that there is an ontology in space and time that evolves according to laws that involve the wave function, and here I explore and discuss this possibility.

In non-relativistic quantum mechanics (NRQM), the orthodox formulation of the theory has several problems that have been discussed in many places (e.g. Tumulka 2022), in particular: that the theory does not clearly state what happens in reality; that it fails to apply the same laws to a measurement apparatus as to every other system of quarks and electrons, or else runs into inconsistencies known as the quantum measurement problem; that it remains unclear how the non-locality required by Bell's theorem (Bell 1987; Goldstein *et al.* 2011) gets reconciled with relativity; that it remains unclear where in the fundamental laws the randomness comes in; and that the physical meaning of observables remains unclear in view of the 'no-hidden-variable' theorems that tell us they cannot be thought of as physical quantities. The merit of Bohmian mechanics is to provide a full, clear, and satisfactory solution of these problems (Bohm 1952, Dürr and Teufel 2009).

These problems persist in QED, not surprisingly, as NRQM should be contained in QED as a limiting case. On top of that, QED has problems that were absent in NRQM, which are as follows.

13.2 Problems of QED

13.2.1 Ultraviolet divergence

The Hamiltonian (and thus the unitary time evolution) is ill defined. The problem is milder if we only consider the time evolution from $t = -\infty$ to $t = +\infty$ (the S matrix), but even that is ill defined; S can heuristically be written as a series (a Dyson series), but the series does not converge. Even the terms in the series are ill defined to begin with, but there are proposals for how to change each term so as to make them well defined (renormalization and similar techniques, Scharf 1995); while some of these changes may appear 'ad hoc', other problems are worse than that.

13.2.2 Born rule for photon positions

NRQM provides, apart from a Hilbert space $L^2(\mathbb{R}^{3N})$ and a Hamiltonian, a Born rule $\rho(q) = |\psi(q)|^2$ for particle positions. In contrast, QED does not provide a Born rule for photon positions (see Tumulka (2022b), Section 7.3.9 for discussion) although experimentalists can measure photon positions. To be sure, we have a Born rule for photon quantum states that are plane waves or locally plane waves, and since quantum states tend to become locally plane waves as $t \to \pm\infty$, we can predict position probabilities for $t = \pm\infty$ but not in between. Some proposals in the literature (such as Good 1957) may be good approximations but do not seem convincing as fundamental laws.

13.2.3 Consequences of excluded negative energy states

While the one-particle Dirac Hamiltonian in $\mathscr{H}_1 = L^2(\mathbb{R}^3, \mathbb{C}^4)$ has spectrum $(-\infty, -m] \cup [m, \infty)$ leading to a decomposition $\mathscr{H}_1 = \mathscr{H}_{1+} \oplus \mathscr{H}_{1-}$ into the contribution of positive and that of negative energies, we usually regard only states without contributions of negative energy as physical. But $\psi \in \mathscr{H}_{1+}$ cannot be concentrated in a bounded region $A \subset \mathbb{R}^3$ (Thaller 1992, Cor. 1.7), which leads to the question how ψ collapses after the particle was detected in A. If the collapsed wave function is not strictly localized in A (while detection probability density is still given by $|\psi|^2$), then the particle could be detected shortly afterwards far away from A, so the same particle could be detected in two space-like separated regions, which sounds not believable. In particular, it apparently follows that either interaction locality (i.e. the Hamiltonian contains no interaction terms between space-like separated space-time points) is violated (so the detector could be triggered by a particle far away), or propagation locality (i.e. wave functions propagate no faster than at the speed c of light) is violated (so the particle's wave packet could travel from one detection region to the other), or both.

Variants of this problem are provided by the Hegerfeldt–Malament theorem (Borchers 1967; Hegerfeldt 1974; Malament 1996; Halvorson and Clifton 2002; Beck 2023), the Reeh–Schlieder theorem (Reeh and Schlieder 1961; Beck 2023), and a related problem that I described in Tumulka 2021, Section 6 paragraph 4. While some authors (e.g. Soffer 2011) think that propagation locality is only approximate and not strict, I find that hard to believe.

13.2.4 Status of the Dirac sea

Everybody has an intuitive understanding of the Dirac sea, but for a physical theory we need more precision. Specifically, given that actual detectors can detect individual electrons (as shown in Fig. 13.1 for a double-slit experiment) but do not detect any background of 'sea electrons', we need a Born rule for the positions of visible electrons (and positrons).

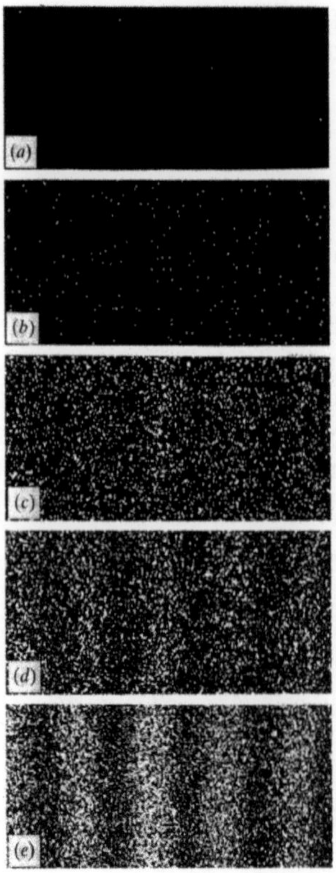

Fig. 13.1: A representation of actual results of a double-slit experiment of Tonomura *et al.*, after (a) 10, (b) 100, (c) 3,000, (d) 20,000, and (e) 70,000 electrons. (Reprinted from Tonomura *et al.* (1989) with the permission of AIP Publishing.)

Standard presentations of QED do not provide such a Born rule. An obvious Born rule is to take, for any configuration $x \in \mathbb{R}^{3n}$ of n electrons and any configuration $\overline{x} \in \mathbb{R}^{3\overline{n}}$ of $\overline{n}$ positrons,

$$\rho(x, \overline{x}) = |\psi^{(n,\overline{n})}(x, \overline{x})|^2 , \tag{13.1}$$

where ψ lies in the standard Hilbert space of the quantum Dirac field,

$$\mathscr{H} = \mathscr{F}(\mathscr{H}_{1+}) \otimes \mathscr{F}(C\mathscr{H}_{1-}) \tag{13.2}$$

with the first factor representing electrons and the second positrons; here,

$$\mathscr{F}(\mathscr{H}) = \bigoplus_{n=0}^{\infty} \mathscr{H}^{\wedge n} \tag{13.3}$$

is the fermionic Fock space, C is the anti-unitary charge conjugation operator, and $C\mathscr{H}_{1-}$ equals in fact $\mathscr{H}_{1+}$. However, this obvious Born rule does not seem convincing (Tumulka (2022a)), also in part because of the Hegerfeldt–Malament theorem and related problems mentioned in the previous sub-section. Alternative candidates are described in Tumulka (2022a, 2024) and in Section 13.5.

Again, the problem is milder at $t = \pm\infty$, essentially because contributions to ψ with different momenta tend to separate in space as $t \to \pm\infty$, so that the distribution of positions at $t = \pm\infty$ coincides with the momentum distribution, which is easier to define than the position distribution. For many experiments, including the one of Figure 13.1, it is a good approximation to set $t = \infty$, but I am asking here for the correct position distribution at finite t: for the distribution to which the approximation is an approximation.

Some authors (e.g. Malament 1996) assert that all these problems indicate that the concept of particles does not make sense in QED. However, the problem here does not originate from any theoretical definition or concept, but it arises from the detection events. When I look at Fig. 13.1, I see individual, localized detection events as well as an obvious statistical distribution of these events, and I wonder which formula governs this distribution.

Moreover, if we lack a convincing Born rule for positions (or, equivalently, position operators), then we have a problem not just with detection probabilities, but also with how to link the quantum state to probabilities even for positions of macroscopic objects. For example, in a situation in which the needle of a measurement instrument can point either to the left or to the right, it is clear in NRQM how to translate the condition ‘points to the left’ into a subset of the configuration space $\mathbb{R}^{3N}$ and thus into a sub-space of the Hilbert space, a projection P, and a probability $\langle\psi|P|\psi\rangle$. But in QED, if we do not have position operators to begin with, then on what grounds could we associate with the condition ‘points to the left’ a sub-space or operator or probability?

13.2.5 The Shale–Stinespring–Ruijsenaars problem

As a step towards defining the time evolution of the Dirac quantum field interacting with a quantized electromagnetic field, it is reasonable to first try to define the time evolution of the Dirac quantum field in the presence of an external, classical electromagnetic field A_μ. As discovered by Shale and Stinespring (1965) (see also Thaller 1992, Section 10.3.3), the latter is well defined in the standard Hilbert space (13.2) if and only if the contribution from A_μ to the one-particle Hamiltonian, $V := \gamma^0\gamma^\mu A_\mu$, has off-diagonal part $V_{+-} := P_{1+}VP_{1-}$ that is a Hilbert-Schmidt operator, $\mathrm{tr}(V_{+-}^\dagger V_{+-}) < \infty$. This happens only, as shown by Ruijsenaars 1977a, 1977b, when the magnetic field vanishes. Of course, that condition is very restrictive and generically violated. And if the time evolution is not well defined in an external A_μ, there is little hope it could be well defined in a quantized A_μ.

One answer might be that (13.2) is not the right Hilbert space. After all, Fock space corresponds to a configuration space containing all configurations of *finitely* many particles, and if configurations of *infinitely* many particles occurred in nature, we might have to use corresponding Hilbert spaces instead of Fock space.

Alternatively, Deckert *et al.* (2010) pointed out that if we change the 'sea space' (i.e. the sub-space of $\mathscr{H}_1$ filled by the Dirac sea) by only finitely many dimensions, then the resulting Hilbert space as in (13.2) would, in a natural sense, stay the same; they were able to construct a Hilbert space for the quantum Dirac field if given, instead of a sea space, a suitable equivalence class of sea spaces. They went on to propose (Deckert *et al.* (2010); Deckert and Merkl (2016)) using different Hilbert spaces for different times or, more generally, for different Cauchy surfaces Σ and showed that their choice of $\mathscr{H}_\Sigma$, which depends only on the magnetic field on Σ, leads to a unique (up to phase) time evolution of the quantum Dirac field, thus solving the Shale–Stinespring–Ruijsenaars problem. However, this approach does not lead to position operators or a Born rule for Dirac particles, as the obvious choice (13.1) would still require the choice of a sea space, for which there is no naturally selected candidate in the presence of an external A_μ.

13.2.6 Curved space-time

In curved space-time $(\mathscr{M}, g)$, the situation becomes worse. It becomes even more problematic than in Minkowski space-time to choose a sub-space of $\mathscr{H}_{1\Sigma}$ as the sea space, as the Hamiltonian depends not only on the Cauchy surface Σ but also on the lapse function, i.e., on how we push Σ towards the future. A cousin of the Shale–Stinespring–Ruijsenaars problem returns, and I do not see how the strategy of Deckert *et al.* (2010) (and Deckert and Merkl 2016) could be applied in this case.

The problem is milder in asymptotically flat space-times, if we consider only very early and very late Cauchy surfaces. However, if we are after the fundamental laws of nature, then we also have to deal with times other than $\pm\infty$. Not to speak of the circumstance that our universe does not seem to be asymptotically flat, but singular in the past and asymptotically de Sitter in the future.

Likewise, the definition of position operators and the Born rule, at least for the analogue of the 'obvious' definition (13.1) in Minkowski space-time, depends on the choice of a sea space, so likewise it becomes even more problematic in curved space-time. Also in this context of quantum field theory in curved space-time, it is often said that the concept of particles breaks down, but again, the problem occurs regardless of whether particles are fundamental, and it occurs as a problem concerning the detectors that we do have and, as pointed out at the end of Section 13.2.4, concerning positions of macroscopic objects.

13.3 And on top of that

Apart from the things about QED that plainly fail to work, there are further oddities.

13.3.1 Complexity

The most striking is perhaps that QED becomes extraordinarily and excessively complex when we try to clearly specify the laws of the theory. This is often hidden in a notation that specifies the Lagrangian in symbols such as $A_\mu(x)$ and $\Psi(x)$ for the field operators, but becomes painfully visible when we try to make explicit or define mathematically what exactly the theory asserts. For example, this is visible in Deckert *et al.* (2010) for just the problem of defining the time evolution of the quantum Dirac field in an external, classical A_μ or in the book of Cohen-Tannoudji *et al.* (1997) that spends the majority of its pages on just defining the theory, without even aiming at mathematical rigor.

This complexity is not only unpleasant, but also odd as we usually expect that fundamental laws are simple. In contrast, I would readily expect that the computation of *predictions* of quantities observed in experiment can be complicated; that is because the experiment itself may involve complicated details that need to be taken into account, but more importantly that computing the prediction may involve *solving* fundamental equations of the theory, such as time evolution equations. But that is different from *stating* the fundamental equations, which should be simple. I suspect that the positivistic attitude of the orthodox view of quantum physics contributes to this difficulty: for a realist, there is a big and clear difference between stating the laws that govern the fundamental physical reality and making predictions for experiments; for a positivist, in contrast, the distinction is blurred, and making predictions is the only activity worthy of a scientist. So, what realists like me would like to see is a set of simple fundamental laws that could be regarded as the definition of the theory and as the ingredients from which to derive empirical predictions.

13.3.2 Wave functions

Another confusing aspect is that it is common in QED to talk about field operators, but not about wave functions. Of course, this is in a way a matter of representation, as one could say that wave functions are included in the field operator picture because the field operators act on some Hilbert space $\mathscr{H}$, and vectors in $\mathscr{H}$ should be the analogue of wave functions (or positive functionals on operator algebras should be the analogues of density matrices). And yet, something is odd. Wave functions or state vectors are central elements of quantum physics, and some arguments such as the Pusey–Barrett–Rudolph theorem (Pusey *et al.* 2012) support that they exist in reality. For comparison, when, in courses on classical physics, the electromagnetic field gets introduced, we pay much attention to how to represent it in a covariant way and to what kind of tensor field it is, such as $F_{\mu\nu} = -F_{\nu\mu}$ and the consequences. But in QED, all attention is paid to the field operators, none to the wave function. I will outline in Section 13.4 a proposal for how to represent the wave function of QED in a covariant way.

13.3.3 Fields or particles

Let us come back to the problems that QED shares with NRQM and that are related to the lack of a clear ontology. The fact that Bohmian mechanics solves the problems for NRQM is why we would like to do something similar for QED. A basic question we face is whether to choose a field ontology or particle ontology. Setting up a particle ontology requires a Born distribution (such as given by position operators) over particle configurations, which leads to the problems mentioned above; but the distributions we see in detections such as Fig. 13.1 suggest that these problems are solvable. Bohm himself already in 1952 proposed (Bohm 1952) to use a particle ontology for fermions and a field ontology for bosons. In my humble opinion, this proposal is not entirely convincing (see also Tumulka 2022b, Section 6.5.2 for discussion). The asymmetry in treating fermions and bosons does not seem justified from the difference between permutation anti-symmetry and permutation symmetry. Then, the field ontology itself seems unnatural in view of boson detection events. Next, it is not clear how to formulate a Born rule for field configurations because the space of field configurations is ∞-dimensional, and there are problems with obtaining a notion of volume in ∞-dimensional space, but a notion of volume is needed when we want to take the function $|\psi|^2$ to be a probability *density*. For fermions, it is unclear how a field ontology could even be set up, as the field operators cannot simply be multiplication by field variables since fermionic field operators anti-commute whereas multiplication operators commute. Furthermore, it is unclear for fermions how to identify the macro-configurations corresponding to any particular field configurations. In view of all of these reasons together, I am more inclined to expect that the final Bohmian version of QED will use a particle ontology for both the fermions and the bosons.

13.4 Particle position picture, starting from the work of Landau and Peierls

Let us turn to some considerations about what a Bohm-style QED could look like. Lev Landau and Rudolf Peierls published a remarkable paper (Landau and Peierls (1930)) in which they proposed a formulation of QED in the particle-position representation for both electrons and photons. I think that if this approach can be carried to the end, then this representation is a particularly natural way of framing QED. Landau and Peierls themselves did not follow up later on this approach, perhaps in part because they did not have a Born rule for photons and perhaps also because their equations were not manifestly Lorentz or gauge invariant. However, they become invariant (Lienert and Tumulka 2024) when formulated in terms of multi-time wave functions (Lienert *et al.* 2020), which here are of the form

$$\Psi = \Psi^{(m,n)}_{s_1 \dots s_m, \mu_1 \dots \mu_n}(x_1 \dots x_m, y_1 \dots y_n)\,, \tag{13.4}$$

where m is the number of electrons (positrons are left out of the model), n is the number of photons, both m and n are variable as in Fock space, $x_j \in \mathbb{R}^4$ is a space-time point considered as the location of the jth electron, $y_k \in \mathbb{R}^4$ a space-time point for the kth photon, x_j and y_k are required to be mutually spacelike separated but otherwise arbitrary, $s_j \in \{1, 2, 3, 4\}$ is the index (associated with the jth electron) for the 4 components of a Dirac spinor, and $\mu_k \in \{0, 1, 2, 3\}$ the index (associated with the kth photon) for the 4 components of a space-time vector as in A_μ. The wave function of a single photon is taken to satisfy the complexified Maxwell equation, up to particle creation and annihilation terms. In fact, the appropriate multi-time evolution equations, one for each particle, read (Lienert and Tumulka 2024):

$$(i\gamma_j^\mu \partial_{x_j,\mu} - m_x)\Psi^{(m,n)}(x_1 \dots x_m, y_1 \dots y_n) = e_x\sqrt{n+1}\,\gamma_j^\rho\,\Psi^{(m,n+1)}_{\mu_{n+1}=\rho}(x_1 \dots x_m, y_1 \dots y_n, x_j)\,, \tag{13.5a}$$

$$2\partial^{\mu}_{y_k}\partial_{y_k,[\mu}\Psi^{(m,n)}_{\mu_k=\nu]}(x_1...x_m,y_1...y_n) =$$

$$\frac{e_x}{\sqrt{n}}\sum_{j=1}^{m}\delta^3_\mu(y_k - x_j)\,\gamma^\mu_j\gamma_{j\nu}\,\Psi^{(m,n-1)}_{\widehat{\mu_k}}(x_1...x_m,y_1...y_{k-1},y_{k+1}...y_n)\,. \quad (13.5b)$$

Here, m_x and $-e_x$ are the mass and charge of the electron, γ^μ_j means γ^μ acting on the index s_j, $[\mu\nu]$ denotes anti-symmetrization in the index pair as in $S_{[\mu\nu]} = \frac{1}{2}(S_{\mu\nu}-S_{\nu\mu})$, δ^3_μ is the Lorentz-covariant 3d delta function (Lienert and Tumulka 2024), and $\widehat{\mu_k}$ means that the index μ_k is omitted. The first equation (13.5a) is the Dirac equation with an additional term on the right-hand side resembling the electromagnetic potential A_μ acting on Ψ, but now provided by the wave function of the next photon, which is part of Ψ itself. The second equation (13.5b) is the Maxwell equation for $A_\mu(y)$ given by Ψ as a function of $y = y_k$ with index $\mu = \mu_k$, and with an additional term on the right-hand side representing the emission of photon k by electron j. Although the notation that I have used in order to make all variables and indices visible leads to long expressions, these equations are really of a remarkable simplicity. Actually, their degree of simplicity would seem quite appropriate for fundamental laws of nature. (But keep in mind that nothing has been done here to deal with the ultraviolet divergence or to incorporate pair creation or positrons at all.) This form of the wave function and the equations appears so natural that I am inclined to think that this Ψ must be what quantum states really are in nature.

13.5 Positrons

Landau and Peierls did not have in mind Bohmian trajectories, but such trajectories can be introduced in standard ways (Tumulka 2022b) once we know how to compute the probability current 4-vector field, that is, once we have the full space-time version of the Born rule. Since we do not have that for photons, I leave them aside and turn to Dirac particles. For a single Dirac particle, the current is given by

$$j^\mu = \overline{\psi}\gamma^\mu\psi\,. \quad (13.6)$$

The question arises how to treat positrons. One possibility that has been considered by several authors (Bohm and Hiley 1993; Colin and Struyve 2007; Deckert *et al.* 2019) is to take the Dirac sea literally and introduce an infinitude of Bohmian particles as electrons of negative energy (or possibly introduce a cut-off that will make the dimension of $\mathscr{H}_{1-}$ finite). It would be of interest to study whether a Bohmian theory of an actual infinitude of particles can work mathematically. Even heuristically, it is not very clear how to think of such a particle configuration. Is it a countably infinite set of points in $\mathbb{R}^3$? Why not a set whose complement is countably infinite? Some heuristics (Tumulka 2022a, Section 5) suggest that the Dirac sea should have exactly two particles at every space point $\boldsymbol{x} \in \mathbb{R}^3$ except for a countable set of $\boldsymbol{x}$'s that each host 0, 1, 3, or 4 particles. The character of sea configurations is also relevant to the question whether it would somehow be visible from the configuration at which places there is a particle missing or a particle in excess, and thus which places should light up in a detector image such as Fig. 13.1.

Let me turn to two alternative candidates for treating positrons, corresponding to alternatives for the Born rule and the position operators.

13.5.1 A new possibility

The first (Tumulka 2022a) is based on defining, for every region $A \subset \mathbb{R}^3$ in physical 3-space, a charge operator $Q(A)$ by

$$Q(A) = -e_x \int_A d^3\boldsymbol{x} \sum_{s=1}^{4} :\Psi_s^\dagger(\boldsymbol{x})\,\Psi_s(\boldsymbol{x}): \quad , \tag{13.7}$$

where $-e_x$ is again the electron charge, $\Psi_s(\boldsymbol{x})$ is the Dirac field operator associated with location $\boldsymbol{x}$ and spin component s, and colons mean normal ordering. According to heuristic reasoning, $Q(A)$ has spectrum in the set $e_x\mathbb{Z}$ of integer multiples of e_x, and the canonical anti-commutation relations for $\Psi_s(\boldsymbol{x})$ imply that $Q(A)$ and $Q(A')$ commute for any A and A', so all of these charge operators can be diagonalized simultaneously. The joint spectrum should then, at least when $\mathbb{R}^3$ is replaced by a box of finite volume, consist of signed configurations, i.e. configurations of finitely many points, of which some are occupied by a positive charge and some by a negative charge. Since the 'sea state', i.e. vector $\Omega = |0\rangle \otimes |0\rangle$ in (13.2) usually called the 'vacuum state', has $\langle\Omega|Q(A)|\Omega\rangle = 0$ for all A but is not an eigenvector of $Q(A)$, the corresponding Born distribution in this picture is actually not concentrated on the empty configuration; rather, in this picture $|\Omega\rangle$ corresponds to a random number of electron and positron particles in random places.

However, an attempt to rigorously define the operators $Q(A)$ has failed (Costa and Tumulka, 2024), so the viability of this approach is questionable.

13.5.2 Another new possibility

A second proposal (Tumulka 2024) can be summarized very briefly by saying that, contrary to what Dirac proposed, *a positron is a Dirac particle with negative energy, rather than a hole among them.* Here are some remarks:

- We may first note that according to the one-particle Dirac equation, a wave packet consisting purely of negative energy contributions will move in an external electromagnetic field exactly like a wave packet of purely positive contributions but with the opposite charge. That is because the charge conjugation operator maps negative energy contributions to positive ones and vice versa, and maps the Dirac Hamiltonian to the one with opposite charge.
- An objection that comes to mind immediately is that if particles of negative kinetic energy existed, so that the free Hamiltonian was unbounded from below, a particle could repeatedly emit radiation with positive energy at the expense of pushing its own energy towards $-\infty$. Then again, if negative energy is allowed, the particle could also emit radiation of negative energy and thereby push its own energy towards $+\infty$, which suggests that, as in a random walk, the actual energy might not grow quickly to either $+\infty$ or $-\infty$. Although at first, this situation may appear rather unstable, it looks like it would take an infinite time before an infinite amount of energy gets radiated off; on top of that, there is an argument

(Tumulka 2024), based on a mapping (that I call W) translating the negative energy states into holes in the Dirac sea, to the effect that the situation is more stable than it may seem, in fact no less stable than the usual Dirac sea picture.

- For a Hamiltonian H not bounded from below, there exists no normalized thermal equilibrium density matrix

$$\rho = \frac{1}{Z} e^{-\beta H} \tag{13.8}$$

  for any inverse temperature $\beta > 0$. On the other hand, in QED we seek a theory of the universe (or a hypothetical universe without non-electromagnetic interactions), and we would not consider the whole universe in thermal equilibrium anyway. In addition, since the mentioned mapping W translates into the Dirac sea picture, which possesses density matrices such as (13.8), there presumably remains some type of universal thermal equilibrium state.
- Another objection that quickly comes to mind is that an external electromagnetic field may lift a negative energy state to positive energy. In the Dirac sea picture, that sounds just right: it creates a hole among the negative energy states, along with a particle of positive energy; in other words, it creates an electron–positron pair. In a model in which positrons are just Dirac particles with negative energy, this process would seem to entail that a positron has disappeared and an electron has appeared, in violation of charge conservation. However, that changes if we make the additional hypothesis that *in the time evolution equations, there are explicit terms for the creation and annihilation of electron–positron pairs.* That is, two further terms are introduced in (13.5), coupling $\Psi^{(m,n)}$ to $\Psi^{(m+2,n-1)}$ and $\Psi^{(m-2,n+1)}$.
- Two nice traits of the proposal are that it ensures interaction locality and propagation locality (circumventing the Hegerfeldt–Malament theorem because H is not bounded from below), and that it extends in a straightforward way to curved space-time, even if not asymptotically flat. If positrons are just Dirac particles of negative energy, and if the concept of positive or negative energy becomes ill-defined in some cases, then in those cases there would not be a fundamental fact about whether a given particle is an electron or a positron—it is just a Dirac particle. Which particles we call electrons and which positrons may thus be a categorization that is meaningful and useful in flat but not in curved space-time, or only at $t = \pm\infty$ but not in between. That is, according to this possibility, it is fundamentally not defined whether a given Dirac particle is an electron or a positron, and only in certain regimes can particles be consistently labeled as electrons or positrons. (This situation is similar to the possibility explored in Goldstein *et al.* (2005) that particles of different species are no different in the fundamental ontology.)

The question whether such a model can be empirically adequate requires further analysis.

13.6 Synthesis

The Bohm-type theory of QED that I am aiming at has particle trajectories for fermions and photons, although satisfactory equations for photon trajectories have not been identified yet. It uses the Landau–Peierls formulation of QED in the particle position representation in the form (13.5) for a multi-time wave function (13.4), with further terms governing pair creation and annihilation. The Bohm-like trajectories depend on a preferred foliation of space-time into space-like hypersurfaces (Dürr *et al.* 1999), while the wave function does not. The creation of Bohmian particles corresponds to stochastic jumps in the configuration (Dürr *et al.* 2020); the jump rates are determined by the probability current. The possibility to treat positrons as Dirac particles of negative energy seems worth considering. The problem of ultraviolet divergence might affect all of the theories considered here in the same way; then again, interior-boundary conditions (Teufel and Tumulka (2021)) have improved this problem in other models and might also help here.

I expect that the empirical predictions of the theory I have outlined coincide with what is usually regarded as the predictions of QED; after all, both are essentially based on the same time evolution expressed in different ways, so that the same series expansions, etc. are applicable. Still, the theory I have outlined would provide a clear ontology, and it would seem much simpler than previous attempts at a fundamental formulation of QED and at the same time better suited to provide an evolution between finite times, also in curved space-time. That would be a satisfactory formulation of QED.

References

Beck, C. (2023). Space-Time-Matter: Some Notes on the Localization Problem in Relativistic Quantum Theory. In Bassi, A., Goldstein, S., Tumulka, R. and Zanghì, N. (eds.) *Physics and the Nature of Reality: Essays in Memory of Detlef Dürr*, pp. 169–185. Heidelberg: Springer. `http://arxiv.org/abs/2305.18118`

Bell, J.S. (1987). *Speakable and Unspeakable in Quantum Mechanics*. Cambridge: Cambridge University Press.

Bohm, D. (1952). A suggested interpretation of the quantum theory in terms of “hidden” variables. I & II. *Physical Review*, **85**(2): 166–193.

Bohm, D. and Hiley, B.J. (1993). *The Undivided Universe: An Ontological Interpretation of Quantum Theory*. London and New York: Routledge.

Borchers, H.J. (1967). A remark on a theorem of B. Misra. *Communications in Mathematical Physics*, **4**: 315–323.

Cohen-Tannoudji, C., Dupont-Roc, J. and Grynberg, G. (1997). *Photons and Atoms*. Weinheim: Wiley.

Colin, S. and Struyve, W. (2007). A Dirac sea pilot-wave model for quantum field theory. *Journal of Physics A: Mathematical and Theoretical*, **40**: 7309–7342. `http://arxiv.org/abs/quant-ph/0701085`

Costa, P. and Tumulka, R. (2024). On the Problem of Defining Charge Operators for the Dirac Quantum Field. In preparation.

Deckert, D.-A., Dürr, D., Merkl, F. and Schottenloher, M. (2010). Time evolution of the external field problem in QED. *Journal of Mathematical Physics*, **51**: 122301. `http://arxiv.org/abs/0906.0046`

Deckert, D.-A., Esfeld, M.A. and Oldofredi, A. (2019). A Persistent Particle Ontology for Quantum Field Theory in Terms of the Dirac Sea. *British Journal for the Philosophy of Science*, **70(3)**: 747–770. `http://arxiv.org/abs/1608.06141`

Deckert, D.-A. and Merkl, F. (2016). External Field QED on Cauchy Surfaces for Varying Electromagnetic Fields. *Communications in Mathematical Physics*, **345**: 973–1017. `http://arxiv.org/abs/1505.06039`

Dürr, D., Goldstein, S., Münch-Berndl, K. and Zanghì, N. (1999). Hypersurface Bohm–Dirac models. *Physical Review A*, **60**: 2729–2736. `http://arxiv.org/abs/quant-ph/9801070`

Dürr, D., Goldstein, S., Teufel, S., Tumulka, R. and Zanghì, N. (2020). Bohmian Trajectories for Hamiltonians with Interior-Boundary Conditions. *Journal of Statistical Physics*, **180**: 34–73. `http://arxiv.org/abs/1809.10235`

Dürr, D. and Teufel, S. (2009). *Bohmian Mechanics: The Physics and Mathematics of Quantum Theory.* Heidelberg: Springer.

Goldstein S., Taylor, J., Tumulka, R. and Zanghì, N. (2005). Are all particles identical?. *Journal of Physics A: Mathematical and General*, **38**: 1567–1576. `http://arxiv.org/abs/quant-ph/0405039`

Goldstein S., Norsen, T., Tausk, D.V. and Zanghì, N. (2011). Bell's theorem. *Scholarpedia*, **6**(10): 8378. `http://www.scholarpedia.org/article/Bell%27s_theorem`

Good, R.H. (1957). Particle Aspect of the Electromagnetic Field Equations. *Physical Review*, **105**(6): 1914–1919.

Halvorson, H. and Clifton, R. (2002). No place for particles in relativistic quantum theories?. *Philosophy of Science*, **69**: 1–28. `http://arxiv.org/abs/quant-ph/0103041`

Hegerfeldt, G. (1974). Remark on causality and particle localization. *Physical Review D*, **10**: 3320–3321.

Landau, L. and Peierls, R. (1930). Quantenelektrodynamik im Konfigurationsraum. *Zeitschrift für Physik*, **62**: 188–200. English translation: Quantum electrodynamics in configuration space. In Dalitz, R.H. and Peierls, R. (eds.) *Selected Scientific Papers of Sir Rudolf Peierls With Commentary*, pp. 71–82. Singapore: World Scientific 1997.

Lienert, M., Petrat, S. and Tumulka, R. (2020). *Multi-Time Wave Functions: An Introduction.* Heidelberg: Springer.

Lienert, M. and Tumulka, R. (2024). Multi-Time Version of the Landau–Peierls Formulation of Quantum Electrodynamics. In preparation.

Malament, D. (1996). In defense of dogma: Why there cannot be a relativistic quantum mechanics of (localizable) particles. In Clifton, R. (ed.) *Perspectives on Quantum Reality: Non-Relativistic, Relativistic, and Field-Theoretic*, pp. 1–10. Dordrecht: Kluwer.

Pusey, M.F., Barrett, J., and Rudolph, T. (2012). On the reality of the quantum state. *Nature Physics*, **8**: 475–478. `http://arxiv.org/abs/1111.3328`

Reeh, H. and Schlieder, S. (1961). Bemerkungen zur Unitäräquivalenz von Lorentzinvarianten Feldern. *Il Nuovo Cimento*, **22**: 1059–1068.

Ruijsenaars, S.N.M. (1977a). Charged particles in external fields. I. Classical theory. *Journal of Mathematical Physics*, **18**(4): 720–737.

Ruijsenaars, S.N.M. (1977b). Charged particles in external fields. II. The quantized Dirac and Klein-Gordon theories. *Communications in Mathematical Physics*, **52**(3): 267–294.

Scharf, G. (1995). *Finite Quantum Electrodynamics*, 2nd edn. Heidelberg: Springer.

Shale, D. and Stinespring, W.F. (1965). Spinor Representations of Infinite Orthogonal Groups. *Journal of Mathematics and Mechanics*, **14**: 315–322.

Soffer, A. (2011). The Maximal Velocity of a Photon. Preprint. `http://arxiv.org/abs/1103.3031`

Teufel, S. and Tumulka, R. (2021). Hamiltonians without ultraviolet divergence for quantum field theories. *Quantum Studies: Mathematics and Foundations*, **8**: 17–35. `http://arxiv.org/abs/1505.04847`

Thaller, B. (1992). *The Dirac Equation*. Berlin: Springer.

Tonomura, A., Endo, J., Matsuda, T., Kawasaki, T. and Ezawa, H. (1989). Demonstration of single-electron buildup of an interference pattern. *American Journal of Physics*, **57**(2): 117–120.

Tumulka, R. (2022a). Positron Position Operators. I. A Natural Option. *Annals of Physics*, **443**: 168988. `http://arxiv.org/abs/2111.12304`

Tumulka, R. (2022b). *Foundations of Quantum Mechanics*. Heidelberg: Springer.

Tumulka, R. (2024). Positron Position Operators. II. A Simple Option. In preparation.

14
On the prospects of a de Broglie–Bohm–Barbour–Bertotti theory

Antonio Vassallo and Pedro Naranjo

14.1 Introduction

The modern perspective of fundamental physics is dominated by quantum theories. Non-relativistic quantum mechanics is used to describe low-energy systems, while quantum field theory is the standard framework for high-energy physics. Quantum theories are so empirically successful that, nowadays, there is a wide consensus about the fact that the fundamental description of the gravitational field must be quantum as well.

However, all quantum theories – whether relativistic or not – have to face a large number of conceptual problems, irrespective of their predictive success. In particular, there are two questions that the quantum formalism used by physicists is unable to address: Firstly, how can the *stable* and *determinate* features of the world, such as macroscopic material structures, be accounted for by a dynamics that just deals with the linear and unitary evolution of the wave function? Secondly, what is the nature of the wave function itself? Is it a concrete object, like a field, or just a compact way to describe how quantum systems behave (including our expectations about measurement outcomes)?

The consequence of these issues is that the picture of the physical world entailed by quantum theories is somewhat obscure, which is a severe drawback if we want to take these theories at face value in a scientific realist spirit. Quoting a recent overview of the foundations of quantum mechanics:

> [M]any physicists have come to believe that fundamental physics is in a state of stagnation, with little meaningful progress having been made in the last few decades [reference omitted]. It seems entirely possible that this has arisen because physicists never properly got to grips with what quantum mechanics tells us about the world, and therefore all subsequent physics has been based on an improper understanding of the earlier theory, leading us into a dead end (Adlam 2021, p. 2).

Usually, this 'improper understanding' is taken to imply the need for an *interpretation* – which, simply speaking, means supplying the quantum formalism with a clear metaphysics (see Maudlin (2019) for a recent textbook on this topic). In this vein, the work of Louis de Broglie (de Broglie 1928, 1960) and David Bohm (Bohm 1952a, 1952b) contributed a great deal to overcome the conceptual issues that concern the 'orthodox' take on quantum mechanics. In doing so, de Broglie and Bohm shed a new light on quantum physics, especially as far as the problem of bridging the conceptual gap between the Schrödinger-like dynamics and the existence of stable and determinate features of the world is concerned. The way the de Broglie–Bohm (dBB) approach to quantum physics achieves that is nowadays very well known and understood (see, e.g. Passon's and Bricmont's chapters in this volume). In a nutshell, the key insight is that quantum physics is not just about the wave function: there are extra degrees of freedom that account for the physical development of material structures in space-time – e.g. the spatial degrees of freedom of material particles. In this sense, the dBB approach comes to grips with the first question mentioned above. However, the question concerning the nature of the wave function remains open in this context. This is because the dBB approach just introduces a 'guidance principle' according to which the wave function *qua* mathematical object determines the equations of motion for the material degrees of freedom, but it leaves the question open as to what kind of physical mechanism this 'pilot wave' picture refers to.

Indeed, two different broad views on the wave function can be recognized in the dBB context. Some people think that this piece of the formalism is best understood as referring to some kind of field-like entity that physically determines in a straightforward sense the evolution of the material degrees of freedom (Bohm 1952a, 1952b is the most famous example; more recent proposals include Albert (1996), and Romano (2020)), while others argue that the wave function is just a way to codify the physical information representing the evolution of the material degrees of freedom (Bohm and Hiley 1993; Goldstein and Zanghì 2013). In this chapter, we argue that, in order to better understand what the nature of the wave function is in a dBB setting, a further step away from 'orthodox' physics is needed. This different perspective on physics is represented by the modern Leibnizian/Machian take on dynamics.

As the designation suggests, this approach to dynamics is rooted in age-old relational and empiricist ideas, which have been revived and adapted to a modern physical framework starting with the seminal work of Julian Barbour and Bruno Bertotti (Barbour 1974; Barbour and Bertotti 1977, 1982). The main realization behind the framework is that an empirically adequate ontology of the physical world should avoid structures whose variation makes no physical difference or which lack explanatory power to account for said differences. The immediate consequence of this metaphysical tenet is that the dynamical description of a closed system should be given in *intrinsic* terms, namely, in terms of relative changes between the degrees of freedom of the system only. This is because Leibnizian/Machian relationalism avoids reference structures external to the system, which are there only to account for some sort of absolute change. The obvious examples of such structures are Newtonian absolute space and time.

This abhorrence for external reference structures further implies that each and every part of a closed system acquires a physical characterization by virtue of standing in some type of physical relations to the rest of the system. A straightforward example of this is the Machian take on inertia: according to Mach, the origin of inertia is not local, that is, inertia is not inherent in the object of which this property is predicated. Rather, inertia accounts for how a material body is dynamically related to the rest of the universe. Note how this 'holistic' perspective nicely fits the dBB setting, where the fundamental dynamical description concerns the universal material system – with a description in terms of sub-systems that is attained only as an appropriate approximation.

In the present context, a Leibnizian/Machian approach can help in figuring out to what extent the wave function can be seen as external to a quantum system (and, thus, a 'foreign' structure to be dispensed with in favour of a fully intrinsic description) or, in some way, part of the quantum system itself (meaning there is something physically tangible about it). The chapter attempts at answering this question by introducing the Leibnizian/Machian approach to physics (Section 14.2.1) and discussing some possible quantum implementations of the framework that exploit the dBB approach (Section 14.2.2). In particular, the latest version of a de Broglie–Bohm–Barbour–Bertotti model is discussed in detail (Section 14.3.1), highlighting some interesting metaphysical morals regarding the nature of the wave function (Section 14.3.2). Finally, some intriguing future lines of research are presented in Section 14.4.

14.2 The modern Leibnizian/Machian approach to physics

14.2.1 Classical mechanics

Modern relational dynamics may be summarized in the following two tenets (Vassallo *et al.* 2022a):

Tenet I. (Spatial relationalism) *The only physically objective spatial information of a physical system is encoded in its shape, intended as its dimensionless and scale-invariant relational configuration.*

Tenet II. (Temporal relationalism) *Temporal structures, such as chronological ordering, duration, and temporal flow, must be defined only in terms of changes in the relational configurations of physical systems.*

The motivation for tenets I and II stems from the relationalist desire to eliminate redundant structures from physical theories, which typically arises because of the presence of various symmetries in the representation of a given physical system. This distaste for redundant structures may be dressed with heavy empiricist overtones by claiming that *all structures whose variation amounts to no empirically observable difference should be shunned from physics.*

The work, pioneered by Barbour and Bertotti, on a theory that fully complies with the relational tenets I and II, eventually led to shape dynamics (SD; see, e.g. Mercati 2018). According to this theory, any closed (or, universal) material system should be stripped of the redundant degrees of freedom resulting from its being embedded in an external space, which means first of all that rigid translations and rotations should be avoided from the picture. This attitude is reminiscent of the Leibnizian static shift argument: taking the material content of the universe and rigidly moving or rotating it in an external Euclidean absolute space would make no empirically observable difference, so why postulating the existence of such a 'metaphysically inert' space?

SD, however, makes a step beyond this Leibnizian attitude. Such a step is motivated by the observation that every measurement simply amounts to the comparison of two physical systems. Formally, this means that only *ratios* of physical quantities carry objective information. Going back to the case of space and time, this comparativist attitude implies that the relationalist should also let go of any notion of size. What remains is simply the *shape* of the system.

Mathematically, the procedure for systematically getting rid of the redundant structure associated with some symmetry is commonly known as *quotienting out*. Schematically, if $\mathcal{Q}$ is the relevant configuration space of a given system and $\mathcal{G}$ is the symmetry group, the quotienting out yielding the space carrying truly physical information is $\mathcal{S} := \mathcal{Q}/\mathcal{G}$, whose dimension is simply $\dim(\mathcal{S}) = \dim(\mathcal{Q}) - \dim(\mathcal{G})$. In the case of N classical particles, $\mathcal{Q}$ is standard configuration space, $\mathcal{G} = \mathsf{Sim}(3)$, the joint group of Euclidean translations T, rotations R, and dilatations S (called *similarity group*), and $\mathcal{S} = \mathcal{Q}/\mathsf{Sim}(3)$ is referred to as the *shape space* of the system.

Although dynamical geometry is not considered in this chapter, let us for completeness indicate the relevant symmetry group and the associated shape space. Let us denote Riem the set of Riemannian 3-geometries and $\mathsf{Diff}(3)$ the group of spatial diffeomorphisms. Then, $\mathsf{Superspace} = \mathsf{Riem}/\mathsf{Diff}(3)$. Further, let $\mathcal{S}_c$ be conformal superspace $\mathcal{S}_c = \mathsf{Superspace}/\text{conformal transformations}$. Finally, the physical space of SD is $\mathcal{S}_\mathcal{V} \equiv \mathsf{Superspace}/\mathrm{VPCT} = \mathcal{S}_c \times \mathbb{R}^+$, where VPCT is the group of *volume-preserving* conformal transformations and $\mathbb{R}^+$ represents the spatial volume or its conjugate variable, the so-called *York time*. The need to restrict the theory to the subset of conformal transformations that keep the volume constant arises from the empirically falsified predictions of a theory taking the full group of conformal transformations: such a theory is unable to account for the expansion of the universe (see Mercati 2018 for details).

A precise mathematical articulation of tenets I and II yields what is today known as the *Mach–Poincaré principle*:

Tenet III. (Mach–Poincaré principle – classic version) *Physical, i.e. relational, initial configurations and their first derivatives alone should uniquely determine the dynamical evolution of a closed system.*

Physically speaking, tenet III merges Mach's idea that the fully relational character of the dynamics can be attained only when the whole universe is considered with Poincaré's insistence on the amount of initial data needed to describe dynamics. A distinction should be drawn between tenets I and II and tenet III: the former establishes an ontological commitment towards spatial and temporal notions, whereas the latter refers to the least possible amount of initial data required to describe the dynamics. But we emphasize that tenet III also takes on an ontological stance (as does tenet IV, which is introduced below). In this regard, the three tenets share the general methodological principle of SD, intended as an ontological commitment, namely the elimination of redundant structure, be it spatial and temporal notions (tenets I and II) or unnecessary initial data (tenets III and IV).

The use of 'first' derivatives is problematic because it places a very strict restriction on the initial data formulation and makes the appearance of an external parameter unavoidable, w.r.t. which *extrinsic* derivatives are introduced, to account for the dynamics, be it the ratio of dilatational momenta in particles models or York time in dynamical geometry. These non-shape degrees of freedom seem to imply a commitment to physical structures other than the relational configuration itself and, thus, mark a conceptual tension with the Leibnizian/Machian spirit (see Vassallo *et al.* 2022b, Section 2.2 for a discussion on this).

The need to overcome this conceptual tension led to the latest refinement of the general framework of relationalism, called *pure shape dynamics* (PSD; see Koslowski *et al.* (2022) for a general technical introduction to the framework). In a nutshell, the qualifier 'pure' means that PSD describes *any* dynamical theory exclusively in terms of the intrinsic geometric properties – which are completely defined in terms of shape space structure alone – of the *unparametrized* curve γ_0 traced out by the physical system in shape space $\mathcal{S}$. This ensures that there are no external reference structures nor clock processes necessary to describe γ_0 in $\mathcal{S}$.

Accordingly, tenet III has to be slightly modified to match PSD's 'intrinsically geometric' nature:

Tenet IV. (Mach–Poincaré principle – modern version) *Physical, i.e. relational, initial configurations and their intrinsic derivatives alone should uniquely determine the dynamical evolution of a closed system.*

The key innovation brought about by tenet IV is to consider in general *higher-order* derivatives of the curve, thereby allowing us to describe the dynamics of a physical system in terms of the curve alone, without the need of any additional non-shape parameters as in standard SD. Clearly, there is a sense in which tenet IV is *weaker* than tenet III, for the former requires more initial data than the latter (higher-order versus first derivatives) to describe a dynamical system. This weakening is anything but a drawback from a modern relationalist standpoint since the upshot is the elimination of any non-shape degree of freedom from the dynamical description of a system—thus delivering a truly relational dynamics cast in terms of the degrees of freedom intrinsic to the system only. This intrinsic character of the framework is achieved through the notion of *intrinsic* derivative, namely an operation that takes the derivative of variables w.r.t. parameters living in the very state space the variables belong to, unlike the extrinsic derivatives appearing in tenet III. A paradigmatic example of intrinsic derivative, which is the one used within PSD, is the derivative w.r.t. the arc length. Interestingly, any derivative w.r.t. an external parameter, be it Newtonian time or otherwise, can be exchanged for an intrinsic derivative (through, for instance, the arc-length parametrization condition; see discussion after equation (14.1)). The reason we care about intrinsic derivatives is because they provide the mathematical structure to account for the ontological requirement imposed by tenet II: dynamics must be given intrinsically, from within the physical system.

Given an already 'quotiented out' physical system, we express the equation of state of the unparametrized curve γ_0 in its associated shape space $\mathcal{S}$ as follows:

$$\begin{aligned} dq^a &= u^a(q^a, \alpha_I^a), \\ d\alpha_I^a &= A_I^a(q^a, \alpha_I^a), \end{aligned} \tag{14.1}$$

and demand that the right-hand side be described in terms of dimensionless and scale-invariant quantities, whose intrinsic change is obtained employing Hamilton's equations of motion. In (14.1), q^a are points in shape space, namely they represent the universal configurations of the system, and u^a is the unit tangent vector defined by the shape momenta p_a:

$$u^a \equiv g^{ab}(q) \frac{p_b}{\sqrt{g^{cd} p_c p_d}},$$

which allows us to define the direction ϕ^A at q^a. It is through the unit tangent vector and the associated direction that the shape momenta enter Hamilton's equations, which are in turn used in the intermediary steps leading to the equation of state (14.1). Finally, α_I^a is the set of any further degrees of freedom needed to fully describe the system. It is this set α_I^a that includes higher-order derivatives of the curve and spares us of the need of additional non-shape degrees of freedom. Among these, one parameter definitely stands out: a measure of the deviation of the curve from geodesic dynamics (cf. κ in Section 14.3.1, equation (14.4)). For consistency, the elements in α_I^a must exhaust the set of all possible dimensionless and scale-invariant quantities that can be formed out of the different parameters entering a given theory.

The derivatives appearing in (14.1) are intrinsic, as anticipated above. Their explicit expression is $dq^a := \frac{dq^a}{ds}$, where s denotes the arc-length parameter of the curve in shape space, i.e. $ds = \sqrt{g_{ab}\, dq^a\, dq^b}$, with g_{ab} being the metric in $\mathcal{S}$. This expression, in turn, enables us to write the relation between intrinsic and extrinsic derivatives by means of the so-called arc-length parametrization condition: $\left(\frac{ds}{dt}\right)^2 = g_{ab}\, \dot{q}^a\, \dot{q}^b$, with $\dot{q}^a$ being the derivative w.r.t. the external, Newtonian time t. Remarkably enough, as shown in Koslowski *et al.* (2022), the equation of state should be taken as a whole and interpreted as giving the relative rates of change of the degrees of freedom of the curve in shape space. Thus, one has $\frac{dq^a/ds}{d\alpha_I^a/ds} = \frac{dq^a}{d\alpha_I^a}$, effectively rendering the dynamics explicitly unparametrized, hence justifying the form adopted by (14.1).

Two crucial features of (14.1) have to be noted. Firstly, by construction, a solution of this set of equations represents the entire history of relational evolution of the universe given certain initial conditions. Indeed, once such initial conditions are selected, (14.1) provides all the physical information needed to fully characterize the universe (i.e. a closed system): all there is to know about the universe is provided by this dynamical evolution. Secondly, (14.1) has an obvious unifying nature. In principle, the whole of relational dynamics, classical, relativistic, and quantum, boils down to the dynamical structure encoded in the above equation of state, which means that all these theories can be formulated in terms of some purely shape variables and their intrinsic derivatives of different orders within their corresponding shape space (Newtonian gravity, which is the example taken in Section 14.2.1, general relativity, which is not discussed here since it requires a paper on its own, and de Broglie–Bohm theory, which we explore in Section 14.3.1). This is one of the most remarkable features of PSD, that is its flexibility to describe an extremely wide range of relational motions: simply speaking, the more structured the system, the more geometric degrees of freedom are needed to describe it.

Finally, it is important to point out how an arrow of time emerges from the dynamics of the curve in shape space: as originally put forward in Barbour *et al.* (2014), the function

$$\mathsf{Com}(q) = -\frac{1}{m_{\text{tot}}^{5/2}} \sqrt{I_{\text{cm}}}\, V_N = \frac{\ell_{\text{rms}}}{\ell_{\text{mhl}}}, \tag{14.2}$$

where I_{cm} is the centre-of-mass moment of inertia, V_N is Newton's potential, and ℓ_{rms} and ℓ_{mhl} account for the greatest and least inter-particle separations, respectively, is a measure of the *complexity* of the N-body system: it accounts for the formation of structures within the system, namely the extent to which particles are clustered. The remarkable feature of (14.2) is its attractor-driven behaviour, whereby the direction of secular growth of complexity *defines* the arrow of time. In general, the recovery of standard temporal and spatial notions in this framework is carried out via the so-called *ephemeris constructions*, which are equations relating said magnitudes to the geometric properties of the curve in shape space. In a sense, these constructions represent the 'inverse' of the quotienting out procedure, in that they explain how and why the impression of there being an external space-time container emerges from a fundamental dynamics that does not feature any spatiotemporal structure properly said.

The above applies to the universe as a whole, as per Mach's insights. What about sub-systems? In the classical N-body case, as is well known, the dynamics features generic solutions that break up the original system into sub-systems, consisting of individual particles and clusters, that become increasingly isolated in the asymptotic regime (Marchal and Saari 1976). Such almost isolated sub-systems will develop approximately conserved charges, namely the energy E, linear momentum $\mathbf{P}$, and angular momentum $\mathbf{J}$. Within the dynamically formed sub-systems, there are pairs of particles that may function as physical rods and clocks. These are referred to as *Kepler pairs* because their asymptotic dynamics tends to elliptical Keplerian motion.

14.2.2 Quantum mechanics

After the short discussion of modern relational theories in the classical case, we now turn our attention to a couple of attempts to extend the programme to the quantum realm by focusing on non-relativistic dBB theory. The suggestion that the dBB approach is a natural choice for constructing a quantum version of SD goes back at least to Vassallo (2015). Indeed, the two frameworks share key insights into physical reality in that both accord a privileged status to 'spatial' descriptions of closed physical systems so that each dynamical stage can be seen as a snapshot of a universal configuration of material bodies. Moreover, given that the standard dBB theory is formulated on a (neo-)Newtonian background structure, the quotienting out required to render the theory relational closely resembles the procedure carried out in classical mechanics. In this section, we are concerned with the two very first attempts at a de Broglie–Bohm–Barbour–Bertotti theory. As shown later, both are incomplete and unsatisfactory on several grounds.

The first model was put forward in Koslowski (2017), and exploited the equivalence between General Relativity (GR) and SD shown in Gomes *et al.* (2011). Koslowski considered a simple quantum toy model whose classical analogue admits two formulations, one possessing key features of the space-time description of GR, and the other sharing key features of SD.

There are two main results of this comparison. Firstly, although both quantum models yield the correct classical limit, the mechanism behind this is markedly different. The timeless quantum constraint equation of the space-time description typical of GR is handled by taking one of its degrees of freedom as an internal clock, w.r.t. which the constrained wave function evolves. The shape dynamics description, instead, exhibits a time-dependent physical Hamiltonian, with this time parameter being the dilatational momentum τ, whereby the shape dynamics analogue of the de Broglie–Bohm 'quantum potential' decays with this time parameter.

Secondly, the shape dynamics analogue of the quantum potential leads to an exponentially large correction to the apparent scale at early times (when $|\tau|$ is small). Remarkably, this inflation of scale is not produced in the constrained space-time formulation. This suggests that inflation could have a natural explanation as semi-classical effects of quantum shape dynamics.

However promising these two features of the SD quantum model may be, there is a deep conceptual flaw, no wonder inherited from the general framework of SD known back then. We are referring to the need for an external parameter to account for dynamics. Closely related to this, this quantum model does not incorporate any measure of complexity to give rise to an arrow of time, *contra* the understanding of physical evolution described in Section 14.2.1.

A second model of quantum shape dynamics has been put forward in Dürr *et al.* (2020), which discusses a dBB dynamics of the N-body system in shape space. This latter model certainly shares some formal similarities with the one we propose in Section 14.3.1, but we claim that such similarities are rather shallow, as shall be argued for presently.

First and foremost, these authors confine themselves to a purely geodesic dynamics in shape space and attempt to reproduce the physics of known regimes through gauge-fixings, effectively restoring the previously-eliminated absolute configuration space. This is physically untenable, for the authors of Dürr *et al.* (2020) do *not* embrace the fundamental insight of SD, namely that shape space is thought of as the fundamental arena where all physics unfolds. According to this insight, laboratory physics is arrived at from the holistic fundamental dynamics in suitably effective regimes within shape space. These effective regimes allow us to write shape space as the Cartesian product of configuration spaces of sub-systems, thereby effectively expressing the fundamental dynamics in terms of separate dynamical sub-systems in each sub-space. In particular, local frames of reference, defined through distant stars and ephemeris time, are local shape sub-structures within the global shape of the universe. This is in stark contrast with the spirit of Dürr *et al.* (2020), where shape space seems to be taken as a convenient mathematical structure wherein physics is simplest (whence their insistence on geodesic dynamics), only to get back standard configuration space physics through suitable liftings and gauge fixings within their fibre bundle structure.

Moreover, the approach of Dürr *et al.* (2020) implies that the wave function is non-normalizable precisely due to their insistence upon describing the interesting physics in configuration space, not shape space. This choice excludes the possibility of taking advantage of the fact that the shape space of the N-body system, the so-called *shape sphere* (Mercati 2018, Section 14.1.7), is *compact*, which guarantees the normalizability of the wave function. This is certainly desirable if one is to endorse the familiar interpretation of $|\Psi^2|$ as a probability distribution. The authors of Dürr *et al.* (2020) argue that the non-normalizability arises from unphysical differences and, further, that it is the universal wave function that fails to be normalizable, while there being no reason why the conditional wave functions, the ones accounting for laboratory physics, should fail to be normalizable. This may be true, but the issue is clouded by their approach, whereas in ours – which primarily relies on the shape space formulation – the matter is readily seen.

Most importantly, these authors, like the model in Koslowski (2017), have not taken into account the attractor-driven behaviour of complexity, required for the emergence of an arrow of time and sub-systems formation. This is the root of the *physical* differences between the model of Dürr *et al.* (2020) and ours: we shall develop a model exhibiting the modified geodesic dynamics typical of PSD, which by construction works within shape space. We then show that this model captures the observable features of the universe – like structure formation and the arrow of time – in appropriate regimes, without betraying the very tenets of SD.

Thus, we claim that Koslowski's and Dürr *et al.*'s models, although arguably do implement relational tenets, fall short of providing a satisfactory account of quantum phenomena because of *conceptual* reasons (that they are "simple" models is a separate matter). We now turn to analysing how PSD fares in this regard.

14.3 A de Broglie–Bohm model of pure shape dynamics

14.3.1 Implementation

Given that the standard dBB theory describes the motion of a universal configuration of particles whose collective behaviour is determined by the wave function associated with the system, a full specification of the dynamics requires two equations. Firstly, the guidance equation, which implements the guidance principle constraining the physically allowed motions based on the wave function's form. Secondly, Schrödinger's equation, which describes the dynamics of the wave function itself. Hence, an appropriate dBB model of PSD is expected to exhibit the same dynamical features of the standard approach, but cast in terms of the geometric properties of unparametrized curves in shape space.

To obtain this, let us first define the relevant shape space, which is simply the same as in the classical N-body system: $\mathcal{S} \equiv Q^N/\mathsf{TRS}$, with Q^N being the configuration space of the standard dBB theory. Obviously, the wave function must be defined on $\mathcal{S}$. Physically, this simply means that all information encoded in Ψ regarding the embedding of the N-body system in an external space must be dropped. Mathematically, this is readily achieved as follows: if Ψ_N is the wave function living in Q^N, its restriction to shape space is $\Psi_{\mathcal{S}} \equiv \Psi_N|_{\mathsf{Sim}}$ (strictly speaking, this reduction works only if the original wave function Ψ_N is single-valued along orbits of Sim; we omit this, considering that it does not affect our analysis). From now on, we omit the label $\mathcal{S}$ from Ψ and work with wave functions on shape space.

With this technical machinery in place, the calculations readily yield (see Farokhi *et al.* (2024) for the technical details of the derivation)

$$\begin{aligned} dQ^a &= u^a(\phi) \\ d\phi_A &= \tfrac{\partial \Phi_A}{\partial Q^a} u^a(\phi) - \tfrac{\partial \Phi_A}{\partial u^a}\left(\tfrac{1}{2} g^{cd}_{,a}(Q) u_c(\phi) u_d(\phi) + \tfrac{1}{\kappa} V_{T,a}(Q)\right) \\ d\kappa &= -2u^a(\phi) V_{T,a}(Q) + \tilde{K}(\kappa, \gamma, Q, \phi) \\ dR(q) &= -\tfrac{1}{\sqrt{\kappa}}\left(g^{ab}(q) R_{,a}(q) S_{,b}(q) + \tfrac{1}{2} R(q) \Delta S(q)\right) \\ dS(q) &= -\tfrac{1}{\sqrt{\kappa}}\left(\tfrac{1}{2} g^{ab}(q) S_{,a}(q) S_{,b}(q) + V_T(q)\right), \end{aligned} \tag{14.3}$$

where $V_T(q) = V(q) - \frac{\Delta R(q)}{2R(q)}$ is the total potential and $\Psi(q) = R(q)\, e^{iS(q)}$. Also, we have used the kinematic metric $g_{ab}(q)$ on shape space to split $S^1_a := \nabla_a S(q)|_{q^a = Q^a}$ into directions ϕ_A and the additional degree of freedom κ, which plays a key role in PSD, namely, a measure of the deviation of the dynamics from geodesic motion:

$$\begin{aligned} u^a(\phi) &= \frac{S^1_a}{\sqrt{g^{ab}(Q) S^1_a S^1_b}} \\ \kappa &= \frac{g^{ab}(Q) S^1_a S^1_b}{R^\alpha}. \end{aligned} \tag{14.4}$$

In the above equation, $u^a(\phi)$ is a unit tangent vector (w.r.t. the kinematic metric g_{ab}) at Q^a that is determined by the direction ϕ_A, and R is the scale variable appearing in a potential homogeneous in R of degree γ, which is required to match the dynamics of the classical N-body system (see Koslowski *et al.* 2022). Finally, the function $\tilde{K}$ is also specified when reproducing the correct classical limit.

To sum up, (14.3) represents the equation of state of a curve obeying the relational counterpart of the dBB laws. The curve gives the succession of instantaneous configurations of the system, which are pairs $(Q^a, \Psi(q))$ of shapes and wave functions on shape space, with Q standing for the actual configuration of the N-body system and q for an arbitrary configuration in shape space.

Note that the second term in the equation for κ in (14.3) violates the equality between the shape momenta $p_a = \sqrt{\kappa} u_a$ and $S_{,a}$ under evolution. This violation represents a substantial departure from the standard theory, but it is not, by itself, a fatal flaw of the relational model: it just means that it is not trivial to find the regime in which the guidance principle holds, i.e. $p_a = S_{,a}$ is recovered. The physical significance of such a departure from the standard theory is easy to grasp: according to (14.3), the emergence of sub-systems satisfying Born statistics, also referred to as 'quantum equilibrium condition', is not always possible – the guidance principle is key to ensuring it, in both dynamical relaxation and typicality arguments; see below. The investigation which conditions may restore Born statistics in this model is still in progress. In other words, we are not postulating the quantum equilibrium condition but trying to derive it from a more general dynamics. In this respect, some tentative arguments for the recovery of the guidance equation can already be given.[1]

To this extent, let us consider two conditions on sub-systems: (i) isolation, whereby the full configuration space may be written as $\mathcal{S} = \mathcal{S}_I \times \mathcal{S}_J \times \cdots$ for sub-systems $I, J, \ldots$, and (ii) boundedness, that is, the scale R_I of the sub-system I tends to a constant value. From the classical case, $K_I = -\gamma\, \kappa\, \frac{D_I}{p_I}$, with γ the homogeneity degree of the potential (see below), and $dR_I = R_I\, \frac{D_I}{p_I}$, which yield the desired result $K_I \approx 0$. Thus, in the regime where (i) and (ii) hold, it can readily be shown from (14.3) that the guidance principle $p_a = S_{,a}$ is recovered and we may invoke either some sort of typicality considerations (Dürr *et al.* 1992) or an analogue of the sub-quantum H-theorem (Valentini 1991a, 1991b) to dynamically arrive at an effective dynamics for sub-systems obeying the Born rule $\rho_I = |\psi_I|^2$, for the probability distribution ρ_I and wave function ψ_I. Both approaches are equally viable so, as the theory stands right now, there is no indication of which of the two views should be considered as the better one.

There is however a potential mathematical problem. Clearly, the equation of state couples Q^a with ψ through the quantum potential, so generally, entangled solutions to Schrödinger equation are expected, meaning we will have to find regimes for the existence of disjoint sub-systems, $\mathcal{H} = \mathcal{H}_1 \otimes \mathcal{H}_2 \otimes \cdots$. This is arguably a challenging problem, whose resolution is necessary to provide a sound and robust mechanism for the emergence of sub-systems. This is being actively investigated.

Remarkably enough, the dynamics exhibited by the equation of state (14.3) shows the attractor-driven behaviour in shape space already stressed in the classical case (Barbour *et al.* 2014; Koslowski *et al.* 2022, Sections 3.5 and 3.6). In a nutshell, this means that the direction of secular growth of complexity – which measures the amount of structure or "clustering" inherent in a shape – *defines* the arrow of time (see Farokhi *et al.* (2024), Section 3.4, for the numerical analysis of the 3-body system, which highlights how the attractor-driven behaviour in shape space is independent of Planck's constant). As already stressed earlier, the asymptotic evolution of the classical N-body system gives rise to the formation of stable substructures, among which Kepler pairs play a major role as physical rods and clocks. However, as already mentioned above, a full technical characterization of the quantum counterpart of these Kepler pairs, being part of the subtle question of the formation of sub-systems satisfying effective Schrödinger equations in the quantum model, is still in progress.

Finally, the analysis of the classical limit of (14.3) is conceptually as simple as in the standard dBB case. In a nutshell, whenever the evolution of the quantum potential $V_Q = -\frac{\Delta R}{2R}(Q)$ along the curve becomes negligible, the evolution of the curve in shape space is effectively described by classical equations of motion determined by the full potential $V_T = V + V_Q$. Hence, the classical regime is attained when V_Q goes to zero. In this case, (14.3) reduces to the equation of state associated with the classical N-body system:

[1] The gist of these arguments is due to joint work with Pooya Farokhi and Tim Koslowski.

$$
\begin{aligned}
dq^a &= u^a(\phi)\,, \\
d\phi_A &= \tfrac{\partial \Phi_A}{\partial q^a} u^a(\phi) - \tfrac{\partial \Phi_A}{\partial u^a}\left(\tfrac{1}{2} g^{bc}_{,a}(q) u_b(\phi) u_c(\phi) + \tfrac{1}{\kappa} V_{,a}(q)\right)\,, \\
d\kappa &= -2u^a(\phi) V_{,a}(q) \mp \gamma\kappa \sqrt{-\left(1 + \tfrac{2\,V(q)}{\kappa}\right)}\,,
\end{aligned}
\tag{14.5}
$$

where $\kappa := p^2/R^\gamma$, with R^γ being the scale component of the potential.

14.3.2 Discussion

The dBB model of PSD just presented is a substantial step forward towards a de Broglie–Bohm–Barbour–Bertotti theory. For starters, it is not a 'shape space analogue' of standard dBB theory, *contra* the spirit of the model by Dürr *et al.* (2020). Indeed, the fact that the guidance principle in the generalized PSD model is not automatically fulfilled indicates that this model is not a mere relational rendition of the standard dBB dynamics. The novelty of the dynamical laws (14.3) resides exactly in the fact that a dBB-like dynamics is not assumed *ab initio*, but it is reached when specific physical conditions are met. This means that the dBB model of PSD has (i) a broader physical significance than a standard dBB one and (ii) it promises to shed light on the very nature of a dBB-like dynamics once a good understanding of the physical conditions under which the guidance principle comes into effect is reached.

What does (14.3) tell us about the nature of the wave function in a "truly" relational setting? The first thing that has to be noted is that the dynamics depicted in this system of equations features both geometrical and quantum degrees of freedom. The quantum part boils down to the phase and amplitude of the relational wave function. We immediately observe that these two groups of degrees of freedom are coupled together, with none of them being mathematically reducible to the other. In this sense, the wave function in the dBB model of PSD is still a fundamental, irreducible piece of formalism that cannot be 'washed' away. In other words, the dBB dynamics in PSD takes on initial values of the form $(Q^a, \Psi(q))$ that are analogous to the non-relational case. However, the structure of (14.3) is still peculiar enough to suggest some features that narrow down the possible choices for a metaphysics of the wave function and the quantum degrees of freedom it represents.

To see the novel contribution that the PSD perspective on quantum physics brings, it is sufficient to ask the question about *where* the quantum degrees of freedom live. Indeed, much of the debate regarding the nature of the wave function in a dBB setting touches upon the fundamental arena where the physical happenings take place. Realistic views of the wave function as a local field are often associated with a commitment to the existence of a higher-dimensional configuration space as the fundamental physical space. Likewise, stances that regard the wave function as some sort of law-like, information-driven element of the formalism tend to accord a privileged ontological status to three-dimensional space. In this context, the dBB model of PSD represents a third option on the metaphysical table.

At the beginning of the chapter, we clarified that the dynamical laws of PSD are formulated in shape space and that an entire history of a universe governed by (14.1) is encoded in an unparametrized curve in this space, without the need of any information external to such a sequence. This suggests that shape space as a whole has physical meaning only insofar as it is considered as the collection of all physically possible curves according to (14.1). Moreover, each and every relational configuration q^a is not fundamentally placed in *any* external space, be it three-dimensional or higher. Otherwise said, fundamentally, PSD is not committed to the existence of neither configuration space nor three-dimensional space – with shape space being a mere abstraction over all physically possible motions. The consequence for the debate regarding the nature of the wave function is immediate: the quantum degrees of freedom are neither 'external' to the material ones (in the sense of there being a separate object encoding the 'quantumness' of the world) nor 'internal' – in the sense of their being just a way to describe the evolution of the material structures. Instead, the quantum degrees of freedom are *on a par* with the material ones.

What does 'being on a par' mean from a Leibnizian/Machian relational perspective? In discussing how spatial relationalism is implemented in the classical case, we made it clear that being a material particle in that setting amounted to nothing over and above standing in spatial relations with the other *relata* making up a shape. From this point of view, the standard talk of degrees of freedom of Newtonian material particles is relationally translated into talk about differences in the web of spatial relations making up the shapes ordered in a dynamical curve. Hence, there is a conceptual link between the notion of relational degree of freedom and the relations making up a shape. This is the sense in which we claim that material and quantum degrees of freedom are on a par: the fundamental relation gluing together the *relata* in a dBB shape is not just spatial but a quantum/spatial hybrid. These two components are distinct, yet they are two sides of the same coin. We have argued at length elsewhere for a metaphysics of hybrid entanglement/spatial relations in PSD (Vassallo *et al.* 2023), so we refer the reader interested in knowing pros and cons of this choice to that paper. What interests us here is that the dBB model of PSD delivers a new take on the nature of the wave function in that it suggests that this mathematical object is 'diluted' inside a shape.

The metaphysical moral that (14.3) suggests is, hence, that the wave function is neither an object nor a description of how quantum systems behave. Rather, it represents (part of) what it is for a system to be quantum. But is not this going back to the 'orthodox' view that everything there is to know about a quantum system is encoded in its quantum state? Granted, the dBB model of PSD puts an emphasis on quantum relations, not quantum states, but the worry remains that we are back to deal with metaphysically suspicious entities with a feeble connection with the manifest image of material objects inhabiting a three-dimensional space. To put the issue more clearly, we may ask the question: how do we get from a picture of shapes being ordered 'one after another' to a picture where a pointer on a display points in a certain direction as a consequence of measurement-like interaction with a quantum sub-system of the universe?

The answer to this question is twofold. The first part concerns the way we obtain ordinary variety spatial and temporal notions out of a timeless sequence of shapes. The details of such a construction are carried out in Vassallo *et al.* (2022b), Section 4.2. The rough idea is that, in order to recover space and time from a purely relational description, it is sufficient to 'reverse' the quotienting out procedure discussed in Section 14.2.1 by using the ephemeris constructions.

The physical meaning of such constructions relies on the notion of complexity, intended as a measure of the clustering among sub-systems. As we saw, the growth pattern of complexity makes it possible to assign a 'time stamp' to each and every shape in a dynamical curve so that global notions of duration and directed flow of time can be achieved. At the same time, this construction leads to a local notion of rod and clock, as exemplified in the formation of Kepler pairs. As a result, we regain a structured picture of the universe as a space-timeful 'arena' for material happenings. Note how this construction does not display the conceptual obscurity with respect to notions such as location and duration that are instead inherent into standard quantum mechanics. This is for the reasons we already mentioned at the beginning of Section 14.2.2, that is because the dBB framework privileges spatial concepts in the physical description, which makes it possible to accommodate the key notion of Leibnizian shape from the beginning.

The second part of the answer concerns how we can obtain a meaningful description of quantum sub-systems whose statistical behaviour matches the Born's rule of standard quantum mechanics out of 'undivided' shapes. This can easily be carried out in the regime where the guidance principle holds. In this case, the decomposition of the global wave function into effective wave functions associated with sub-systems can be carried out in the same vein as in the standard dBB theory.

Hence, assessing the prospects of a de Broglie–Bohm–Barbour–Bertotti theory boils down to assessing whether and to what extent this twofold answer is viable. For starters, it is still not clear what notion of complexity can be defined in a dBB context. The simple definition (14.2) works well in classical mechanics, but it should be amended to account for the more complex motions exhibited in a dBB context. This is by no means an unsurmountable problem since nothing speaks against the possibility of defining a dBB complexity that serves the same purposes of its classical counterpart in order to implement the ephemeris constructions. Instead, a more delicate issue concerns finding a recipe to define effective wave functions obeying the Born rule. Firstly, at this stage, it is not clear under which physical conditions (14.3) satisfies the guidance principle, and which subset of such physical conditions allows for a rewriting of the wave function in terms of effective terms with mutually disjoint supports. Furthermore, it is still an open question whether the conditions under which the guidance principle is fulfilled will naturally select a probability distribution $\rho = |\Psi|^2$ as a typicality measure or, rather, will point toward a generic starting distribution that quickly evolves towards $|\Psi|^2$ as some sort of relaxation limit.

In conclusion, the model (14.3) is still far from being fully worked out, but it represents a window that is supposed to appear as far as the dBB theory is concerned. As the previous discussion hopefully highlights, by combining the original approach to quantum physics of de Broglie and Bohm with the deep insights into relational physics of Barbour and Bertotti, it is possible to substantially progress the understanding of the quantum world, which goes far beyond the picture originally suggested in Bohm's 1952 seminal papers in terms of material particles moving under the influence of a pilot wave. The project may of course crash and burn at any point, but this is a virtue rather than a vice: it proves that the work on quantum PSD is a legitimate physical endeavour that carries a deep ontological import, and not just an idle exercise in *a priori* metaphysics. From this point of view, the prospects of a de Broglie–Bohm–Barbour–Bertotti theory look as promising and exciting as ever.

14.4 Conclusion: on the 'geometrizing away' of the wave function

One of the key questions that motivated the research into a fully Leibnizian/Machian formulation of the dBB theory concerned the possibility of reducing the wave function to some geometric features of the framework. The essential intuition in this respect was to show that the 'influencing' role of the wave function on the material degrees of freedom of the theory could be rendered as some geometric constraints on the physical curves in shape space. The original suggestion made in the literature (Vassallo 2015; Vassallo and Ip 2016) was to incorporate Bohm's quantum potential into shape space's metric in a vein similar to how the gravitational potential is dispensed with in general relativity (a similar idea in the quantum gravity context dates back to Pitowsky (1991)). The expectation was that the geodesics motions in shape space would exhibit the quantum behaviour typical of the dBB theory. This target slightly changed with the onset of PSD, where the interesting dynamical curves are no longer geodesics (recall the crucial role of the 'curvature' κ). In this new scenario, the geometrized wave function should have been rendered as some additional geometric degree of freedom of the curve besides the direction ϕ^A and κ. This option, however, is cumbersome to implement and, in the end, is not really needed in order to make good sense of the model (14.3). Does it mean that the dream of geometrizing away the wave function has been finally abandoned? Well, not really. Indeed, preliminary research shows that this is the way to go if we want to push the parsimony-driven reductionist spirit of PSD to its extreme consequences.

To better understand the direction we are heading to, recall the crucial role that the quotienting out procedure plays in PSD. This procedure permits to rigorously eliminate from the mathematical formalism any theoretical feature that is deemed redundant or metaphysically suspicious under the Leibnizian/Machian lenses. The most realistic example is of course the elimination of the degrees of freedom related to the embedding of a universal material configuration in an external space, as represented by the quotienting out of translations, rotations, and dilations. Note that this reasoning can be pushed beyond the spatial side, for example by quotienting out permutation transformations to signify that material particles do not possess intrinsic identity. The metaphysical significance of the quotienting out procedure is, hence, clear: a quotiented out formal feature is something that does not belong to the fundamental ontology of the theory, although it can be recovered in the appropriate limit via an ephemeris construction.

Compare this with what happens to the standard universal wave function of the dBB theory in (14.3). In this case, this mathematical object is not strictly speaking quotiented out; rather, it is mathematically restricted to the quotiented out space constituting the dynamical arena of the theory. This restriction is sufficient to guarantee that all and only the relational physical information encoded in the original wave function is translated into the PSD model. However, if we consider (14.3) from this perspective, it is no surprise that there is no straightforward way to reduce it to geometrical degrees of freedom of a dynamical curve: the quantum degrees of freedom inherent into the wave function are rendered relational but they are still 'out there', as opposed to the spatiotemporal degrees of freedom that are entirely omitted from the fundamental ontology via the quotienting out operation.

Here lies a possible key insight: why not quotienting out the wave function itself, instead of just restricting it to shape space? This obviously begs the question as to *what* should be quotiented out in the wave function. Undoubtedly, the wave function contains a tremendous amount of information, not least the statistical information about possible measurement outcomes. So a more useful question to ask may be: what is it that *should not* be quotiented out? There is no definite answer to this question as yet but, intuitively, the most fundamental and indispensable information that we can think of is that encoded in the guidance principle. This is because this principle encodes the fundamental dynamical behaviour of material structures. Everything over and above this – e.g. statistical information about measurement outcomes – may be seen just as a useful description to tie the fundamental behaviour of matter to experimental procedures that make this behaviour empirically observable.

If it should be possible to quotient out the statistical information that leads to Born's rule – and if we agree that the manifest 'quantumness' of the world is encoded in $|\Psi|^2$ – this would mean that the fundamental theory would be pre-quantum in a clear physical sense: it would be a theory that recovers Born's rule in an appropriate limit, thus implying that it is a non-fundamental feature of reality, but something that 'emerges' at a certain point. If, besides this, a complete quotienting out of the spatiotemporal degrees of freedom should be possible, the end result theory would in addition be non-spatiotemporal.

Such a speculative theory would be a game changer in the quest for a theory of quantum gravity. Indeed, the common motivation for the pursuit of the many quantum gravity programmes active nowadays is superseding general relativity – and, possibly, quantum field theory – by providing a quantum treatment of gravitational phenomena. The hidden premise behind this goal is that quantum physics is a window into the fundamental nature of the world. If pursued until its most extreme consequences, the PSD programme would challenge this received view about quantum physics by highlighting how going quantum does not mean reaching the most basic layer of reality – in fact, quantum phenomena would not be fundamental at all. This 'final' PSD model would constitute a radically novel framework for fundamental physics, according to which quantum and gravitational phenomena emerge from a genuinely fundamental domain that is inherently pre-quantum and does not feature the general relativistic space-time. This is, of course, pure speculation at the moment. However, it is exciting to see how far the original ideas of de Broglie, Bohm, Barbour, and Bertotti may lead in the quest for the understanding of the deepest nature of reality.

Acknowledgements

The authors gratefully acknowledge financial support from the Polish National Science Centre under grant no. 2019/33/B/HS1/01772.

References

Adlam, E. (2021). *Foundations of Quantum Mechanics.* Cambridge: Cambridge University Press.

Albert, D. (1996). Elementary quantum metaphysics. In Cushing, J., Fine, A. and Goldstein, S. (eds.) *Bohmian Mechanics and Quantum Theory: An Appraisal*, pp. 277–284. Dordrecht: Springer.

Barbour, J. (1974). Relative-distance Machian theories. *Nature*, **249**: 328–329.

Barbour, J. and Bertotti, B. (1977). Gravity and inertia in a Machian framework. *Il Nuovo Cimento B*, **38**(1): 1–27.

Barbour, J. and Bertotti, B. (1982). Mach's principle and the structure of dynamical theories. *Proceedings of the Royal Society A*, **382**: 295–306.

Barbour, J., Koslowski, T. and Mercati, F. (2014). Identification of a gravitational arrow of time. *Physical Review Letters*, **113**: 181101.

Bohm, D. (1952a). A suggested interpretation of the quantum theory in terms of "hidden" variables. I. *Physical Review*, **85**(2): 166–179.

Bohm, D. (1952b). A suggested interpretation of the quantum theory in terms of "hidden" variables. II. *Physical Review*, **85**(2): 180–193.

Bohm, D. and Hiley, B.J. (1993). *The Undivided Universe. An Ontological Interpretation of Quantum Theory.* London: Routledge.

de Broglie, L. (1928). La nouvelle dynamique des quanta. In *Electrons et photons. Rapports et discussions du cinquième Conseil de physique tenu à Bruxelles du 24 au 29 octobre 1927 sous les auspices de l'Institut international de physique Solvay*, pp. 105–132. Gauthier-Villars.

de Broglie, L. (1960). *Non-linear Wave Mechanics.* Amsterdam: Elsevier B.V.

Dürr, D., Goldstein, S. and Zanghì, N. (1992). Quantum equilibrium and the origin of absolute uncertainty. *Journal of Statistical Physics*, **67**: 843–907.

Dürr, D., Goldstein, S. and Zanghì, N. (2020). Quantum motion on shape space and the gauge dependent emergence of dynamics and probability in absolute space and time. *Journal of Statistical Physics*, **180**: 92–134.

Farokhi, P., Koslowski, T., Naranjo, P. and Vassallo, A. (2024). A de Broglie-Bohm Model of Pure Shape Dynamics: *N*-body system. *Foundations of Phyiscs*, **54**, 40. `https://doi.org/10.1007/s10701-024-00776-4`.

Goldstein, S. and Zanghì, N. (2013). Reality and the role of the wavefunction in quantum theory. In Ney, A. and Albert, D.Z. (eds.) *The Wave Function. Essays in the Metaphysics of Quantum Mechanics*, pp. 91–109. New York: Oxford University Press. `http://arxiv.org/abs/1101.4575`.

Gomes, H., Gryb, S. and Koslowski, T. (2011). Einstein gravity as a 3D conformally invariant theory. *Classical and Quantum Gravity*, **28**: 045005.

Koslowski, T. (2017). Quantum inflation of classical shapes. *Foundations of Physics*, **47**(5): 625–639.

Koslowski, T., Naranjo, P. and Vassallo, A. (2022). Pure shape dynamics: General framework. *Classical and Quantum Gravity*, **39**: 215017. DOI 10.1088/1361-6382/ac9725.

Marchal, C. and Saari, D. (1976). On the final evolution of the n-body problem. *Journal of Differential Equations*, **20**: 150–186.

Maudlin, T. (2019). *Philosophy of Physics: Quantum Theory.* Princeton: Princeton University Press.

Mercati, F. (2018). *Shape Dynamics.* Oxford: Oxford University Press.

Pitowsky, I. (1991). Bohm's quantum potentials and quantum gravity. *Foundations of Physics*, **21**(3): 343–352.

Romano, D. (2020). Multi-field and Bohm's theory. *Synthese*. DOI 10.1007/s11229-020-02737-6.

Valentini, A. (1991a). Signal-locality, uncertainty, and the subquantum H-theorem. I. *Physics Letters A*, **156**(1): 5–11.

Valentini, A. (1991b). Signal-locality, uncertainty, and the subquantum H-theorem. II. *Physics Letters A*, **158**(1): 1–8.

Vassallo, A. (2015). Can Bohmian mechanics be made background independent? *Studies in History and Philosophy of Modern Physics*, **52**: 242–250.

Vassallo, A. and Ip, P. H. (2016). On the conceptual issues surrounding the notion of relational Bohmian dynamics. *Foundations of Physics*, **46**(8): 943–972.

Vassallo, A., Naranjo, P. and Koslowski, T. (2022a). The physics and metaphysics of pure shape dynamics. In Vassallo, A. (ed.) *The Foundations of space-time Physics: Philosophical Perspectives*, pp. 108–132. New York: Routledge.

Vassallo, A., Naranjo, P. and Koslowski, T. (2022b). A proposal for a metaphysics of self-subsisting structures I. Classical physics. *Synthese*, **200**. DOI 10.1007/s11229-022-03865-x.

Vassallo, A., Naranjo, P. and Koslowski, T. (2024). A proposal for a metaphysics of self-subsisting structures II. Quantum physics. *Foundations of Physics*, **54**, 65. https://doi.org/10.1007/s10701-024-00800-7

15

How weak values illuminate the role of 'hidden' variables as predictive tools

Xabier Oianguren-Asua, Albert Solé, Carlos F. Destefani, and Xavier Oriols

15.1 Introduction

In the literature, weak values (Aharonov *et al.* 1988; Tamir and Cohen 2013) have traditionally been discussed either mostly from an 'ontological perspective' (Fankhauser and Dürr 2021), or from an 'experimental perspective' (Dressel *et al.* 2014; Kofman *et al.* 2012).The lack of a clear distinction between the experimental, the ontological, and the formal descriptions of weak values has led many physicists to suspect that weak values are either mere theoretical oddities with no practical import or experimental 'tricks' lacking serious theoretical significance. In this chapter, we provide a comprehensive description of weak values which is suitable for physicists with different interests, ranging from those focused on their operational aspects to those interested in their ontological implications.[1] To achieve this, we explicitly distinguish (i) their *experimental* description in terms of laboratory protocols and statistical post-processing, (ii) their *formal* description within the general state-operator formalism, and (iii) their *ontological* significance according to different quantum theories or interpretations.

By considering the first two descriptions, we draw the first main conclusion of this chapter: weak values and their functions, along with the associated operational techniques, enable numerous new predictions and provide experimental access to a wealth of information about microscopic systems. Typically, this information is not attainable with more conventional characterization tools like expectation values of usual operators. Through an insightful case study, we demonstrate the remarkable characterization ability of weak values, even in situations where conventional expectation values fail to elucidate the relevant properties of a quantum system.

Moreover, this same case study serves as an illustration of the second main conclusion of the chapter: the predictive power of weak values remains unaffected by debates on their ontological status. We should recall that such an emphasis on practicality is not novel. For instance, when dealing with the *wave function*, it has already been

[1]We refer to a physicist who demands a fully fledged description of reality from quantum mechanics as having an '*ontological* attitude'. In contrast, we say that a physicist who uses quantum mechanics merely as a predictive algorithm has an '*empiricist* attitude'.

widely accepted and has proven to be highly productive, as witnessed by the impressive developments in quantum technologies regardless of ongoing debates on the meaning of the wave function. We consider that similar developments can occur if the usefulness of weak values as predictive and characterization tools is duly recognized.

This does not entail, however, that ontological discussions are completely idle in this context. Apart from their obvious interest in the foundations of quantum mechanics, we contend that certain quantum theories or interpretations, even if motivated through ontological arguments, can be of interest for those with an 'empiricist attitude'. There are certain weak values or functions thereof that are singled out as physically relevant by Bohmian mechanics (Bohm 1952; Holland 1993; Oriols and Mompart 2019; Dürr and Teufel 2009) and other 'hidden-variable' theories, whereas they do not hold any particular significance from the standpoint of other approaches. We show by means of the case study that the specific weak values highlighted by some of these 'hidden' variable theories are valuable for characterizing problematic quantum systems, a characterization that can be contrasted in the laboratory. This illustrates that such theories provide heuristically convenient frameworks to analyse quantum systems both theoretically and experimentally. As a consequence, we draw the third main conclusion of the chapter: one may use certain 'hidden' variable theories, such as Bohmian mechanics, as heuristic tools in the experimental characterization of quantum systems, even if one is not necessarily committed to the ontology of such theories.

15.2 On the meaning of weak values

As stated in Section 15.1, in this section we aim to characterize three defining facets of weak values, namely their *experimental*, *formal*, and *ontological* descriptions. In order to clarify this threefold distinction, it is useful to apply it first to standard *expectation values* that are better understood than weak values. Regarding the *experimental* description[2] of expectation values (choosing the kinetic[3] momentum's expectation as a driving example), consider the following statistical laboratory protocol:

(i) A certain microscopic system, called the *object system* or *object*, is prepared identically (with the same wave function[4]) M times in a laboratory.

[2] In the philosophical literature, an operational reduction of a concept is sometimes understood as the *definition* of that concept in purely *observational* terms or by specifying a set of operations in a laboratory. We are *not* attempting to provide such an analysis here. Firstly, the operations involved in the protocols we describe do not solely concern purely observational quantities or methods. They require a significant amount of theory, including bridge principles of the predictive core of quantum mechanics, such as those related with the preparation of systems. Secondly, in our experimental description of both standard expectations and weak values, we resort to notions, such as 'laboratory measurements', that, for lack of space, we cannot fully analyse in terms of specific laboratory operations. Now, even if the experimental description of weak values is theory-laden, the amount of theory involved is much lower than that required for the discussion of the formal and ontological aspects of weak values.

[3] Note that the canonical momentum operator lacks electromagnetic gauge invariance, whereas the kinetic momentum operator, obtained from the commutator of the position and the Hamiltonian operators, is gauge invariant. In the absence of a magnetic field and assuming the Coulomb gauge both operators are equivalent, rendering such distinctions irrelevant. For simplicity, we adopt this scenario throughout the chapter.

[4] In this chapter, we exclusively deal with pure states. Extending the discussion to mixed states would mainly introduce averages of the considered results across the mixture, with no substantial

(ii) For each preparation, a standard or so-called 'strong' laboratory measurement of the momentum is registered, resulting in $\mathfrak{p}^{(k)}$ for the kth repetition. A standard or 'strong' laboratory measurement is performed by first letting the microscopic object interact with a calibrated ancilliary microscopic system, called an *ancilla*, constituting the part of the measurement device that directly interacts with the object. This interaction is designed to correlate in an (approximately) one-to-one fashion the momentum's value of the object with some property of the ancilla, typically its position. In general, the process substantially alters the pre-measurement state of both sub-systems due to the reciprocity of interactions. The position of the ancilla is then coupled in an (approximately) one-to-one fashion with the position of a macroscopic pointer, providing the outcome of the measurement. This last coupling is mediated by the chain of interacting systems that constitute the measurement device.[5]

(iii) Finally, the quantity $\langle\mathfrak{p}\rangle_M := \sum_{k=1}^{M}\mathfrak{p}^{(k)}/M$, known as the *average value of momentum*, is computed. If M is large enough, upon several repetitions of the protocol with different ensembles of identically prepared systems, we expect the obtained average values to converge to a fixed number $\langle\mathfrak{p}\rangle := \lim_{M\to\infty}\langle\mathfrak{p}\rangle_M$. This asymptotic average value is called the *(empirical) expectation value of momentum.*

Then, in order to relate this last quantity to the *formalism* of quantum mechanics, one should consider the state-vector $|\psi\rangle$ and the momentum operator $\hat{P}_s$ that models the preparation procedure mentioned above. As is well known, after assuming the mathematical description of a measurement by Von Neumann (1955), including Born's rule, one obtains that $\langle\mathfrak{p}\rangle = \langle\psi|\hat{P}_s|\psi\rangle$. Thus, we refer to $\langle\hat{P}_s\rangle := \langle\psi|\hat{P}_s|\psi\rangle$ as the *(formal) expectation value of momentum.*

Lastly, the *ontological* significance of expectation values may vary depending on specific ontological attitudes. Some physicists may argue that expectation values can only be understood as properties of the ensembles, while others may suggest that they also characterize each individual system. Due to the unavoidable backaction of the measurement, some may interpret them as properties of the systems after the measurement has occurred. However, as $\langle\hat{P}_s\rangle$ depends solely on the wave function of the systems before the backaction occurs, others may claim that this quantity characterizes the systems' pre-measurement state. What matters to us is that, regardless of ontological disputes, nobody would question the practical utility of expectations in characterizing quantum systems. As shown later, similar conclusions can be inferred regarding weak values.

change in Conclusions 1-4 and 7. We defer this analysis to future research, along with the exploration of open system weak values (derivable from many-body weak values (Destefani and Oriols (2023a))).

[5] As a result, in a standard laboratory measurement, the *macroscopic pointer* gets correlated in an (approximately) one-to-one manner with the measured property of the *object*. In this respect, although the ancilla is always present in experimental terms, one may omit any reference to it, as is generally done. However, the role of the ancilla becomes prominent in what we refer to as 'weak' laboratory measurements in the next sub-section. Thus, we already mention the ancilla here in order to clarify the contrast between standard and 'weak' laboratory measurements.

15.2.1 On the meaning of weak values post-selected in position

15.2.1.1 Weak values as quantities obtained in a laboratory. Let us now proceed with the experimental description of a weak value, similarly to what we have just done with standard expectation values. We take the weak value of momentum post-selected in position as a driving example. As described later, there are two closely related asymptotic-empirical weak values that we denote by $\mathcal{P}_{e,r}(x,t,\tau)$ and $\mathcal{P}_{e,i}(x,t,\tau)$ respectively.[6] Starting with $\mathcal{P}_{e,r}(x,t,\tau)$, this is the quantity that can be approximated after performing the following laboratory protocol and subsequent statistical analysis:

(i) A certain microscopic system, called the *object system* or *object*, is prepared identically (with the same wave function) M times in a laboratory.

(ii) For each preparation, a 'weak' laboratory measurement of the momentum is registered at the object's time t, resulting in a value $\mathfrak{p}_w^{(k)}$ for the kth repetition. In general, a 'weak' laboratory measurement of an observable involves obtaining information about this observable while causing minimal disturbance to the object system. This is usually achieved by coupling the object with the measurement device's *ancilla* through a limited interaction that only slightly correlates the targeted object's information (in our case, the momentum) with some property of the ancilla, typically its position. Next, this property of the ancilla is correlated in an (approximately) one-to-one fashion with the position of a macroscopic pointer, as we have described above for a standard laboratory measurement. In this way, the initial (pre-measurement) state of the object is only slightly modified after the interaction.[7]

(iii) A short time τ after t, a standard laboratory measurement of the position of the object is done, yielding the value $x^{(k)}$ for the kth repetition.

(iv) At this point, M pairs of values $(\mathfrak{p}_w^{(k)}, x^{(k)})$ with k in $\{1, ..., M\}$ have been experimentally determined. Next, the average $\mathfrak{p}_w^{(k)}$, for those cases in which $x^{(k)}$ is a certain number x, is computed. In other words, one computes the average momentum (weakly measured) of those systems found at x a time τ later. We refer to such a conditional average as the *empirical weak value.* Mathematically, it can be written as

$$\mathcal{P}_{e,r}^{M}(x,t,\tau) := \frac{\sum_{k\in\sigma_x} \mathfrak{p}_w^{(k)}}{M_x}, \tag{15.1}$$

where M_x is the number of systems found at x and σ_x is the set of labels for those systems. The limit of this conditional average when $M \to \infty$, known as the *conditional expectation* in the theory of probability, is what we define as the

[6] The subscript 'e' refers to 'empirical', whereas the subscripts 'r' and 'i' respectively refer to the 'real' and 'imaginary' parts of a complex number. The reason for using these subscripts will be clear after deriving the formal counterparts of the quantities.

[7] Specific instructions for a 'weak' measurement are strongly contingent upon the studied system. The reader can find detailed guidelines for electrons in Marian *et al.* (2016) and for photons in Kocsis *et al.* (2011).

asymptotic-empirical weak value $\mathcal{P}_{e,r}(x,t,\tau)$. Note that for a small τ, $\mathcal{P}^M_{e,r}$ is experimentally understood as the *local average of momentum around* x, being $\mathcal{P}_{e,r}$ its expected value.

Following the theory of probability, for any τ, the conditional expectation $\mathcal{P}_{e,r}$ can be rewritten as $\mathcal{P}_{e,r}(x,t,\tau) = \int d\mathfrak{p}_w\, \mathfrak{p}_w\, \mathbb{P}(\mathfrak{p}_w|x,t,\tau)$, where $\mathbb{P}(\mathfrak{p}_w|x,t,\tau)$ is the conditional probability (density) to obtain the value $\mathfrak{p}_w$, given that the system is found at x. That is,

$$\mathbb{P}(\mathfrak{p}_w|x,t,\tau) := \lim_{M\to\infty} \frac{\text{\# cases: momentum (weakly measured) } \mathfrak{p}_w \text{ at time } t, \text{ \& particles at } x, \text{ a time } \tau \text{ after}}{M_x}. \tag{15.2}$$

Finally, in order to relate $\mathcal{P}_{e,r}(x,t,\tau)$ with its counterpart within the quantum formalism, it is useful to express it avoiding probabilities conditioned on x. For this, note that the probability to measure $\mathfrak{p}_w$ as momentum *and* x as position, $\mathbb{P}(\mathfrak{p}_w, x|t,\tau)$, is defined by equation (15.2) but replacing M_x for M. Then, given the definition of conditional probability, $\mathbb{P}(\mathfrak{p}_w|x,t,\tau) = \mathbb{P}(\mathfrak{p}_w, x|t,\tau)/\mathbb{P}(x|t,\tau)$, with the marginalized probability density $\mathbb{P}(x|t,\tau) = \int d\mathfrak{p}_w \mathbb{P}(\mathfrak{p}_w, x|t,\tau)$, one obtains that

$$\mathcal{P}_{e,r}(x,t,\tau) = \frac{\int d\mathfrak{p}_w\, \mathfrak{p}_w\, \mathbb{P}(\mathfrak{p}_w, x\,|\,t,\tau)}{\int d\mathfrak{p}_w\, \mathbb{P}(\mathfrak{p}_w, x\,|\,t,\tau)}. \tag{15.3}$$

Now consider the asymptotic-empirical weak value $\mathcal{P}_{e,i}$. It is defined after a nearly identical protocol, with the exception that, in step (ii), instead of the position of the ancilla, it is its momentum (i.e. the conjugate property) that is correlated in an (approximately) one-to-one manner with the macroscopic pointer.

One can, of course, follow the same protocols but perform a 'weak' laboratory measurement of another observable instead of the momentum in step (ii). In the general case, for an observable G we can define the asymptotic-empirical weak values $\mathcal{G}_{e,r}(x,t,\tau)$ and $\mathcal{G}_{e,i}(x,t,\tau)$, after changing $\mathfrak{p}_w$ by the weak measurement of G in the so-far written formulas. At this juncture, we arrive at a clear initial conclusion.

- **CONCLUSION 1:** It is possible to asymptotically determine the weak values $\mathcal{G}_{e,r}(x,t,\tau)$ and $\mathcal{G}_{e,i}(x,t,\tau)$ in a laboratory. For this, averages of 'weak' laboratory measurements concerning G are necessary, using sufficiently large ensembles of identically prepared systems found to be in x a time τ after the weak measurement.

15.2.1.2 Weak values as elements of the quantum formalism. We are now interested in exploring, within the formalism of non-relativistic[8] quantum mechanics, what the protocols described in the last sub-section correspond to, regardless of the preferred interpretation or ontology of the theory. Among other basic principles, we will assume the unitary Schrödinger evolution between measurements, the Born's rule, and the

[8]Note that, in more general quantum theories, wherein for instance the Schrödinger equation is no longer applicable, the same empirical weak values might correspond to different formal weak values.

von Neumann mathematical description of measurements (von Neumann 1955). These principles are widely accepted as either fundamental or effective[9] in practically any non-relativistic quantum theory (von Neumann 1955; Dürr and Teufel 2009; Oriols and Mompart 2019).

Let us begin with the formal characterization of the protocol associated with the asymptotic-empirical weak value $\mathcal{P}_{e,r}$. Recall that the procedure involves M repetitions of identically prepared experiments (with $M \to \infty$). Each repetition comprises, in turn, three steps to be formalized: (a) a weak coupling of the two microscopic systems that we have called the object system, s, and the ancilla, a; (b) a subsequent standard laboratory measurement of the ancilla's position; and, (c) a time τ after, a standard laboratory measurement of the position of the object. Let x be the position coordinate of the object (with corresponding position operator $\hat{X} = \int x|x\rangle\langle x|dx)$ and p_s its momentum coordinate (with corresponding momentum operator $\hat{P}_s = \int dp_s p_s |p_s\rangle\langle p_s|)$. We assume with generality that the object's initial state (at time t) is $|\psi(t)\rangle = \int \psi(p_s)|p_s\rangle dp_s$. Similarly, let y be the position coordinate of the ancilla (with corresponding position operator $\hat{Y} = \int y|y\rangle\langle y|dy)$, p_a its momentum coordinate (with corresponding momentum operator $\hat{P}_a = \int p_a|p_a\rangle\langle p_a|dp_a)$, and its initial state $|f(t)\rangle = \int f(y)|y\rangle dy$.

Following Von Neumann (1955), we model standard laboratory measurements in (b) and (c) through the *collapse* law and Born's rule, while the ancilla–object interaction in (a) is modelled through the Hamiltonian $\hat{H}_{sa} = \gamma \hat{P}_s \otimes \hat{P}_a$, which operates during an interval T, negligible if compared to the object system's dynamics. This Hamiltonian correlates the object's momentum with the displacement of the ancilla. As is well known, the strength of this interaction is given by $\gamma > 0$, where larger values of γ result in a stronger correlation between the ancilla's displacement (and eventually the macroscopic pointer's displacement) and the object's momentum. Due to the reciprocity of the interaction, the strength γ is positively correlated with the disturbance of both systems' pre-measurement states. Therefore, the weak coupling can be properly modelled assuming that γ is small enough. Finally, in order for the macroscopic pointer in (b) to indicate a hypothetical measured value $\mathfrak{p}_w$ when the ancilla gets collapsed at $y = \gamma T \mathfrak{p}_w$, it is typical to assume that the ancilla's state is calibrated to be centred around a reference position $y = 0$ (satisfying $\int y|f(y)|^2 dy = 0$) with a precision on the order of σ (which, formally, implies a support of the order of σ for $f(y)$).

Let us now find the formal correlate of $\mathcal{P}_{e,r}$, given the indications made above. Note that the initial state of the ancilla–object composite system is given by the separable wave function $|\Psi(t)\rangle_{sa} = |\psi(t)\rangle \otimes |f(t)\rangle = \int dp_s\, \psi(p_s)|p_s\rangle \otimes \int dy\, f(y)|y\rangle$. Next, during time T, the von Neumann interaction among both sub-systems takes place, described by the unitary operator $\hat{U}_{sa} = exp(-iT\hat{H}_{sa}/\hbar)$. This interaction leads to the following entangled state of the composite system:

$$|\Psi_{pre}\rangle_{sa} := \hat{U}_{sa}|\Psi(t)\rangle_{sa} = \iint \psi(p_s) f(y - \gamma T p_s)\, |p_s\rangle\, |y\rangle dp_s dy. \tag{15.4}$$

[9] For instance, in Bohmian mechanics, the *collapse* law is understood to be a consequence of deterministic particle dynamics that 'select' an effective wave function during the ancilla's channelization process of a measurement (Dürr *et al.* 1992).

This result holds for any value of γ, meaning we have not yet derived any consequence of assuming a specific strength for the interaction. At this point, the read-out of the ancilla's position occurs, wherein a measured displacement $y = y_{\mathfrak{p}_w} := \gamma T \mathfrak{p}_w$ corresponds to an object momentum $\mathfrak{p}_w$. This causes the collapse operation $\hat{I}_s \otimes |y_{\mathfrak{p}_w}\rangle\langle y_{\mathfrak{p}_w}|$ on the ancilla–object composite system, which leaves this system in the state $\int \psi(p_s) f(y_{\mathfrak{p}_w} - \gamma T p_s)|p_s\rangle \otimes |y_{\mathfrak{p}_w}\rangle dp_s$, and the object system in the perturbed state

$$|\tilde{\psi}_{\mathfrak{p}_w}(t)\rangle_s := \hat{I}_s \otimes \langle y_{\mathfrak{p}_w}|\Psi_{pre}\rangle_{sa} = \int \psi(p_s) f(y_{\mathfrak{p}_w} - \gamma T p_s)|p_s\rangle dp_s. \tag{15.5}$$

This is the object's state after the measurement in (b), for any value of the coupling strength γ. Thus, the read-out perturbs the object system's wave function differently as a function of the read-out value $y_{\mathfrak{p}_w}$, even if we choose γ to be small enough for a weak coupling. The assumption that γ is small and thus that we are modelling a "weak" laboratory measurement finally allows to neglect the terms of second order (and *not* of first order, as misleadingly[10] suggested by some authors (Vaidman, 2017; Cohen, 2017)) in the Taylor expansion[11]

$$f(y - \gamma T p_s) = f(y) - \gamma T p_s \frac{df(y)}{dy} + O((\gamma T)^2). \tag{15.6}$$

In order to formalize step (iii) in the experimental protocol, we assume that the (slightly perturbed) object system freely evolves for a time τ, according to its time-evolution operator $\hat{U}_s(\tau) = \exp\left(-i\hat{H}_s\tau/\hbar\right)$, where $\hat{H}_s$ is the Hamiltonian governing the evolution of the object alone. Finally, after τ, a standard laboratory measurement of the system's position x is performed. According to Born's rule, since we did not normalize the state (15.5), the joint probability of obtaining the read-out $y = y_{\mathfrak{p}_w}$ in the weak measurement and x in this final standard measurement is $\mathbb{P}(\mathfrak{p}_w, x \,|\, t, \tau) = |\langle x|\hat{U}_s(\tau)|\tilde{\psi}_{\mathfrak{p}_w}(t)\rangle_s|^2$. Expanding, we have

[10] For a sufficiently small $\gamma T > 0$, it might seem reasonable to neglect the first-order term as well, assuming that $f(y - \gamma T p_s) \simeq f(y)$. This would imply that the object's state would not be perturbed at all by the weak measurement, $|\tilde{\psi}_{\mathfrak{p}_w}\rangle \simeq f(y)|\psi\rangle$ (since $f(y)$ is a factor that disappears through normalization), apparently allowing the possibility to extract information about the object without the individual repetitions experiencing any significant backaction. We argue that such a backaction-free weak measurement is impossible if one wishes to obtain the weak value. The contribution of the zeroth-order term to the probability density (15.7) vanishes in the conditional expectation (15.8), meaning the weak value emerges strictly from the first-order term. Therefore, since the first-order term is the key element in the emergence of the weak value, it is also necessary to consider it in the post-weak-measurement state of the system. That is, the correct lowest order approximation for this state is $|\tilde{\psi}_{\mathfrak{p}_w}\rangle \simeq f(y)|\psi\rangle - \gamma T df(y)/dy \hat{P}_s|\psi\rangle$, which is a different state as a function of the weakly measured y; i.e. the system's perturbation is not negligible.

[11] Note that we assumed $1/T\gamma$ to be significantly larger than the range of p_s that the device is measuring.

$$\mathbb{P}(\mathfrak{p}_w, x \,|\, t, \tau) =$$

$$\left| f(y_{\mathfrak{p}_w}) \langle x|\hat{U}_s(\tau) \int \psi(p_s)|p_s\rangle dp_s - \gamma T \frac{\mathrm{d}f(y_{\mathfrak{p}_w})}{\mathrm{d}y} \langle x|\hat{U}_s(\tau) \int \psi(p_s) p_s |p_s\rangle dp_s + O((\gamma T)^2) \right|^2 = \tag{15.7}$$

$$= |f(y_{\mathfrak{p}_w})|^2 |\langle x|\hat{U}_s(\tau)|\psi(t)\rangle|^2 - 2\gamma T \,\mathrm{Re}\Big\{ f^*(y_{\mathfrak{p}_w}) \frac{\mathrm{d}f(y_{\mathfrak{p}_w})}{\mathrm{d}y} \langle \psi(t)|\hat{U}_s^\dagger(\tau)|x\rangle \langle x|\hat{U}_s(\tau)\hat{P}_s|\psi(t)\rangle \Big\} + O((\gamma T)^2).$$

If the ancilla's rest state $f(y) = R(y)exp(iS(y)/\hbar)$ has an odd phase $S(y)$ and an odd or even modulus $R(y)$, implying that $\int y f^*(y) \frac{\mathrm{d}f(y)}{\mathrm{d}y} dy = -1/2$ and $\int y |f(y)|^2 dy = 0$, we obtain

$$\int \mathfrak{p}_w \, \mathbb{P}(\mathfrak{p}_w, x \,|\, t, \tau) d\mathfrak{p}_w = \frac{1}{\gamma T} |\langle x|\hat{U}_s(\tau)|\psi(t)\rangle|^2 \mathrm{Re} \frac{\langle x|\hat{U}_s(\tau)\hat{P}_s|\psi(t)\rangle}{\langle x|\hat{U}_s(\tau)|\psi(t)\rangle} + O(1). \tag{15.8}$$

At this point, one should realize the significance of the first-order term in (15.6), as it constitutes the dominant term.

Finally, for the marginalized probability density $\int \mathbb{P}(\mathfrak{p}_w, x \,|\, t, \tau) d\mathfrak{p}_w$, we obtain

$$\int \mathbb{P}(\mathfrak{p}_w, x \,|\, t, \tau) d\mathfrak{p}_w = \frac{1}{\gamma T} |\langle x|\hat{U}_s(\tau)|\psi(t)\rangle|^2 + O(1), \tag{15.9}$$

from which we deduce that, to leading order, the asymptotic-empirical weak value $\mathcal{P}_{e,r}$ in (15.3) within the formalism of quantum mechanics yields

$$\mathcal{P}_{e,r}(x, t, \tau) \simeq \mathrm{Re}\left[\frac{\langle x|\hat{U}_s(\tau)\hat{P}_s|\psi(t)\rangle}{\langle x|\hat{U}_s(\tau)|\psi(t)\rangle} \right]. \tag{15.10}$$

In order to derive the formal counterpart of the laboratory protocol defining $\mathcal{P}_{e,i}$, one should adopt the same assumptions about the interaction strength γ and the ancilla's wave packet. However, in this case, it is the Fourier transform of the ancilla's wave packet that is used both in the expression for the state of the object after the weak measurement's read-out and in the preceding Taylor expansion. Through an otherwise analogous derivation, one finally obtains that

$$\mathcal{P}_{e,i}(x, t, \tau) \simeq \mathrm{Im}\left[\frac{\langle x|\hat{U}_s(\tau)\hat{P}_s|\psi(t)\rangle}{\langle x|\hat{U}_s(\tau)|\psi(t)\rangle} \right]. \tag{15.11}$$

This strongly suggests the definition of a *formal weak value* encoding the two asymptotic-empirical weak values,

$$\mathcal{P}_f(x, t, \tau) := \frac{\langle x|\hat{U}_s(\tau)\hat{P}_s|\psi(t)\rangle}{\langle x|\hat{U}_s(\tau)|\psi(t)\rangle} \simeq \mathcal{P}_{e,r}(x, t, \tau) + i\, \mathcal{P}_{e,i}(x, t, \tau). \tag{15.12}$$

In general, if in the laboratory protocol we had weakly measured another observable G with an associated operator $\hat{G}_s$, and defined the asymptotic-empirical functions

$\mathcal{G}_{e,r}(x,t,\tau), \mathcal{G}_{e,i}(x,t,\tau)$ analogously, the resulting formal weak value would be given by

$$\mathcal{G}_f(x,t,\tau) := \frac{\langle x|\hat{U}_s(\tau)\hat{G}_s|\psi(t)\rangle}{\langle x|\hat{U}_s(\tau)|\psi(t)\rangle} \simeq \mathcal{G}_{e,r}(x,t,\tau) + i\,\mathcal{G}_{e,i}(x,t,\tau). \tag{15.13}$$

Note that formal weak values do not depend on any particular measurement context and contain information about the state of the system before its measurement, $|\psi(t)\rangle$, similarly to what occurs with expectation values. Thus, we reach the following conclusion:

- **CONCLUSION 2:** The experimental conditional expectations of Conclusion 1, for a weak enough measurement, can be predicted *within any non-relativistic* quantum theory (where the von Neumann measurement protocol and the Born rule are accepted), through the formal weak value (15.13), which only depends on the state $|\psi(t)\rangle$ of the pre-measurement system and its propagator $\hat{U}_s(\tau)$.

Weak values for which τ is very small can be understood as *expectation values around the position* x, both formally and experimentally. We have already seen that this is the case experimentally. To show this is also true formally, let us define the limit weak value $\mathcal{G}_\ell(x,t) := \lim_{\tau\to 0} \mathcal{G}_f(x,t,\tau)$, which is a simultaneously well-defined field in configuration and time for any observable G (even for those not commuting with $\hat{X}$). This field satisfies

$$\int |\psi(x,t)|^2\, \mathcal{G}_\ell(x,t)\, dx = \langle\psi(t)|\hat{G}_s|\psi(t)\rangle = \langle\hat{G}_s\rangle(t), \tag{15.14}$$

where we have defined $\psi(x,t) := \langle x|\psi(t)\rangle$. That is, the standard expected value of an operator $\hat{G}_s$ can be recovered through the spatial average of its limit weak value $\mathcal{G}_\ell(x,t)$. This is why it is appropriate to refer to $\mathcal{G}_\ell(x,t)$ as the *local expectation value of* G *(around* x*)*. It provides *information* on how the expectation value $\langle\hat{G}_s\rangle(t)$ is distributed along the configurations x. Note that we use the ambiguous term *'information'* here deliberately because we do not want to attach any ontological significance to the local expectations, beyond their experimental and formal interpretations. Moreover, since $\hat{G}_s$ is Hermitian, even the real part of the local expectation field alone, $\mathcal{G}_\ell^r(x,t) := \mathrm{Re}[\mathcal{G}_\ell(x,t)]$, behaves itself as an expectation density: $\int |\psi(x,t)|^2\, \mathcal{G}_\ell^r(x,t)\, dx = \langle\hat{G}_s\rangle(t)$. This implies that the imaginary part of the weak value, with generality, satisfies $\int |\psi(x,t)|^2 \mathrm{Im}[\mathcal{G}_\ell(x,t)]dx = 0$.[12] Summing up, we can conclude that:

- **CONCLUSION 3:** From both an empirical perspective and a formal perspective *within any non-relativistic quantum theory*, weak values define expectation values around certain configurations x when $\tau \to 0$, known as *local expectation values.*

[12]The attentive reader may have noticed that, when referring to $\mathcal{G}_\ell(x,t)$ as local expectations, we have used the terminology introduced by Holland (1993). We have willingly done so, but two differences should be noted. Firstly, only $\mathcal{G}_\ell^r$ is defined to be a 'local expectation' in Holland (1993). Secondly, Holland introduces local expectation values within an ontological discourse, interpreting them as properties of individual Bohmian particles and without a clear reference to any laboratory protocol.

Either the complex field $\mathcal{G}_\ell(x,t) := \lim_{\tau\to 0}\mathcal{G}_f(x,t,\tau)$ or the real field $\mathcal{G}_\ell^r(x,t) := \lim_{\tau\to 0}\mathrm{Re}[\mathcal{G}_f(x,t,\tau)]$ are possible local expectation values.

Assuming the polar decomposition of the object system's wave function, $\psi(x,t) = R(x,t)e^{iS(x,t)/\hbar}$, local expectation values can be further analysed. For instance, the local expectation value of the momentum can be developed as

$$\mathcal{P}_\ell(x,t) = \frac{\langle x|\hat{P}_s|\psi(t)\rangle}{\langle x|\psi(t)\rangle} = \frac{\partial S(x,t)}{\partial x} - i\frac{\hbar}{R(x,t)}\frac{\partial R(x,t)}{\partial x}. \tag{15.15}$$

It turns out that this example clearly highlights the informational richness of weak values. If the real part of this local expectation is known, which is experimentally given by $\lim_{\tau\to 0}\mathcal{P}_{e,r}(x,t,\tau)$, it is possible to compute the phase of the wave function $S(x,t)$ by its integration, up to a constant (a global phase[13]). Moreover, in the laboratory protocol to obtain it, we also determine the probability density $R^2(x,t)$ through the marginal density $\mathbb{P}(x\,|\,t,\tau)$ of equation (15.9). This implies that the laboratory protocol for $\mathcal{P}_{e,r}$ (with a small enough τ), reveals the wave function of the systems as it was prior to any measurement back-action, $|\psi(t)\rangle$.[14,15] It follows that weak values and their functions can be as informative regarding the characterization of quantum systems as the wave function itself is. Thus, we can conclude that:

- **CONCLUSION 4:** The weak values of G and their associated protocols provide new ways to characterize quantum systems, beyond the corresponding operator's expectation values.

Conclusions 1–4 suffice to convey one of the main messages of this chapter, particularly aimed at physicists leaning towards an empiricist attitude: weak values are useful tools that are worth considering and further developing, regardless of discussions about their ontological significance. Let us make, next, a brief incursion into interpretive matters, addressing the ontological significance of weak values according to different quantum theories.

15.2.1.3 Weak values as elements of reality. Presumably, each quantum theory provides a distinctive view on the ontology of weak values. Let us first consider what we refer to as the 'standard' or 'orthodox' view, which attributes properties to microphysical systems via the so-called '*eigenvalue–eigenstate link*'. As it is well known, this link establishes that a system has a property G with determinate value g if and only if the system's wave function is an eigenstate of the operator representing the property, $\hat{G}_s$, with eigenvalue g. As a consequence of this interpretive tenet, if the wave function

[13] In general, the phase of the wave function or its gradient are *not* physically determinable because both depend on the arbitrarily selectable electromagnetic gauge. Nonetheless, when magnetic fields are absent, in the Coulomb gauge, the gradient of the phase turns out to match a gauge invariant quantity and hence becomes determinable.

[14] Again, we here assume a single-particle quantum system in a pure state.

[15] In fact, this is merely one among numerous ways in which weak values enable the experimental determination of the wave function of a system. For example, in the pioneering experimental work by Lundeen *et al.* (2011), the authors determined the wave function through *momentum post-selected* local expectations (a class of weak value that is elucidated in Section 15.2.3).

of the system is a superposition of eigenstates of $\hat{G}_s$ associated with different eigenvalues, simply, the system does not determinately possess the property G. This view also includes the collapse-upon-measurement as one of its postulates. Given that the collapse ensures that, after a measurement of property G, the wave function transforms into an eigenstate of the operator $\hat{G}_s$, it follows from the standard view that, *after a measurement of property* G, this property is always a determinate, well-defined property of the system with the value corresponding to the measured outcome. Yet this is *not* the case, in general, *before* the measurement. So, the standard view has the consequence that, in general, measurement processes not only disturb the measured system but 'create' or make determinate the measured property (Mermin 1993).

Regarding weak values, when the state of the system $|\psi\rangle$ is already an eigenstate of the observable $\hat{G}_s$, with $\hat{G}_s|\psi\rangle = g|\psi\rangle$, its associated formal weak value gives $\mathcal{G}_f(x, t, \tau) = g$ for all x and τ. In light of the above explanation, within the standard view, such a system determinately has the property G with value g, regardless of its position or anything else. Therefore, even if $\mathcal{G}_f(x, t, \tau)$ is computed through an average, the standard view attributes the weak value, in such a case, to each individual system of the ensemble (as the well-defined property G with value g).

However, in general, the state $|\psi\rangle$ is *not* an eigenstate of $\hat{G}_s$. In such a case, according to the eigenvalue–eigenstate link, the determinate properties of the system would correspond to a set of operators *different than* $\hat{G}_s$. Given the scarcity of properties available to the advocate of the standard approach, we can conclude that:

- **CONCLUSION 5:** Quantum theories assuming the eigenstate–eigenvalue link as a fundamental interpretive postulate cannot generally relate weak values with properties determinately possessed by the system (nor with their averages) and thus these theories have a too sparse ontology to attribute an ontological meaning to weak values.

One may wonder what would happen if we understood that the average observable G (such as momentum or energy) around a position x was actually a property of any of the systems found at x. This is what happens according to the standard view, *but only for systems whose state is an eigenstate* of $\hat{G}_s$. Now we want to drop this restriction. In such an alternative view, the local expectation value of the momentum would imply a well-defined momentum for the particle system at all times (determined by the system's position). Consequently, the particle of the system would trace well-defined trajectories in x space. A minimal version of this theory would focus on the real local expectations $\mathcal{G}_\ell^\tau(x, t)$, which already yield the operator expectations. According to such a minimal version, the momentum of the particle when it is at x would be equal to the gradient of the phase of the wave function, as shown in equation (15.15). For well-behaved wave functions, this would result in non-crossing deterministic trajectories, each ascribable to individual realizations of the system (Berndl *et al.* (1995)). This, in turn, would imply that if we measured at time t the position of the system with trajectory $x^k(t)$, the result would correspond with $x = x^k(t)$ because that would have been the actual position of the particle. However, after the measurement, the trajectory would be governed by the phase of a wave function that would have been uncontrollably perturbed,[16] so the entire individual trajectories, although determinate at the ontological level, would not be epistemically accessible. Hence, a stochastic description of positions would still be necessary epistemologically, to have an empirically adequate theory. Similarly, the other observables would also be ontologically determined through the position of the particles. In this theory, the real local expectation of G, $\mathcal{G}_\ell^r(x,t)$, would give the value of G corresponding to the particle when found to be in x at time t. It is worth noting that, in this theory, an eigenstate of an operator is a state in which all possible trajectories exhibit the same value for the corresponding property. This would explain why, when we measure an observable in an eigenstate, we always obtain the same value with a probability equal to 1.

Such a 'literal interpretation of local expectations' is indeed very close to Bohmian mechanics (Bohm 1952; Dürr and Teufel 2009; Holland 1993; Oriols and Mompart 2019), with the primary difference being the terminology used for certain functions. For instance, in both cases, the momentum of the particle, denoted as $p_B(x,t)$, is equal to (the real part of) the momentum weak value. However, the kinetic energy is given by $p_B^2(x,t)/(2m)$ in Bohmian mechanics, while, in the 'literal interpretation of local expectations', it is the real part of the weak value of $\hat{P}_s^2/(2m)$ that is referred to as the kinetic energy. As shown in the next sub-section, these quantities do *not* coincide. In the Bohmian theory, the difference between them is the so-called 'quantum potential'.

Given that, in Bohmian mechanics, $p_B = \frac{\partial S(x,t)}{\partial x}$ is the pre-existent value (previous to any interaction) of the momentum of the particles, and we have demonstrated that this quantity is equivalent to the real local expectation of momentum, $p_B = \mathrm{Re}(\mathcal{P}_\ell)$, it follows that the protocol to obtain the weak value $\mathrm{Re}(\mathcal{P}_\ell)$ described in the subsection 15.2.1.1 serves as an experimental protocol for accessing this property within the framework of Bohmian mechanics. Now, it is essential to emphasize that, according to alternative quantum theories, the same experimental protocol may be interpreted differently because, among other things, it involves averaging and post-selecting data obtained from an ensemble of different experiments, rather than from an individual experiment. Furthermore, there exist non-standard but empirically equivalent versions of Bohmian mechanics that propose different velocity fields (Deotto and Ghirardi 1998). In these theories, it is evident that the protocol associated with $\mathrm{Re}(\mathcal{P}_\ell)$ does *not* constitute a measurement in any sense of the momentum of the particles. This shows that, to the extent that one can talk of 'measurement' in this context, it is a heavily theory-laden notion. Moreover, since all these theories are empirically equivalent, one can, at most, empirically establish what they have in common but not where they diverge.[17] It is the ontology of each quantum theory that determines the ontic import of each weak value.

Finally, let us remark that there is a theory where local expectations correspond to *averages* of ontologically determined properties that may vary from preparation to preparation. According to this theory, the particles follow well defined trajectories

[16]We previously observed that the perturbation is significant even for the weakest of the measurement couplings.

[17]Thus, the actual existence of the Bohmian trajectories cannot be experimentally proved. For a more in-depth discussion on the 'undetectability' of Bohmian trajectories, please refer to Fankhauser (2025) in this volume.

that evolve stochastically, and the Bohmian trajectories represent average paths. This theory is Nelson's stochastic mechanics where the expected (but not the individual) particle momentum is equal to the Bohmian momentum p_B (accompanied by an *osmotic momentum* component describing a diffusive process, given by $p_O := \mathrm{Im}\mathcal{P}_\ell$). Of course, as it happens in Bohmian mechanics, there might be a difference among Nelson's theory and a 'literal interpretation of weak values' that naively takes the weak value of any operator, $\hat{G}$, to represent the average of the property G in the ensemble.

We can therefore conclude that:

- **CONCLUSION 6:** Bohmian mechanics *allows* to interpret weak values post-selected in position (when $\tau \to 0$) as properties of individual systems at specific positions.[18] In Nelson's stochastic trajectory theory, these weak values *can* instead be understood as averages of individual system properties.

Note that other quantum theories may result in different, alternative ontological readings of weak values. One may just claim, for instance, that the momentum's local expectation value, (15.10), is merely 'the gradient of the phase of the wave function', and that this is what is measured when performing the corresponding laboratory protocol (Fankhauser and Dürr 2021). We do not intend to object to such analyses, but in our view, they remain incomplete in terms of characterizing the ontological meaning of the local expectation value unless they are accompanied by a characterization of the *ontological* meaning of the wave function itself. This can potentially give rise to further challenges, particularly considering the multi-dimensional nature of the wave function field.[19] However, due to space constraints, we have limited our discussion of the ontological meaning of weak values to the alternatives already presented.

15.2.2 On the meaning of a function of a weak value

Weak values are interesting, among other reasons, because usually they cannot be rewritten as the expectation of a Hermitian operator, thus offering an additional characterization tool for quantum systems. It turns out that a *function* of weak values (such as p_B^2, that is, the square of the Bohmian momentum $p_B = \mathrm{Re}\mathcal{P}_\ell$ in equation (15.15)), in addition to not being expressible as a standard expectation either, generally cannot be written as a weak value of some Hermitian operator, not at least of the most obvious choice of operator (for example $\hat{P}_s^2$). And yet, experimentally, *functions* of weak values can still be obtained in a laboratory simply by applying those functions

[18] We have pointed out that within Bohmian mechanics, in general, one can think of the real local expectation $\mathcal{G}_\ell^\tau(x,t)$ associated with the operator $\hat{G}_s$ as representing some property (not necessarily G) of the systems located at position x at time t. This interpretation is reinforced by prominent Bohmians, such as Bohm and Hiley (1993) or Holland (1993). Other prominent Bohmians, like Bell (1990) or Dürr *et al.* (2004), consider that the only property attributed to the Bohmian particles is their position, because it suffices to explain all experiments in the laboratory (as they ultimately involve only the positions of measurement apparatus pointers). Moreover, advocating for a minimal set of properties eliminates the possibility of drawing misleading conclusions. Nonetheless, a large number of properties provide practical heuristic power in characterizing quantum systems, as shown in the case study.

[19] Note that Bohmian-like theories avoid this difficulty, at least, when it comes to specifying the nature of weak values. For these theories, weak values unambiguously represent properties of the *particles*, even though they formally depend on the wave function.

to the numbers resulting from the previously explained weak value protocols. This means that *functions* of weak values constitute a vast source of new information to characterize quantum systems, even beyond weak values themselves. As shown later in the case study, far from being arbitrary operations performed on the experimental results, they can possess very interesting features.

Let us illustrate this with a particular example (that relates with the discussion of Section 15.2.1.3 and the case study). In terms of the polar decomposition of the wave function, the local expectation of the kinetic energy operator $\hat{P}_s^2/2m$ (the *weak value of the square of the momentum operator*) reads:

$$\frac{1}{2m}\frac{\langle x|\hat{P}_s^2|\psi(t)\rangle}{\langle x|\psi(t)\rangle} = \frac{1}{2m}\left(\frac{\partial S(x,t)}{\partial x}\right)^2 - \frac{\hbar^2}{2m}\frac{1}{R(x,t)}\frac{\partial^2 R(x,t)}{\partial x^2} - i\frac{1}{2m}\left[\hbar\frac{\partial^2 S(x,t)}{\partial x^2} + 2\frac{\hbar}{R(x,t)}\frac{\partial R(x,t)}{\partial x}\frac{\partial S(x,t)}{\partial x}\right]. \tag{15.16}$$

If one re-examines the momentum local expectation $\mathcal{P}_\ell = p_B + i\,p_O$ developed in (15.15) (which encodes the Bohmian momentum $p_B = \partial S(x,t)/\partial x$ and the Nelsonian osmotic momentum $p_O = -(\hbar/R)\partial R(x,t)/\partial x$), it becomes evident that the *square of the momentum operator's weak value*, $\mathcal{P}_\ell^2/2m$, yields a different thing. The real part of (15.16), which is enough to recover the kinetic energy's *standard expectation*, indeed contains the *square of the momentum operator's real weak value*, $\text{Re}\{\mathcal{P}_\ell\}^2/2m$, namely, the Bohmian kinetic energy, but there is an additional term that depends on the osmotic momentum p_O. This term is called the *quantum potential* in Bohmian mechanics,

$$Q(x,t) := -\frac{\hbar^2}{2m}\frac{1}{R(x,t)}\frac{\partial^2 R(x,t)}{\partial x^2} = \frac{1}{2m}\left(\hbar\frac{\partial p_O(x,t)}{\partial x} - p_O(x,t)^2\right). \tag{15.17}$$

Such a development shows that it is possible to determine $p_B^2/2m$ and Q in a laboratory as *functions* of the local expectation of momentum $\mathcal{P}_\ell$. Our point is, as we will show in the case study, that the individual quantities $p_B^2/2m$ and Q separately provide a richer source of information than their sum, the kinetic energy's real local expectation.

- **CONCLUSION 7:** In general, *functions of weak values of operators* are not equivalent to *weak values of functions of operators.* However, *functions* of weak values are still experimentally determinable by applying those functions to empirically determined weak values (e.g., when post-processing). Therefore, they constitute predictive tools that provide an additional contrastable source of information about quantum systems.

It is important to note that we have not addressed any ontological implication of the *functions* of weak values. Therefore, Conclusion 7 remains significant for an empiricist attitude reader.

15.2.3 Generalization to other post-selections

If we keep the same state preparation procedure for the empirical weak value protocol and perform a weak measurement of the observable $\hat{G}_s$ as previously described, but instead of measuring the position of the system particles x after time τ, we measure another observable B with associated operator $\hat{B} = \sum_b b|b\rangle\langle b|$, we can define additional weak values. These are known as weak values 'post-selected in B'. Empirically, they correspond to the average of weak measurements for the replicas in which a specific eigenvalue b is obtained in the subsequent measurement. The derivation of the corresponding formal weak values is analogous to the one that we have discussed, and it leads to the following expression for the generalized formal weak values:

$$\mathcal{G}_f^B(b,t,\tau) = \frac{\langle b|\hat{U}_s(\tau)\hat{G}_s|\psi(t)\rangle}{\langle b|\hat{U}_s(\tau)|\psi(t)\rangle}. \tag{15.18}$$

When the limit $\tau \to 0$ is considered, both empirically and formally, the standard expectation $\langle\hat{G}_s\rangle(t)$ can still be recovered by an average of the weak values over all possible b, weighted by their probability density $|\psi(b,t)|^2 := |\langle b|\psi\rangle|^2$. Thus, both empirically and formally, Conclusions 1–4 (even Conclusion 7) can be generalized, since $\mathcal{G}_\ell^B(b,t) = \lim_{\tau\to 0}\mathcal{G}_f^B(b,t,\tau)$ is still a local expectation value, but now in the 'locality' of $B = b$, whatever this might mean.

The standard approach with the eigenvalue–eigenstate link has the same difficulties to generally attribute an ontological meaning to weak values post-selected in B (or their corresponding 'local' expectation values) as it has in the case of weak values post-selected in position. In fact, even position-based 'hidden' variable theories, such as Bohmian mechanics, do not attach a natural ontological meaning to them. Following the same 'literal' reading of weak values however, one can define a so-called 'modal' theory for each post-selection B, which would do so. These theories would have a 'hidden' variable ontology in the usual sense but with well-defined material trajectories in the b-space instead of x-space. According to them, quantities such as $\mathcal{G}_\ell^B(b,t)$ would constitute determinate properties of the individual systems 'located' at $B = b$, and the corresponding laboratory procedures would be the natural protocols to determine those properties.

While general weak values may be valuable in characterizing quantum systems, we consider that position post-selected weak values hold a privileged status. Position is epistemically privileged since one can consider that all measurements are ultimately measurements of position (the position of a pointer, pixel on a screen, or similar indicators). But position is special in several other respects. For instance, in the position basis, the Schrödinger equation has a natural decoupling into a continuity equation and the well-known Hamilton–Jacobi equation (or even into a version of Newton's second law) (Bohm 1952; Holland 1993). The resemblance of these equations with the classical ones, allows to extend the intuitions and methods of classical fluid and particle dynamics to the quantum domain in a *guided* and rigorous manner.

15.3 Exemplifying Bohmian-like theories as heuristic tools

Finally, let us present a case study to illustrate our principal theses: the utility of weak values and their *functions* as novel ways to characterize quantum systems, the fact that this utility is independent of assessments of their ontological import, and how their natural incorporation into Bohmian-like theories make these theories practically valuable tools. Our case study delves into a cornerstone issue in quantum thermodynamics: the thermalization of a closed quantum system. To maintain clarity, we provide a qualitative overview of the problem and highlight the relevant results for our argument. A more detailed technical analysis of this case study can be found elsewhere (Destefani and Oriols 2023b).

Let us consider two[20] identical electrons (with respective degrees of freedom x_1 and x_2) confined in a disordered harmonic trap with Coulomb and exchange interactions between them. In general, after multiple scattering events with the speckles of a random pattern in the (disordered) potential field, an initially localized wave packet unitarily evolves towards a *fully dispersed* state. This time evolution is illustrated in Fig. 15.1. It is natural to infer that an *equilibrium* state is eventually reached. Indeed, if we calculate the standard expectations of observables such as the kinetic energy or the velocity, we will observe that they relax over time, converging to steady-state values. The characteristic time constant that quantitatively describes this equilibration is called *the thermalization time*, t_{eq}.

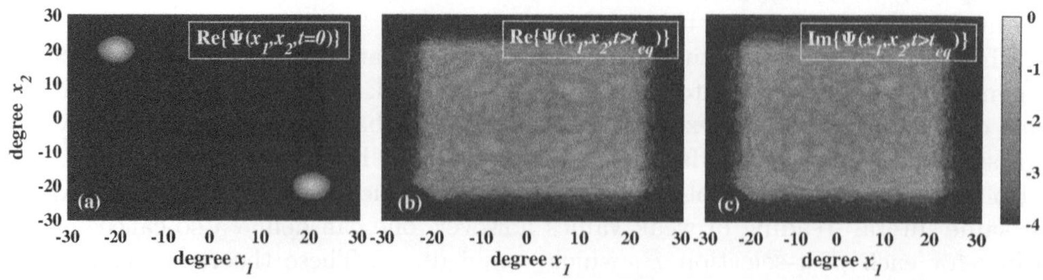

Fig. 15.1: Time evolution of a two-particle wave function $\Psi(x_1, x_2, t)$ in a disordered harmonic potential. (a) Real part of the wave function at $t = 0$. (b) Real part of the wave function at a representative time $t > t_{eq}$. (c) Imaginary part at the same time $t > t_{eq}$ (note that the imaginary part at $t = 0$ is zero for this particular scenario). All vertical axes and colourmaps represent the same ranges. Atomic units are used. The colourmap is the $log_{10}(\cdot)$ value of the modulus of each component.

In Fig. 15.2, three different two-particle simulations are presented to exemplify various scenarios in such a quantitative study of thermalization. The left column depicts a case where both particles have initial wave packets with central positions close to zero

[20] For this, one needs to deal with many-body weak values, as introduced in Destefani and Oriols 2023a. Let us outline their key features. In the case of N *distinguishable* particles it is possible to weakly measure the momentum of the 'jth particle', which if followed by a strong measurement of the system's configuration, $\vec{x} \in \mathbb{R}^N$, allows the definition of a conditional expectation $\mathcal{P}_{e,r}^{(j)}(\vec{x}, t, \tau)$ and its formal counterpart, $\mathcal{P}_f^{(j)} = \langle \vec{x}|\hat{U}_s \hat{P}_j|\psi\rangle / \langle \vec{x}|\psi\rangle$. *Indistinguishable* particles however, only allow a 'particle-agnostic' weak measurement of the momentum, followed by the determination of, say, the position of one of the particles. One can still find that the corresponding formal weak value is a function of the different $\mathcal{P}_f^{(j)}$ which are always formally well-defined quantities.

and identical central momenta. In the middle column, the initial wave packets have zero central momentum but opposite central positions (this is the situation depicted in Fig. 15.1). In the right column, the initial central positions are opposite and the central momenta are initially large.

The process of thermalization is revealed in Fig. 15.2 through the time evolution of various expectation values. In the top row, Fig. 15.2(a) and (d) depict the gradual evolution of kinetic and potential energy expectations towards thermalization. In the middle row, Fig. 15.2(b) and (h) display the thermalization of position and velocity expectations.

Our argument revolves around the observation that in certain configurations, many standard expectation values fail to capture the thermalization time due to conservation symmetries. This is illustrated in Fig. 15.2(e), where position and velocity expectations become 'blind' to thermalization, or in Fig. 15.2(g), where standard energy expectations are the 'blind' ones. Fortunately, there exist decompositions of standard expectations into expectations of *weak value functions* that still convey the thermalization signature for most such scenarios. This already highlights the practicality of weak values beyond conventional quantum metrics and any interpretative or ontological discussion. Yet, not all decompositions into weak value functions successfully characterize the thermalization time. And here is our main point: far from needing to look for informative weak values by trial and error, Bohmian-like theories naturally suggest appropriate decompositions due to their close connection to some classical notions.

Let us illustrate this through particular decompositions of the *kinetic energy* that naturally emerge from Bohmian-like theories. In particular, we present the analyses suggested by Bohmian mechanics, the hydrodynamic quantum theory (Madelung 1927), and Nelson's stochastic mechanics.

Within Bohmian mechanics, along the direction j, all the energy contributions apart from the classical potential energy are attributed to the Bohmian kinetic energy, $K_{B,j} = p_{B,j}^2/2m$, and the quantum potential Q_j (as defined in (15.17) for a single-particle system). In terms of standard quantum operators, the energy contribution along j that is not due to the classical potential is given by the kinetic energy operator $\hat{P}_j^2/2m$. As we found in equation (15.16), the (real) local expectation of this operator is equal to $K_{B,j} + Q_j$, which implies the decomposition $\langle \hat{P}_j^2/(2m)\rangle = \langle K_{B,j}\rangle + \langle Q_j\rangle$. We also found that the two quantities in the right-hand-side, although natural in Bohmian mechanics, are not individually associated with Hermitian operator expectations. They are *functions of weak values*, for which we still devised a laboratory protocol. Namely, by following the empirical protocol for local expectations, one can determine both the local expectation $p_{B,j}$ and the probability density $R^2(x,t)$, to subsequently evaluate $K_{B,j}$ and Q_j employing their respective definitions in terms of the former quantities.

This decomposition remains natural in hydrodynamic versions of quantum mechanics, where possible Bohmian trajectories can be viewed as elements of a fluid. In these formulations, the quantum potential Q_j arises from the internal pressure of the fluid, and thus, it is treated in analogy to classical fluid mechanics as part of the total kinetic energy of the fluid elements, along with their momentum-associated energy $K_{B,j}$.

A similar, yet alternative decomposition is also natural in Nelson's theory, where

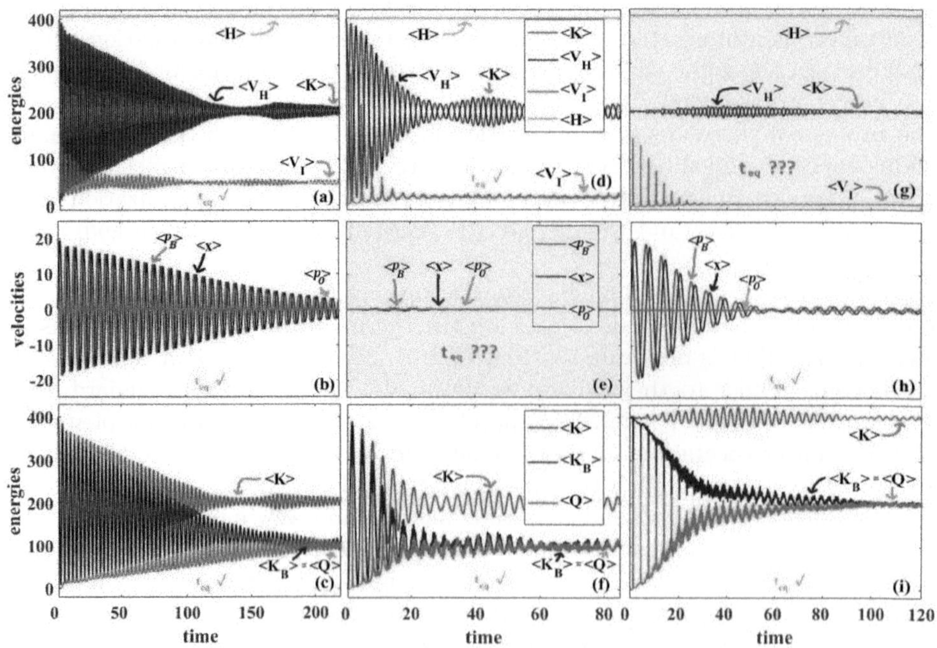

Fig. 15.2: In the first column, we show a representative initial condition where energy, position, and momentum expectations serve to identify thermalization. In the second column, a scenario in which position and momentum are 'blind' to the onset of thermalization. In the third column, we show a scenario where kinetic and potential energies are unable to reflect thermalization times. The first row shows the time evolution of standard energy expectations: the standard kinetic $\langle K\rangle$, harmonic $\langle V_H\rangle$, Coulomb repulsion $\langle V_I\rangle$, and total $\langle H\rangle$ energies ($\langle V_I\rangle$ is magnified by 100). The second row shows the time evolution of standard momentum and position expectations $\langle p_B\rangle, \langle x\rangle$, together with the osmotic momentum expectation $\langle p_O\rangle$. The third row shows the time evolution of Bohmian $\langle K_B\rangle$ and osmotic $\langle Q\rangle$ kinetic energy expectations, relative to $\langle K\rangle$. Values in (g) are divided by two. Legends in left and right columns follow the ones in the middle column. Vertical axis in middle and right columns are the same as in the left column. All in atomic units.

the real and imaginary parts of the momentum local expectation are the two momentum components of the stochastic trajectories, respectively, the current momentum $p_{B,j}(\vec{x},t)$ and the osmotic (or diffusion) momentum $p_{O,j}(\vec{x},t)$. Two momentum contributions imply two kinetic energy contributions, one being $K_{B,j}$ and the other one being $K_{O,j} := p^2_{O,j}/2m$. It is shown in Destefani and Oriols (2023b) that these functions are related to the standard kinetic energy expectation by $\langle \hat{P}^2_j/(2m)\rangle = \langle K_{B,j}\rangle + \langle K_{O,j}\rangle$.[21]

[21]This means that $\langle K_{O,j}\rangle = \langle Q_j\rangle$, although $K_{O,j} \neq Q_j$, as discussed in Destefani and Oriols (2023b).

Hence, we can alternatively decompose the (standard) average kinetic energy into these two functions of weak values, which are likewise experimentally determinable.

Now, these 'hidden' variable theories not only serve as a means to derive such decompositions, but within their conceptual frameworks, it is natural that they are valid decompositions for identifying the thermalization time. It is inherent to any of the three approaches that, following the classical concept of *energy equipartition*, a system in thermal equilibrium distributes its energy equally among all its energetic components. And indeed, as it is demonstrated in Destefani and Oriols (2023b), after the thermalization time t_{eq}, the expected current and osmotic kinetic energies (and expected quantum potential), all become equal to each other and each equal to half of the standard kinetic energy expectation. This implies that the times $t > t_{eq}$ (after thermalization), are characterised by the condition $\langle \hat{P}_j^2/(2m)\rangle \simeq 2\langle K_{B,j}\rangle \simeq 2\langle K_{O,j}\rangle \simeq 2\langle Q_j\rangle$. Fig. 15.2(c), (f) and (i) demonstrate that the 'hidden' variables $\langle K_{B,j}\rangle$ and $\langle Q_j\rangle$ (or $\langle K_{O,j}\rangle$) converge to equal values after t_{eq}, effectively revealing the onset of thermalization. Importantly, as illustrated in the figure, this holds true even when Hermitian position or velocity expectations, as well as the expectation of kinetic energy itself, are insensitive to the thermalization process.

Anyone familiar with any of the aforementioned theories, *regardless of their ontological preferences,* could potentially arrive at this conclusion. From our viewpoint, this success of certain functions of weak values in characterizing a seemingly pathological quantum system clearly illustrates the *predictive* power and physical interest of weak values. Moreover, this case study also shows the *heuristic* potential inherent in Bohmian-like theories that can serve as a guide to find the relevant functions.

15.4 Final remarks

In this chapter, we have shown how weak values are very useful predictive and characterization tools, regardless of ontological disputes, and how certain hidden variable theories provide heuristic frameworks that guide us towards physically informative weak values. To illustrate these conclusions, as a case study, it is shown how the equilibration time in certain scenarios is better determined from functions of weak values taken from the Bohmian machinery rather than from standard operators. To properly acknowledge the merits of our work, we emphasize that all results shown in Section 15.3 in terms of (functions of) *Bohmian-inspired* weak values are not merely theoretical embellishments but also outcomes directly accessible in the laboratory following the protocol outlined in sub-section 15.2.1.1. This connection between theoretical and experimental *Bohmian-inspired* weak values opens up new and unexplored possibilities for characterizing quantum systems, with quantum thermalization discussed in this chapter serving as just the first example.

Let us conclude with a final, more speculative remark. It is relevant to question how one should approach elements of a theory that prove to be useful in enhancing its predictive power. Should they be considered merely as useful fictions or as insights into the inner workings of reality? Consider, for example, the Copernican system with its heliocentric hypothesis. As it is well known, in the preface of *De Revolutionibus* written by Andreas Osiander, this hypothesis was deemed a mere useful fiction that would allow astronomers to perform certain calculations regarding planetary movements with more ease. However, the heliocentric hypothesis ended up considered not merely useful but true. In 1808, John Dalton used the 'imaginative' and 'hidden' notion that matter consists of atoms to explain fixed ratios in chemical combinations. For several decades, the mainstream resisted the ontological implications of this calculation method ('who [has] ever seen a gas molecule or an atom?', was the objection by Marcelin Berthelot still in 1877 (Herbert 2011)). Yet, as we know, the story concluded with the atomic hypothesis being accepted as true. Although the atomic hypothesis has undergone significant refinement, and our current conception of what an atom is substantially differs from what Dalton believed, we still acknowledge the validity of his insight. Now, we ponder whether a similar narrative could unfold with weak values and theories like Bohmian mechanics, not as the ultimate truths (being fully aware that we are dealing with non-relativistic quantum phenomena) but as indicative of the right path to follow. May we be experiencing a '*déjà vu*' in this regard? Only time will tell...

References

Aharonov, Y., Albert, D.Z. and Vaidman, L. (1988). How the result of a measurement of a component of the spin of a spin-1/2 particle can turn out to be 100. *Physical Review Letters*, **60**(14): 1351.

Bell, J. (1990). Against "measurement". *Physics World*, **3**(8): 33.

Berndl, K., Dürr, D., Goldstein, S., Peruzzi, G. and Zanghì, N. (1995). On the global existence of Bohmian mechanics. *Communications in Mathematical Physics*, **173**: 647–673.

Bohm, D. (1952). A suggested interpretation of the quantum theory in terms of "hidden" variables. I. *Physical Review*, **85**(2): 166–179.

Bohm, D. and Hiley, B.J. (1993). *The Undivided Universe: An Ontological Interpretation of Quantum Theory*. London: Routledge.

Cohen, E. (2017). What weak measurements and weak values really mean: Reply to Kastner. *Foundations of Physics*, **47**(10): 1261–1266.

Deotto, E. and Ghirardi, G.C. (1998). Bohmian mechanics revisited. *Foundations of Physics*, **28**(14): 1–30.

Destefani, C.F. and Oriols, X. (2023a). Assessing quantum thermalization in physical and configuration spaces via many-body weak values. *Physical Review A*, **107**(1): 012213.

Destefani, C.F. and Oriols, X. (2023b). Kinetic energy equipartition: A tool to characterize quantum thermalization. *Physical Review Research*, **5**: 033168.

Dressel, J., Malik, M., Miatto, F.M., Jordan, A.N. and Boyd, R.W. (2014). Colloquium: Understanding quantum weak values: Basics and applications. *Reviews of Modern Physics*, **86**(1): 307.

Dürr, D. and Teufel, S. (2009). *Bohmian Mechanics: The Physics and Mathematics of Quantum Theory*. Berlin: Springer.

Dürr, D., Goldstein, S. and Zanghi, N. (1992). Quantum equilibrium and the origin of absolute uncertainty. *Journal of Statistical Physics*, **67**: 843–907.

Dürr, D., Goldstein, S. and Zanghì, N. (2004). Quantum equilibrium and the role of operators as observables in quantum theory. *Journal of Statistical Physics*, **116**: 959–1055.

Fankhauser, J. (2025). The (un)detectability of Bohmian trajectories. In Oldofredi, A. (ed.) *Guiding Waves in Quantum Mechanics: One Hundred Years of de Broglie-Bohm Pilot-Wave Theory*. Oxford: Oxford University Press.

Fankhauser, J. and Dürr, P.M. (2021). How (not) to understand weak measurements of velocities. *Studies in History and Philosophy of Science Part A*, **85**: 16–29.

Herbert, N. (2011). *Quantum Reality: Beyond the New Physics*. New York: Anchor.

Holland, P. R. (1993). *The Quantum Theory of Motion: An Account of the de Broglie-Bohm Causal Interpretation of Quantum Mechanics*. Cambridge: Cambridge University Press.

Kocsis, S., Braverman, B., Ravets, S., Stevens, M.J., Mirin, R.P., Shalm, L.K. and Steinberg, A.M. (2011). Observing the average trajectories of single photons in a two-slit interferometer. *Science*, **332**(6034): 1170–1173.

Kofman, A.G., Ashhab, S. and Nori, F. (2012). Nonperturbative theory of weak pre- and post-selected measurements. *Physics Reports*, **520**(2): 43–133.

Lundeen, J.S., Sutherland, B., Patel, A., Stewart, C. and Bamber, C. (2011). Direct measurement of the quantum wave function. *Nature*, **474**(7350): 188–191.

Madelung, E. (1927). Quantum theory in hydrodynamical form. *Zeitschrift für Physik*, **40**: 322.

Marian, D., Zanghì, N. and Oriols, X. (2016). Weak values from displacement currents in multiterminal electron devices. *Physical Review Letters*, **116**(11): 110404.

Mermin, N.D. (1993). Hidden variables and the two theorems of John Bell. *Reviews of Modern Physics*, **65**(3): 803.

Oriols, X. and Mompart, J. (2019). *Applied Bohmian Mechanics: From Nanoscale systems to Cosmology*. Boca Raton, Florida, USA: CRC Press.

Tamir, B. and Cohen, E. (2013). Introduction to weak measurements and weak values. *Quanta*, **2**(1): 7–17.

Vaidman, L. (2017). Weak value controversy. *Philosophical Transactions of the Royal Society A: Mathematical, Physical and Engineering Sciences*, **375**(2106): 20160395.

Von Neumann, J. (1955). *Mathematical Foundations of Quantum Mechanics*. Princeton: Princeton University Press.

16
Spin-aware movement of electrons and time-of-flight momentum spectroscopy

Siddhant Das

16.1 Introduction

The de Broglie–Bohm (dBB) world-view is grounded in the mantra *matter moves*. It is commonly expressed in terms of a dynamical theory for particles or fields – the elements of matter – and their quantum mechanical wave functions. The role of the wave functions is to steer matter in motion, analogous to, yet distinct from, the role played by electromagnetic or gravitational fields in classical mechanics. The most compelling, empirically successful version of the dBB (pilot-wave or Bohmian) theory features point-like particles pursuing continuous trajectories in space-time, determined by their wave function (see Bohm 1952a, 1952b; Bohm and Hiley 1993; Dürr and Teufel 2009; Holland 1993; Goldstein 2021; Bricmont 2016; Oriols and Mompart 2012; Struyve 2005).

For a single particle with position $\mathbf{R}$, the (Bohmian) trajectories solve the *guiding equation*

$$\frac{\mathrm{d}}{\mathrm{d}t}\mathbf{R}(t) = \mathbf{v}^{\Psi}\big(\mathbf{R}(t), t\big) \tag{16.1}$$

a first-order equation of motion for the particle's velocity, where $\mathbf{v}^{\Psi}(\mathbf{r}, t)$ is a vector field that depends on the wave function Ψ.
The Ψ function satisfies Schrödinger's equation

$$i\hbar\frac{\partial\Psi}{\partial t} = \mathcal{H}\Psi \tag{16.2}$$

at *all* times. (So in this theory 'wave-particle duality' simply means that there is both a wave and a particle.) The type of wave function, the Hamiltonian $\mathcal{H}$, and the velocity field $\mathbf{v}^{\Psi}$ depend on the nature of particles, the presence of external potentials, and whether the motion is Galilei- or Lorentz-covariant. Their exact forms are educated guesses motivated by symmetry arguments, simplicity, empirical sufficiency, the anticipated classical limit, and arguably even operational input (Wiseman 2007). Once specified, the dBB dynamics is determined by Equations (16.1) and (16.2), and only these. The particle's trajectory $\mathbf{R}(t)$ is thus uniquely defined for all t, given suitable initial conditions $\mathbf{R}(0)$ and $\Psi(\mathbf{r}, 0)$.[1]

An example that is most commonly examined is a non-relativistic spin-0 particle of mass m moving in an external potential $V(\mathbf{r}, t)$, for which Ψ is a complex-valued function of $\mathbf{r}$ and t,

$$\mathbf{v}^{\Psi}(\mathbf{r}, t) = \frac{\hbar}{m} \mathrm{Im}\left[\frac{\Psi^* \boldsymbol{\nabla}\Psi}{\Psi^*\Psi}\right] \tag{16.3}$$

or, alternatively, $\mathbf{v}^{\Psi}(\mathbf{r}, t) = \boldsymbol{\nabla} S/m$ writing $\Psi = |\Psi| \exp(iS/\hbar)$, and

$$\mathcal{H} = -\frac{\hbar^2}{2m}\nabla^2 + V. \tag{16.4}$$

The particle's motion is generally non-Newtonian, readily illustrated in the problem of quantum tunnelling or barrier penetration (see Norsen 2013; Dewdney and Hiley 1982). Now widely accessible, particle trajectories derived from the velocity field (16.3) offer an intelligible, paradox-free, and persuasive picture of various quantum phenomena. A notable recurring example is that of the double-slit experiment with (or without) monitoring (cf. Philippidis *et al.* 1979; Tastevin and Laloë 2018).

Of course, the dBB theory does not stop at comforting pictures. As is widely recognized, it yields the same experimental predictions as the textbook/standard quantum formalism (Bohm 1952b; Dürr *et al.* 2004) – thankfully, without glorifying the role of observation. An illustration of how this theory proceeds from pictures to predictions in the context of laboratory momentum measurements is provided below.

In this chapter, we focus on the dBB equations for a single spin-1/2 particle, which are based on spinor-valued wave functions. Early on, it was thought that incorporating spin into the repertoire of the dBB theory would prove to be an impasse,[2] to wit: David Bohm reported in a letter quoted in Freire Junior (2019), p. 97:

> von Neumann thinks my work correct, and even 'elegant', but he expects difficulties in extending it to spin.

But within a year, following his 1952 papers, Bohm demonstrated how the theory could naturally account for the behaviour of spin-1/2 particles (Bohm 1953). Many physicists contributed to this subject in the following years, including Takehiko Takabayasi, Ralph Schiller, Jayme Tiomno, John S. Bell, and especially Peter R. Holland – it would be a worthy undertaking for historians to chronicle these developments.

Spin-dependent guiding equations have received relatively little attention in the dBB literature, outside of discussions of the EPRB–Bell or Stern–Gerlach experiments (see, e.g. Norsen 2014; Dewdney *et al.* 1988). Of greater concern, in almost

[1] Berndl *et al.* (1995) and Teufel and Tumulka (2005) establish the existence of the (N-particle) dBB dynamics under very general assumptions on the external potentials and initial conditions.

[2] This may have been partially influenced by Bohm's presentation of Equation (16.1) in a second-order, Newtonian form (Bohm 1952a), since it is generally considered that spin is a type of 'non-classical two-valuedness' and a definitive symbol of the break from classical modes of thought; in part, this view may have also been reinforced by the 'no hidden variable' theorem due to Kochen and Specker, which forbids assigning pre-existing values (such as the particle position $\mathbf{R}$) to all the spin components of a particle; see Mermin (1993).

all instances where spin-1/2 particles are discussed, the analysis is based on an inadequate non-relativistic velocity field,[3] which contradicts the relativistic velocity in the non-relativistic limit (see below). This oversight, while usually not compromising much conceptual accuracy, is regrettable because atoms are composed entirely of spin-1/2 particles, which are most relevant to chemistry[4] and, for that matter, nearly everything. The subject surely deserves better treatment.

What better way is there of getting acquainted with the dBB dynamics of fermions than the simple setting of a single spin-1/2 particle trapped within a box (one of the first problems a student encounters in quantum mechanics)? The particle-in-a-box problem is certainly of historical significance considering the interpretive debates between Born, Einstein, and Bohm, the specifics of which are not addressed here (but see Myrvold 2003, Section 3.1; de Broglie 1960, Chapter XI, Section 3; Holland 1993, Section 6.5.3).

The motion of a spin-1/2 particle (henceforth referred to as an electron) prepared in the ground state of a spherical box is investigated here in both non-relativistic (Section 16.2) and relativistic (Section 16.3) versions of the dBB theory. In both cases, the Bohmian trajectories are explicitly computable in a coordinate-free way – a rare luxury. The presentation assumes some familiarity with Pauli and Dirac's equations and is at a level of quantum mechanics and mathematical methods typical of courses taken by first-year graduate students in physics.

Switching gears, in Section 16.4, we consider the response of the electron to a sudden removal of the confining potential. While the Bohmian trajectories are no longer analytically solvable, it is shown that the escaping electron exhibits free-Newtonian behaviour in the far field – a hallmark of unbounded dBB motion (Daumer *et al.* (1996, 1997)).

This problem offers a segue to examine a widely used technique for determining the momentum of an abruptly freed bound electron by measuring its arrival (or flight) time on a distant detector (Section 16.5).[5] Section 16.6 repurposes the electron-in-a-box example to illustrate this experimental technique, recovering the expected quantum mechanical result *without reference to a quantum observable*. Here, we outline the typical steps involved in applying the dBB theory to experiments, explaining how seemingly random experimental outcomes can arise within a deterministic theory.

Our analysis highlights a crucial point: the dBB theory not only makes it possible to predict experimental results but also validates or limits the assumptions that experimenters invariably make when transforming their raw data into final reported results. This is an important nuance that is often overlooked when comparing different 'interpretations' of quantum mechanics.

Final concluding thoughts are offered in Section 16.7.

[3] Perhaps following an early proposal of John S. Bell, (Bell 1971, pp. 171–181).

[4] See Lombardi and Fortin's contribution in this volume.

[5] This is in contrast to the dBB treatments of momentum measurements discussed, e.g. in Holland (1993), Section 6.8; Dürr and Teufel (2009), Section 15.1.2; Bricmont (2016), Section 2.A.3; or Dürr *et al.* (2004), Section 5.5, which involve a position measurement at a late but *fixed time* rather than a time-of-flight measurement at a specified detection surface.

16.2 Electron-in-a-box: non-relativistic treatment

For an electron of mass m moving at non-relativistic speeds, the wave function is a two-component spinor (or two-spinor) $\Psi = \big(\psi_+(\mathbf{r},t), \psi_-(\mathbf{r},t)\big)^\top$. The dBB velocity field and Hamiltonian are given, respectively, by (Bohm and Hiley 1993, Chapter 10 and Struyve 2005, Chapter 2)

$$\mathbf{v}^\Psi(\mathbf{r},t) = \frac{\hbar}{m}\mathrm{Im}\left[\frac{\Psi^\dagger\nabla\Psi}{\Psi^\dagger\Psi}\right] + \frac{\hbar}{2m}\frac{\nabla\times\left(\Psi^\dagger\boldsymbol{\sigma}\Psi\right)}{\Psi^\dagger\Psi}, \tag{16.5}$$

and

$$\mathcal{H} = -\frac{\hbar^2}{2m}\left(\boldsymbol{\sigma}\cdot\nabla\right)^2 + V(\mathbf{r},t)\,\mathbb{1}. \tag{16.6}$$

Here, V is an external potential, $\mathbb{1}$ denotes the 2×2 unit matrix (often omitted, but to be understood), and $\boldsymbol{\sigma} = \sigma_x\hat{\mathbf{x}}+\sigma_y\hat{\mathbf{y}}+\sigma_z\hat{\mathbf{z}}$ is Pauli's spin-vector matrix. Incorporating (16.6) into (16.2) gives the Schrödinger–Pauli equation.[6] The first (second) term on the right-hand side of (16.5) is also called the *convective* or *osmotic* (*spin* or *Gordon*) velocity. The convective term is a natural generalization of the spin-0 dBB velocity (16.3) to spinor-valued wave functions.

A word of caution: The spin velocity term is omitted in most discussions of spin in the dBB literature. As emphasized in Section 16.1, without it, the resultant velocity field would not be recovered from the relativistic velocity field (Eq. (16.30), below) in the non-relativistic limit. The latter is, in fact, *uniquely pinned down* by covariance considerations Holland (1999, 2003), as is, therefore, its non-relativistic limit given by (16.5).[7]

In dropping the spin term (or by inadvertently using the spin-0 equations), one faces the unexpected consequence that an electron in the hydrogen ground state is stationary. This mistake has practically become lore, inspiring novel proposals for modifying the dBB guiding equations to cure the 'stationary electron malady', e.g. Finley (2021). Nevertheless, as demonstrated in Colijn and Vrscay 2002, 2003a, 2003b (see also footnote 11), if the full velocity field (16.5) is taken into account, the electron is in fact *not* stationary in the hydrogen ground state.

We now turn to a discussion of an electron trapped in an infinite spherical well (or spherical box) of radius a. For this, assume

$$V(r) = \begin{cases} 0, & r\le a \\ \infty, & r>a \end{cases}, \qquad r = \|\mathbf{r}\|, \tag{16.7}$$

in (16.6); the Hamiltonian simplifies to

[6] If a magnetic field $\mathbf{B}$ is present, the gradient in Equations (16.5) and (16.6) must be replaced by the gauge covariant derivative involving the vector potential (minimal coupling prescription). The Pauli interaction term $\propto \boldsymbol{\sigma}\cdot\mathbf{B}$ is automatically generated by this and need not be put in by hand.

[7] One can arrive at (16.5), or equivalently the current density $\mathbf{j} = (\Psi^\dagger\Psi)\mathbf{v}^\Psi$, using the non-relativistic Schrödinger–Pauli equation; see Wilkes (2020).

$$\mathcal{H} = -\frac{\hbar^2}{2m}\nabla^2\mathbb{1} + V(r)\mathbb{1}, \tag{16.8}$$

in view of the identity $(\boldsymbol{\sigma}\cdot\boldsymbol{\nabla})^2 = \nabla^2\mathbb{1}$—a special case of the spin-matrix identity

$$(\mathbf{a}\cdot\boldsymbol{\sigma})(\mathbf{b}\cdot\boldsymbol{\sigma}) = (\mathbf{a}\cdot\mathbf{b})\mathbb{1} + i(\mathbf{a}\times\mathbf{b})\cdot\boldsymbol{\sigma}. \tag{16.9}$$

The Schrödinger–Pauli equation thus reduces to two uncoupled Schrödinger equations for the spinor components $\psi_\pm$, allowing for simple *space-spin-separated* or *spin-polarized* two-spinor solutions of the form

$$\Psi(\mathbf{r},t) = e^{-itE/\hbar}\psi(\mathbf{r})\chi, \tag{16.10}$$

where ψ is a solution of the time-independent Schrödinger equation $\hbar^2\nabla^2\psi = 2m(V - E)\psi$, and χ is an otherwise arbitrary, constant two-spinor. (ψ is commonly referred to as the coordinate or orbital wave function, or simply the 'spatial-part' of Ψ, while χ is called the 'spin-part' of Ψ.) The ground-state wave function, given by (Griffiths and Schroeter, 2018, p. 130), is distinguished by its simple dependence on r (in particular, it exhibits no zeros or nodes within the spherical box):

$$\psi(\mathbf{r}) = \frac{1}{\sqrt{2\pi a}\,r}\sin\left(\frac{\pi r}{a}\right) \tag{16.11}$$

for $r < a$ and zero otherwise. It corresponds to

$$E = \frac{\hbar^2\pi^2}{2ma^2}. \tag{16.12}$$

For spin-polarized wave functions, the guiding equation simplifies to

$$\frac{\mathrm{d}}{\mathrm{d}t}\mathbf{R}(t) = \frac{\hbar}{m}\mathrm{Im}\left[\frac{\boldsymbol{\nabla}\psi}{\psi}\right](\mathbf{R}(t),t) + \frac{\hbar}{2m}\left[\frac{\boldsymbol{\nabla}\times\left(|\psi|^2\hat{\mathbf{s}}\right)}{|\psi|^2}\right](\mathbf{R}(t),t), \tag{16.13}$$

introducing the constant *unit* vector

$$\hat{\mathbf{s}} = \frac{\chi^\dagger\boldsymbol{\sigma}\chi}{\chi^\dagger\chi} \tag{16.14}$$

– the so-called unit spin-vector associated with Ψ.[8] The numerator of the last term in Equation (16.13) is equivalent to $\left(\boldsymbol{\nabla}|\psi|^2\right)\times\hat{\mathbf{s}}$ since $\boldsymbol{\nabla}\times\hat{\mathbf{s}} = \mathbf{0}$ (the components of $\hat{\mathbf{s}}$ do not depend on the spatial coordinates). We thus have

[8]For any two-spinor $\chi = (\chi_+,\chi_-)^\top$, $\chi^\dagger\boldsymbol{\sigma}\chi = 2\mathrm{Re}[\chi_+^*\chi_-]\,\hat{\mathbf{x}} + 2\mathrm{Im}[\chi_+^*\chi_-]\,\hat{\mathbf{y}} + (|\chi_+|^2 - |\chi_-|^2)\,\hat{\mathbf{z}}$; consequently

$$\begin{aligned}\left\|\chi^\dagger\boldsymbol{\sigma}\chi\right\|^2 &= (\chi^\dagger\sigma_x\chi)^2 + (\chi^\dagger\sigma_y\chi)^2 + (\chi^\dagger\sigma_z\chi)^2 = 4|\chi_+^*\chi_-|^2 + (|\chi_+|^2 - |\chi_-|^2)^2\\ &= |\chi_+|^4 + 2|\chi_+|^2|\chi_-|^2 + |\chi_-|^4 = (\chi^\dagger\chi)^2 \Rightarrow \|\hat{\mathbf{s}}\| = 1.\end{aligned}$$

$$\frac{\mathrm{d}}{\mathrm{d}t}\mathbf{R}(t) = \frac{\hbar}{m}\,\mathrm{Im}\left[\frac{\boldsymbol{\nabla}\psi}{\psi}\right](\mathbf{R}(t), t) + \frac{\hbar}{m}\,\mathrm{Re}\left[\frac{\boldsymbol{\nabla}\psi}{\psi}\right](\mathbf{R}(t), t) \times \hat{\mathbf{s}}. \tag{16.15}$$

In the present case, the spatial wave function ψ, Equation (16.11), is a function of r only; therefore,

$$\frac{\boldsymbol{\nabla}\psi}{\psi} = \psi^{-1}\frac{\mathrm{d}\psi}{\mathrm{d}r}\hat{\mathbf{e}}_r = \left[\frac{\pi}{a}\cot\left(\frac{\pi r}{a}\right) - \frac{1}{r}\right]\hat{\mathbf{e}}_r, \qquad r < a, \tag{16.16}$$

with $\hat{\mathbf{e}}_r = r^{-1}\mathbf{r}$. Since this quantity is real-valued, the first term on the right-hand side of (16.15) vanishes. Consequently, only the second nonzero term, i.e. the spin or Gordon term, determines the electron's velocity within the box. Incorporating (16.16) into Equation (16.15), we obtain

$$\dot{\mathbf{R}}(t) = \frac{\hbar}{mR(t)}\left[\frac{\pi}{a}\cot\left(\frac{\pi R(t)}{a}\right) - \frac{1}{R(t)}\right]\mathbf{R}(t) \times \hat{\mathbf{s}}, \tag{16.17}$$

where $R(t) = \|\mathbf{R}(t)\|$ is the distance of the electron from the centre of the box at time t. (The overdot indicates a derivative taken w.r.t. t.) Equation (16.17) holds for $R(t) < a$; the electron's velocity is undefined for $R(t) \geq a$, i.e., outside and on the surface of the box (where its wave function is identically zero).

However, not surprisingly, *no* solution of (16.17) starting within the volume of the box allows the particle to escape. To establish this, note that

$$\frac{\mathrm{d}}{\mathrm{d}t}R^2(t) = 2\,\mathbf{R}(t)\cdot\dot{\mathbf{R}}(t) \propto \mathbf{R}(t)\cdot\big(\mathbf{R}(t)\times\hat{\mathbf{s}}\big) = 0, \tag{16.18}$$

i.e., $R(t)$ is a constant of the motion with $R(t) = R(0)$. The electron's trajectory thus lies on the surface of a sphere of radius $R(0)$, concentric with the spherical box. In particular, $R(0) < a$ implies that $R(t) < a$ for *all* t.

Furthermore, since the component of velocity along the spin-vector $\hat{\mathbf{s}}$ is zero, we have

$$\frac{\mathrm{d}}{\mathrm{d}t}\big(\mathbf{R}(t)\cdot\hat{\mathbf{s}}\big) = \dot{\mathbf{R}}(t)\cdot\hat{\mathbf{s}} = 0. \tag{16.19}$$

This implies that $\mathbf{R}(t)\cdot\hat{\mathbf{s}} = \text{const.} = \mathbf{R}(0)\cdot\hat{\mathbf{s}}$, or equivalently,

$$\big(\mathbf{R}(t) - \mathbf{R}(0)\big)\cdot\hat{\mathbf{s}} = 0. \tag{16.20}$$

That is, $\mathbf{R}(t)$ lies in a plane *perpendicular* to $\hat{\mathbf{s}}$ that passes through the point $\mathbf{R}(0)$.[9] It can thus be inferred that the electron trajectory is part of a circle (formed by the intersection of this plane with the spherical surface of radius $R(0)$) whose centre is situated at a distance $\|\mathbf{R}(0)\cdot\hat{\mathbf{s}}\|$ from the centre of the box.

[9] Recall that the equation of a plane in the point-normal form is given by $(\mathbf{r} - \mathbf{r}_0)\cdot\hat{\mathbf{n}} = 0$.

In fact, the motion is *periodic*. To demonstrate this, we solve Equation (16.17) with initial condition $\mathbf{R}(0) = \mathbf{R}_0$, assuming $R_0 < a$. Since we have established that $R(t) = R(0) = R_0$, Equation (16.17) can be written as

$$\dot{\mathbf{R}}(t) = -\,\omega_0\, \mathbf{R}(t) \times \hat{\mathbf{s}}, \tag{16.21}$$

where

$$\omega_0 = \frac{\hbar}{mR_0^2}\left[1 - \frac{\pi R_0}{a}\cot\left(\frac{\pi R_0}{a}\right)\right], \tag{16.22}$$

is a positive[10], trajectory-specific constant, determined by the initial condition. (Note that Equation (16.21) is the familiar Bloch equation for the precession of a magnetic moment $\mathbf{M}(t)$ in a uniform magnetic field $\mathbf{B}$, with Larmor frequency ω_0.) Differentiating (16.21) w.r.t. t, we have

$$\ddot{\mathbf{R}}(t) = -\,\omega_0\, \dot{\mathbf{R}}(t) \times \hat{\mathbf{s}} \overset{(16.21)}{=} \omega_0^2\left(\mathbf{R}(t)\times\hat{\mathbf{s}}\right)\times\hat{\mathbf{s}} = \omega_0^2\left[\left(\mathbf{R}(t)\cdot\hat{\mathbf{s}}\right)\hat{\mathbf{s}} - \|\hat{\mathbf{s}}\|^2\,\mathbf{R}(t)\right], \tag{16.23}$$

applying the triple vector cross product identity $\mathbf{a}\times(\mathbf{b}\times\mathbf{c}) = (\mathbf{a}\cdot\mathbf{c})\,\mathbf{b} - (\mathbf{a}\cdot\mathbf{b})\,\mathbf{c}$. Since $\|\hat{\mathbf{s}}\| = 1$ and $\mathbf{R}(t)\cdot\hat{\mathbf{s}}$ is a constant of the motion, Equation (16.23) reduces to

$$\ddot{\mathbf{R}}(t) + \omega_0^2\,\mathbf{R}(t) = \omega_0^2\left(\mathbf{R}_0\cdot\hat{\mathbf{s}}\right)\hat{\mathbf{s}}, \tag{16.24}$$

the (classical) equation of motion for a forced simple harmonic oscillator. Its general solution is the familiar one:

$$\mathbf{R}(t) = \left(\mathbf{R}_0\cdot\hat{\mathbf{s}}\right)\hat{\mathbf{s}} + \cos(\omega_0 t)\,\mathbf{J} + \sin(\omega_0 t)\,\mathbf{K}, \tag{16.25}$$

where $\mathbf{J}$ and $\mathbf{K}$ are arbitrary constant vectors. Evidently, the motion is periodic with period $2\pi/\omega_0$, characteristic of an individual trajectory.

The variation with radial distance of the angular velocity ω_0, Equation (16.22), reflects the profile of the wave function.[11] Electrons orbiting closer to the centre of the box ($R_0 \ll a$) have an angular velocity

$$\omega_0 \sim \frac{\pi^2}{3}\frac{\hbar}{ma^2} + \mathcal{O}\left(\frac{R_0}{a}\right)^2, \tag{16.26}$$

whereas for those moving close to the boundary of the box,

[10] $x\cot x < 1$ for $0 < x < \pi$.

[11] Electron orbits for the hydrogen ground-state wave function are similar in all respects to the present ones, except that $\omega_0 = \hbar/(ma_0R_0)$ (Colijn and Vrsacy 2002, Equation (7)), where a_0 is the Bohr radius. Those for an isotropic harmonic potential of trapping frequency ω have $\omega_0 = \omega$, independent of R_0.

$$\omega_0 \sim \frac{\hbar}{ma(a - R_0)}, \tag{16.27}$$

as R_0 approaches a. The latter suggests that the electron is moving at an astronomically high speed near the box's edge. Such a feature is not surprising in a non-relativistic treatment. However, it is absent in the relativistic account presented below, where the dBB velocity $\mathbf{v}^{\Psi}$ does not exceed c (the speed of light in vacuum).

To complete the trajectory calculation, note that $\mathbf{R}(0) = \mathbf{R}_0 \overset{(16.25)}{\Rightarrow} \mathbf{J} = \mathbf{R}_0 - (\mathbf{R}_0 \cdot \hat{\mathbf{s}})\,\hat{\mathbf{s}}$. To determine $\mathbf{K}$, differentiate (16.25) w.r.t. t and use the initial velocity condition $\dot{\mathbf{R}}(0) \overset{(16.21)}{=} -\omega_0\,\mathbf{R}_0 \times \hat{\mathbf{s}}$. This yields $\mathbf{K} = -\,\mathbf{R}_0 \times \hat{\mathbf{s}}$. The explicit solution for the Bohmian trajectory is thus given by

$$\mathbf{R}(t) = \cos(\omega_0 t)\,\mathbf{R}_0 + \sin(\omega_0 t)\,\hat{\mathbf{s}} \times \mathbf{R}_0 + \big[1 - \cos(\omega_0 t)\big]\,\big(\mathbf{R}_0 \cdot \hat{\mathbf{s}}\big)\,\hat{\mathbf{s}}, \tag{16.28}$$

which is reminiscent of Rodrigues' rotation formula. Geometrically, this means that the position of the particle at time t is obtained by rotating its initial position vector $\mathbf{R}_0$ by an angle $\omega_0 t$ about the spin-vector-axis counterclockwise, as illustrated in Fig. 16.1.

16.3 Electron-in-a-box: relativistic treatment

In what follows, we present a brief relativistic account of the electron-in-a-box problem. For this, Dirac's Hamiltonian

$$\mathcal{H} = -\,i\hbar c\,\boldsymbol{\alpha} \cdot \boldsymbol{\nabla} + \beta mc^2 + V(r) \tag{16.29}$$

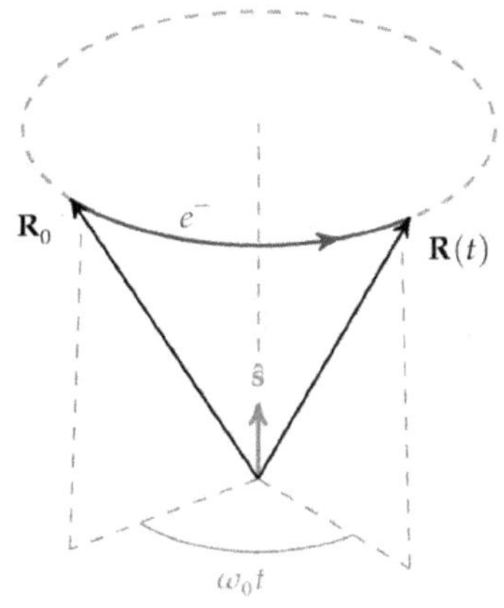

Fig. 16.1: A de Broglie–Bohm electron trajectory for the spin-polarized ground-state wave function of a spherical box.

replaces the Schrödinger–Pauli Hamiltonian (16.6) of the non-relativistic theory, and the dBB velocity field is given by (Bohm 1953; de Broglie 1960, Chapter XVI; Holland 1992)

$$\mathbf{v}_{\mathrm{R}}^{\Psi}(\mathbf{r},t) = c\frac{\Psi^{\dagger}\boldsymbol{\alpha}\Psi}{\Psi^{\dagger}\Psi}, \tag{16.30}$$

where $\Psi(\mathbf{r},t)$ is a four-spinor wave function. In these equations,

$$\boldsymbol{\alpha} = \begin{pmatrix} \mathbb{0} & \boldsymbol{\sigma} \\ \boldsymbol{\sigma} & \mathbb{0} \end{pmatrix} \tag{16.31}$$

is the 4×4 Dirac α-vector matrix containing, in the off-diagonal blocks, the Pauli-spin-vector-matrix introduced earlier, $\beta = \mathrm{diag}(1,1,-1,-1)$, and $\mathbb{0}$ is the 2×2 zero or null matrix. This velocity field reproduces (16.5) in the non-relativistic limit. In fact, $\mathbf{v}_{\mathrm{R}}^{\Psi}$ may be separated into convective and spin components by using Gordon's decomposition identity (Greiner 1992, Section 8.1).

That $\|\mathbf{v}_{\mathrm{R}}^{\Psi}(\mathbf{r},t)\| \leq c$ regardless of Ψ is a consequence of a so-called Fierz identity (Van Oosterhout 1966; Kofink 1937, Section 4):

$$(\Psi^{\dagger}\Psi)^2 - |\Psi^{\dagger}\boldsymbol{\alpha}\Psi|^2 = \mu^2 + \nu^2, \tag{16.32}$$

where $\mu = \Psi^{\dagger}\beta\Psi$, and $\nu = -\Psi^{\dagger}(\alpha_x\alpha_y\alpha_z)\beta\Psi$ are *real-valued* quantities. Clearly, the left-hand side of (16.32) is ≥ 0, establishing our claim.[12]

Next, we address the electron in a spherical box problem. To begin, the relativistic ground-state wave function needs to be found. It turns out that we can resurrect the non-relativistic ground-state wave function to obtain its relativistic counterpart (see Appendix A for details), with the final result

$$\Psi(\mathbf{r},t) = e^{-itE_{\mathrm{R}}/\hbar}\begin{pmatrix} \psi(\mathbf{r})\chi \\ -\dfrac{i\hbar c}{E_{\mathrm{R}} + mc^2}(\boldsymbol{\sigma}\cdot\nabla)\,\psi(\mathbf{r})\chi \end{pmatrix}, \tag{16.33}$$

where $\psi(\mathbf{r})$ is the non-relativistic spatial wave function (16.11), χ is an arbitrary, constant two-spinor, and

$$E_{\mathrm{R}} = mc^2\sqrt{1 + \frac{2E}{mc^2}}, \tag{16.34}$$

with $E =$ (16.12). ($E_{\mathrm{R}} \approx mc^2 + E$ for $E \ll mc^2$.) Note that (16.33) is not space-spin separated as its non-relativistic counterpart (16.10). In what follows, we set

$$\frac{\hbar c}{E_{\mathrm{R}} + mc^2} = \gamma \tag{16.35}$$

for brevity (γ has the physical dimensions of a length).

[12] Actually, the very form of the velocity field (16.30) forces $\|\mathbf{v}_R^{\Psi}\|$ to be *bounded* for any Ψ. This is a consequence of a theorem in linear algebra, which states that for any nonzero $\Xi \in \mathbb{C}^n$, and $n \times n$-Hermitian matrix A, the ratio $(\Xi^{\dagger}A\Xi)/(\Xi^{\dagger}\Xi)$ (the so-called Rayleigh quotient) is real-valued and $\leq \lambda_{\mathrm{max}}$—the largest eigenvalue of A (Horn and Johnson (2013), Theorem 4.2.2., p. 234). In light of this result and observing that $\lambda_{\mathrm{max}} = 1$ for α_x, α_y, or α_z (since each α matrix squares to the identity matrix), it follows that every component of (16.30) is $\leq c$, hence via the triangle inequality, $\|\mathbf{v}_R^{\Psi}\| \leq \sqrt{3}\,c$. However, to establish the tighter bound $\|\mathbf{v}_R^{\Psi}\| \leq c$, additional input about the α matrices is needed.

Since ψ is real and $\nabla\psi = \mathrm{d}\psi/\mathrm{d}r\,\hat{\mathbf{e}}_r$, we obtain

$$\begin{aligned}\Psi^\dagger\Psi &= \psi^2\,(\chi^\dagger\chi) + \gamma^2\left(\frac{\mathrm{d}\psi}{\mathrm{d}r}\right)^2(\hat{\mathbf{e}}_r\cdot\boldsymbol{\sigma}\chi)^\dagger(\hat{\mathbf{e}}_r\cdot\boldsymbol{\sigma}\chi)\\ &= \left(\psi^2 + \gamma^2\left(\frac{\mathrm{d}\psi}{\mathrm{d}r}\right)^2\right)(\chi^\dagger\chi),\end{aligned} \tag{16.36}$$

using $\boldsymbol{\sigma} = \boldsymbol{\sigma}^\dagger$ and the identity (16.9). Furthermore, following some simplifications,

$$\Psi^\dagger\boldsymbol{\alpha}\Psi = -i\gamma\left(\psi\frac{\mathrm{d}\psi}{\mathrm{d}r}\right)\chi^\dagger\Big[\boldsymbol{\sigma}\,(\hat{\mathbf{e}}_r\cdot\boldsymbol{\sigma}) - (\hat{\mathbf{e}}_r\cdot\boldsymbol{\sigma})\,\boldsymbol{\sigma}\Big]\chi. \tag{16.37}$$

The terms enclosed in square brackets evaluate to $2i\,\hat{\mathbf{e}}_r\times\boldsymbol{\sigma}$. Thus, in terms of the unit spin-vector (16.14) introduced earlier, we have

$$\mathbf{v}_\mathrm{R}^\Psi(\mathbf{r},t) = c\,F\left(\frac{\gamma}{\psi}\frac{\mathrm{d}\psi}{\mathrm{d}r}\right)\hat{\mathbf{e}}_r\times\hat{\mathbf{s}}, \tag{16.38}$$

where

$$F(\xi) = \frac{2\xi}{1+\xi^2}.$$

Even without substituting the known expression for ψ, it follows quite explicitly from (16.38) that $\|\mathbf{v}_\mathrm{R}^\Psi\| \le c$, given $\|\hat{\mathbf{e}}_r\times\hat{\mathbf{s}}\| \le 1$, and $|F(\xi)| \le 1$ for any real ξ. And, in the non-relativistic limit as $c\to\infty$,[13] equivalently $\gamma\to 0$, such that $\gamma c\to\hbar/(2m)$ (recalling (16.35)), the right-hand side of (16.38) obligingly reproduces the non-relativistic velocity field, Equations (16.16-16.17). The *absolute necessity* of the spin term of Equation (16.5) cannot be overemphasized here. Without it, the expected non-relativistic limits would fail to follow.

It is a fortunate coincidence that the relativistic guiding equation obtained by incorporating (16.38) into (16.1) has precisely the same form as its non-relativistic counterpart, Equation (16.17). This implies that the relativistic electron orbits are circular as well, given by the closed-form expression (16.28), except that ω_0 must be replaced by

$$\omega_{0\mathrm{R}} = \frac{c}{R_0}F\left(\frac{mR_0}{\hbar}\gamma\omega_0\right), \tag{16.39}$$

where ω_0 is the R_0-dependent non-relativistic angular velocity of Equation (16.22). Given that F is a bounded function with linear behaviour around zero, $\omega_{0\mathrm{R}}$ is bounded for all $0\le R_0\le a$. In particular,

$$\omega_{0\mathrm{R}} \sim \frac{2c}{\gamma}\left(1-\frac{R_0}{a}\right) + \mathcal{O}(a-R_0)^3, \tag{16.40}$$

as $R_0\to a^-$, thus taming the divergence encountered in (16.27).

[13] More appropriately, as the nondimensionalised velocity $(ma/\hbar)\,c \gg 1$.

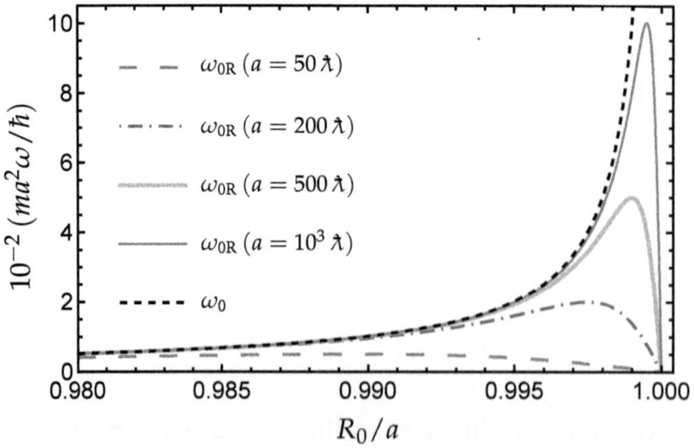

Fig. 16.2: Nondimensionalized relativistic (ω_{0R}) and non-relativistic (ω_0) angular velocities vs. R_0, the electron's distance from the centre of the box. Here, a and $\lambda\!\!\!^{-} = \hbar/(mc)$ denote the box radius and the reduced Compton wavelength of the electron, respectively. Notable differences between ω_0 and ω_{0R} only appear near the edge of the box for $R_0 > 0.9a$. Note that ω_{0R} (ω_0) vanishes (blows up) as $R_0 \to a^-$.

The relativistic and non-relativistic angular velocities are compared in Fig. 16.2 for selected values of a. The (relativistic) orbital velocity increases monotonically with increasing distance from the centre of the box, reaching a maximum value (characteristic of the box radius a), after which it begins to decrease and eventually vanishes asymptotically as R_0 approaches a. In other words, the electrons slow down to rest at the box's boundary. Numerical calculations indicate that ω_{0R} is appreciably different from ω_0 only for $a - R_0 \lesssim \hbar/(mc) \approx 3.86 \times 10^{-13}$ m (the reduced Compton wavelength). Therefore, FAPP, the ground-state electron orbits are adequately described by the non-relativistic theory.

16.4 Electron motion post-deconfinement

Left to itself, the trapped electron will continue orbiting indefinitely, leaving little more to say about this system. To evoke a more dramatic response, one could imagine opening the box suddenly, say, at time t_0, and subsequently investigating the wave function of the particle, and the evolution of its motion, due to such a 'shock'. Specifically, this could aid in analyzing an experiment designed to catch the emerging electron on, e.g. a microchannel plate.

The quantities typically reported in this kind of experiment are either the electron's impact position, its arrival time, or both. These measurements are ubiquitous in particle physics and serve as the basis for measuring momentum, energy, etc. The measurement of momentum is discussed in detail in Section 16.5. In what follows, we present a non-relativistic discussion of the electron trajectories post-deconfinement.

In the absence of the spherical box potential (16.7), the wave function evolves freely[14] with Hamiltonian $\mathcal{H} = -\hbar^2/(2m)\nabla^2\mathbb{1}$ (cf. Equations (16.6) and (16.9)), thus continuing to remain space-spin separated, as (16.10), for $t > t_0$. Over time, the electron is transported to distances $\gg a$ as the wave function spreads isotropically, filling the entire space. The wave function for any t can be expressed as follows:

$$\Psi(\mathbf{r}, t) = \begin{cases} (16.10), & t \leq t_0 \\ \psi_>(\mathbf{r}, t)\chi, & t > t_0 \end{cases}, \tag{16.41}$$

where $\psi_>$ solves the free-particle Schrödinger equation

$$i\hbar\frac{\partial\psi_>}{\partial t} = -\frac{\hbar^2}{2m}\nabla^2\psi_>. \tag{16.42}$$

In order for the wave function and dBB velocity to be continuous at $t = t_0$, we require

$$\psi_>\left(\mathbf{r}, t_0^+\right) = e^{-\,it_0E/\hbar}\,\psi(\mathbf{r}), \tag{16.43}$$

where $\psi(\mathbf{r})$ is the spatial part of the ground-state wave function, Equation (16.11).

The solution of Equation (16.42) with initial condition (16.43) can be found in the literature (Moshinsky 1952b; Godoy 2003, 2007). However, for completeness, we derive it in Appendix B by methodically reducing the three-dimensional calculation to a one-dimensional one that can be solved with ease, obtaining

$$\psi_>(\mathbf{r}, t) = \frac{N_0}{r}\Big[M(r+a, k, \tau) - M(r-a, k, \tau) - M(a-r, k, \tau) + M(-a-r, k, \tau)\Big]. \tag{16.44}$$

Here,

$$N_0 = \frac{i\,e^{-\,iEt_0/\hbar}}{2\sqrt{2\pi a}}, \qquad k = \frac{\pi}{a}, \qquad \tau = \frac{\hbar}{m}(t - t_0), \tag{16.45}$$

and $M(r, k, t)$ is Moshinsky's function, which inevitably appears in problems involving matter waves suddenly released from bounded domains; see Nussenzveig 1992; del Campo *et al.* 2009. Note that the electron wave function ceases to be *stationary* post-deconfinement, i.e. it is no longer separated in $\mathbf{r}$ and t. A few snapshots of $\psi_>$ are plotted in Fig. 16.3.

Besides the general dispersive spreading of the wave function, with its amplitude gradually diminishing over time for $r < a$, $\psi_>$ presents a train of tiny ripples, or wavelets, close to the boundary of the spherical box.[15] With time, these wavelets move farther away, nucleating into distinct, self-similar wave packets. This wave phenomenon, called *diffraction in time* – due to its 'close mathematical resemblance with

[14] We are neglecting, for brevity, the electron's free fall due to gravity in the problem.

[15] Note that $\psi_>$ becomes nonzero at *all* distances from the spherical trap as soon as the trap is removed. A relativistic treatment of the problem expectedly modifies this feature, requiring the wave function to strictly vanish for any $r > a + c(t - t_0)$, at time t (Moshinsky 1952a, Section IV; Godoy and Villa 2016).

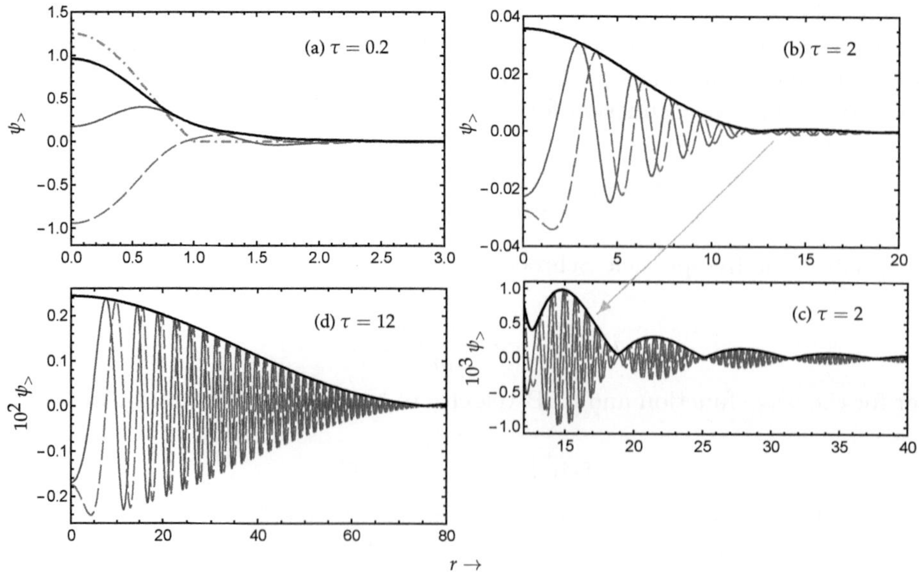

Fig. 16.3: Snapshots of the spatial wave function $\psi_>$ at selected instant of τ: Re$[\psi_>]$ (solid line), Im$[\psi_>$ (dashed line), and $|\psi_>|$ (thick solid line) for $\hbar = m = a = 1$. The dash-dotted curve in panel (a) denotes the spatial ground-state wave function.

the intensity of light in the Fresnel diffraction by a straight edge' (Moshinsky 1952a), arises in response to a sudden change in the boundary conditions of the wave function at some surface (that of the spherical box, $r = a$, in the present case).

Some of these aspects of the wave function are easier understood considering the approximation

$$\psi_>(\mathbf{r}, t) \approx 2\,N_0 \sqrt{\frac{2\pi}{i\tau}}\,\frac{a}{r}\,\frac{\sin(ar/\tau)}{(ar/\tau)^2 - \pi^2}\,e^{i(r^2+a^2)/(2\tau)}, \tag{16.46}$$

which is derived in Appendix A assuming $r \gg a$ and $r \gg \pi\tau/a$. Interestingly, though, except for very small values of τ, it turns out to be a very good approximation to (16.44) for practically all r and τ.

Now, turning to the Bohmian trajectories, the simplified guiding equation (16.15) (applicable for spin-polarized wave functions) continues to be valid for $t > t_0$ except that $\psi \to \psi_>$. As $\psi_>$ is spherically symmetric, we write

$$\frac{\boldsymbol{\nabla}\psi_>}{\psi_>} = \Big[\mathcal{R}(r,t) + i\,\mathcal{I}(r,t)\Big]\,\hat{\mathbf{e}}_r, \tag{16.47}$$

where $\mathcal{R}$ ($\mathcal{I}$) denotes the real (imaginary) part of $\psi_>^{-1}\,\mathrm{d}\psi_>/\mathrm{d}r$. This allows expressing the equation of motion for $t > t_0$ in the form

$$\dot{\mathbf{R}}(t) = \frac{\hbar}{mR(t)}\Big(\mathcal{I}\big(R(t),t\big)\,\mathbf{R}(t) \;+\; \mathcal{R}\big(R(t),t\big)\,\mathbf{R}(t)\times\hat{\mathbf{s}}\Big), \tag{16.48}$$

where $\hat{\mathbf{s}}$ is the unit spin-vector defined in Equation (16.14). (For $t \leq t_0$, Equation (16.17) applies.) Note that the first term within parentheses (originating from the convective velocity term) is now *nonzero*. Also, the velocity field is now explicitly time-dependent, making it much harder to determine Bohmian trajectories.

For convenience, let us use a spherical-polar coordinate system where the polar axis points along $\hat{\mathbf{s}}$. The position and velocity of the particle at time t are given by (cf. Reed 2019, Chapter 2, or Symon 1971, Section 3.5)

$$\mathbf{R} = R\,\hat{\mathbf{e}}_r, \qquad \dot{\mathbf{R}} = \dot{R}\,\hat{\mathbf{e}}_r + R\,\dot{\Theta}\,\hat{\mathbf{e}}_\theta + R\sin\Theta\,\dot{\Phi}\,\hat{\mathbf{e}}_\phi, \tag{16.49}$$

respectively, where $R(t)$, $\Theta(t)$, and $\Phi(t)$ denote the spherical coordinates of the particle at time t. We also have $\hat{\mathbf{s}} = \cos\theta\,\hat{\mathbf{e}}_r - \sin\theta\,\hat{\mathbf{e}}_\theta$, hence $\hat{\mathbf{e}}_r \times \hat{\mathbf{s}} = -\sin\theta\left(\hat{\mathbf{e}}_r \times \hat{\mathbf{e}}_\theta\right) = -\sin\theta\,\hat{\mathbf{e}}_\phi$. Incorporating this into (16.48) and comparing the result with $\dot{\mathbf{R}}$, Equation (16.49), yields the component equations

$$\dot{R}(t) = \frac{\hbar}{m}\,\mathcal{I}\big(R(t),t\big), \qquad \dot{\Theta}(t) = 0, \qquad \dot{\Phi}(t) = -\frac{\hbar}{mR(t)}\,\mathcal{R}\big(R(t),t\big). \tag{16.50}$$

The second equation has the obvious solution $\Theta(t) = \Theta(t_0)$, the value of Θ at time t_0. In fact, Θ being the angle of inclination of the electron's position vector $\mathbf{R}$ to the spin-vector $\hat{\mathbf{s}}$ (see Fig. 16.1) is a constant of the motion for $t < t_0$ as well.

To this point our considerations have been exact. Unfortunately, the remaining two equations in (16.50) cannot be solved analytically. However, we may extract the qualitative behaviour of the Bohmian trajectories in the far field (i.e., $r \gg a$), which is relevant for our subsequent discussion of momentum measurements, below. For this, we utilize (16.46) to obtain without further approximations $\mathcal{I}(r,t) \approx r/\tau$, in turn, $\dot{R} \approx R(t)/(t - t_0)$. The solution is

$$R(t) \approx v_\infty(t - t_0), \tag{16.51}$$

where v_∞ is a constant of integration. The escaping electron thus acquires a *constant radial velocity* at large separations from the spherical box; see Fig. 16.4 (right panel). The constant v_∞ is trajectory-specific and varies with the exact initial position of the electron.[16]

The remaining equation for $\dot{\Phi}$ can be analysed in a similar way using (16.46). One finds in view of (16.51) that $\Phi(t) \to$ const., as $t \to \infty$. Individual Bohmian trajectories therefore become approximately straight lines of constant Θ and Φ in the far field, as shown in Fig. 16.4 (left panel). In other words, they become *asymptotically Newtonian with a well-defined constant velocity*, which is a prominent attribute of Bohmian trajectories in scattering problems (Römer *et al.* 2005; Dürr and Teufel 2009, Chapter 16).

The spin velocity contribution is more pronounced in the near-field (i.e. at distances $\lesssim a$), where the trajectories tend to circulate $\hat{\mathbf{s}}$, as shown in Fig. 16.4 (middle panel).

[16] For an initial Gaussian wave function of width σ_0, the guiding equations can be solved exactly, yielding $v_\infty = \hbar R_0/(2m\sigma_0^2)$ (Holland and Philippidis 2003, Equation (8.7)). That is, the further the electron starts from the centre of the wave packet, the faster it escapes.

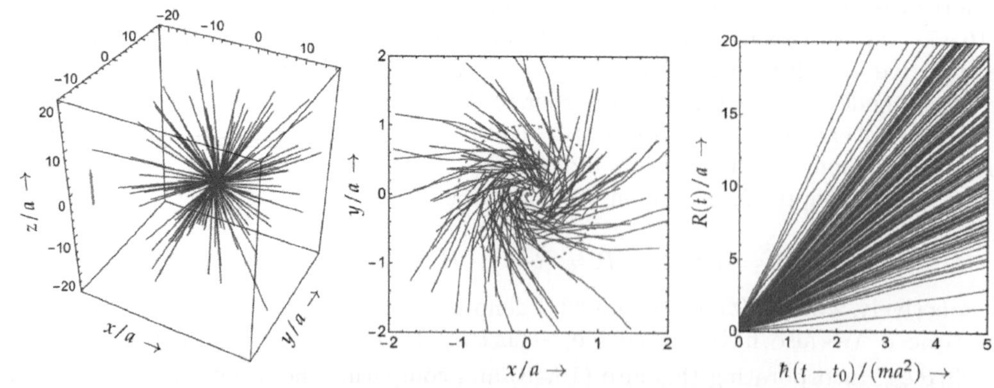

Fig. 16.4: Left panel: A collection of Bohmian trajectories emanating from the spherical trap generated by numerically integrating Equations (16.50) for a random sample of initial positions. The spin-vector $\hat{\mathbf{s}}$ is parallel to the arrow depicted in the figure. Middle panel: Bird's-eye view of the trajectories in the near-field with $\hat{\mathbf{s}}$ pointing out of the page. The dashed circle marks the boundary of the spherical trap that was released at time t_0. Right panel: Displacement vs. time graphs of the same paths demonstrating how quickly the electron's radial velocity approaches a constant.

However, the spin velocity decreases with distance as $\mathcal{R}(r,t)$ becomes subleading to $\mathcal{I}(r,t)$ in the far field. As a result, it is the convective velocity primarily governing the motion of the unbound electron post-deconfinement, whereas the spin velocity alone is responsible for the motion of the bound electron (Section 16.2).[17]

16.5 Momentum spectroscopy: general considerations

For a particle prepared in the wave function $\Psi(\mathbf{r})$, the standard/textbook quantum formalism mandates that the result of a momentum measurement is random with distribution $|\tilde{\Psi}(\mathbf{p})|^2$ (or $\tilde{\Psi}^\dagger\tilde{\Psi}$ for spinor-valued Ψ), where

$$\tilde{\Psi}(\mathbf{p}) = \left(2\pi\hbar\right)^{-3/2}\int_{\mathbb{R}^3} d^3r\ \Psi(\mathbf{r})\exp\left(-\frac{i}{\hbar}\mathbf{p}\cdot\mathbf{r}\right) \tag{16.52}$$

is the momentum-space wave function. It is easily calculated for the ground state of the spherical box (Equation (16.10)):

$$\tilde{\Psi}(\mathbf{p}) = \left(\frac{a}{\hbar}\right)^{3/2}\frac{\sin q}{q\left(\pi^2 - q^2\right)}\,\chi, \qquad q = \frac{a}{\hbar}\|\mathbf{p}\|. \tag{16.53}$$

However, the question of what precisely constitutes a 'momentum measurement', i.e. what experiments could validate the above prediction, is often passed over in silence.

[17] Although this pattern is somewhat paradigmatic (see also Holland and Philippidis 2003), the spin velocity can dominate the electron's motion over very long distances in some special cases, such as when a spin-polarized electron is propagating within a waveguide and its motion is bounded in all but one direction (Das *et al.* 2019).

On the other hand, a plethora of experimental literature exists on *recovering* the momentum distribution (or the closely related Wigner distribution) from directly observable quantities, such as the impact position or arrival times of the particle (Ullrich *et al.* 2003; Wolf and Helm 2000; Kurtsiefer *et al.* 1997, 1995). But since *reconstruction is always theory-dependent*, we must, in the end, turn to a theory to relate the directly measured quantities with the desired ones. To discuss this point further, let us take a closer look at the widely used 'time-of-flight momentum spectroscopy' technique.

The key step of this method involves a measurement of the transit time or time-of-flight (ToF) of a particle prepared in the wave function of interest at a starting time t_0 and allowed to travel freely. When this particle registers at a remote detector at a (random) time t_1, its ToF

$$t_f = t_1 - t_0 \tag{16.54}$$

is used to define the particle's momentum, or, rather, its magnitude p, as follows:

$$p = \frac{mL}{t_f}. \tag{16.55}$$

Here, L is the known distance between the source and the detector (typically much larger than the width of the prepared wave packet). After repeating the experiment numerous times and accumulating the flight times, an empirical distribution of t_f, in turn, p given (16.55), is obtained. The latter is expected to reproduce the quantum mechanical distribution

$$\Lambda_{\mathrm{QM}}(p) = \int_{\mathbb{R}^3} d^3p' \; \delta(\|\mathbf{p}'\| - p)\, \tilde{\Psi}^{\dagger}(\mathbf{p}')\tilde{\Psi}(\mathbf{p}'), \tag{16.56}$$

where δ is Dirac's delta function. In mathematical terms, if $\Pi_{\mathrm{exp}}(t_f)$ is the measured ToF distribution, one expects

$$\Lambda_{\mathrm{QM}}(p) = \frac{mL}{p^2}\, \Pi_{\mathrm{exp}}\left(\frac{mL}{p}\right). \tag{16.57}$$

At the core of this scheme lies Equation (16.55), *which translates the measured* t_f *to the reported* p. It is motivated under the assumption that free Newtonian motion prevails, enabling the particle to approach the detector at a constant speed p/m; with p being the magnitude of the particle's momentum. Werner Heisenberg, for instance, makes just such an assumption

> The momentum of the [bound] electron can most readily be measured by suddenly rendering the interaction of the electron with the nucleus and neighbouring electrons negligible. *It will then execute a straight-line motion* and its momentum can be measured in the manner already explained (Heisenberg 1987, p. 32, emphasis added).

Considering how passionately Heisenberg contended that since motion and particle trajectories are not disclosed to us outside of measurement, talking about them has no meaning or advantage (Aristarhov 2022), it may seem incredibly circular that he refers to motion in the very conception of momentum measurements. Indeed, according to Valentini (2010), Einstein cautioned Heisenberg that his treatment of observation was unduly laden with the outdated theory of classical mechanics – and this would eventually get him into 'hot water'.

However, in all fairness to Heisenberg and other quantum physicists who helped themselves to Newtonian presuppositions in the early stages of theory development, one could argue à la Valentini (2010) pp. 503–504, paraphrasing Einstein, that while

> experiment is theory-laden, and correct measurement procedures must be laden with the correct theory, [...] when new experimental phenomena are discovered—phenomena that require the formulation of a new theory – in practice the old theory is at first assumed to provide a reliable guide to interpreting the observations [...] Note that this is a practical necessity, for the new theory has yet to be formulated. However – and here is the crucial point – once the new theory *has* been formulated, one ought to be careful to use the new theory to design and interpret measurements, and not continue to rely on the old theory to do so. For one may well find that consistency is obtained only when the new laws are found *and applied to the process of observation.*

But today, nearly a century later, many quantum physicists boldly reject the existence of particle trajectories[18], even though trajectory-based considerations are essential to numerous experiments designed to verify the predictions of quantum mechanics, including the ToF momentum spectroscopy method under consideration.

An alternative way to justify the ToF method without using classical trajectories would start with the quantum mechanical prediction $\Pi_{\text{QM}}(t_f)$ for the directly measured ToF distribution and deduce that Equation (16.55) is simply a change of variables that relates it to the momentum distribution via Equation (16.57) (with Π_{QM} replacing Π_{exp}). However, this is easier said than done because calculating the quantum mechanical ToF distribution has been a long-standing issue, in the absence of a recognized arrival-time observable, with various proposed solutions; see, Muga and Leavens 2000; Muga *et al.* 1998; Das and Nöth 2021.[19] Here, we will not go into a case-by-case examination of the existing suggestions to determine which ones validate or contradict the experimental method in question.

But, against this background, it is clear that a trajectory containing quantum theory such as the dBB theory – which is not dependent on quantum observables to describe experiments – would be more appropriate for evaluating this experimental technique. In the following section, we offer a dBB analysis of the ToF momentum spectroscopy method by applying it to the electron-in-a-box example.

16.6 ToF momentum spectroscopy: a dBB account

For mathematical convenience, we assume that the ToF detector has a spherical surface with radius $L \gg a$ surrounding the spherical box. Once the box is abruptly removed at time t_0, as shown in Secton 16.4, the trapped electron veers from its otherwise circular orbit and starts to spiral out, eventually following a rectilinear path. An individual

[18]Echoing, e.g. Landau and Lifshitz: 'In quantum mechanics there is no such concept as the path of a particle' (Landau and Lifshitz 1977, p. 2).

[19]It has even been suggested that the 'time of arrival cannot be precisely defined and measured in quantum mechanics' (Aharonov *et al.* 1998) or 'that wave mechanics cannot accommodate an exact and ideal arrival-time concept' (Allcock 1969). Should these suggestions be taken seriously, Equation (16.55) becomes vacuous.

electron, therefore, contacts the detector (resulting in a detection event) at a definite time t_1 at which

$$R(t_1) = L, \tag{16.58}$$

where $R(t)$ denotes the electron's distance from the centre of the box at time t. Equations (16.58) and (16.54) implicitly define the ToF of the electron as a function of its initial position $\mathbf{R}_0$ within the box (which determines $R(t)$ via the guiding Equations (16.17) and (16.50)) and the known parameters L and t_0.

While we are only able to determine this t_f numerically (see below), for $L \gg a$ one expects

$$t_f \approx \frac{L}{v_\infty}, \tag{16.59}$$

where v_∞ is the asymptotic Bohmian velocity of the electron (cf. Section 16.4); keeping in mind that the brief period spent in the near-field close to the trapping region contributes negligibly to the total ToF. It thus follows from (16.59) that *the p determined via* (16.55) *is simply the asymptotic dBB momentum of the electron acquired in flight, post deconfinement, viz.,* mv_∞.[20] Just this insight, free from implicit Newtonian assumptions, provides a much-needed physical foundation for Equation (16.55) in quantum mechanical contexts.

Although a single electron can be dependably prepared in the ground state of a trap at the start of each experimental run, controlling its position is practically impossible; for this reason, the measured arrival time t_1 randomly varies from one experimental run to the next. The statistical distribution of t_1, and hence t_f, may be calculated thanks to the *quantum equilibrium hypothesis* (QEH), which postulates that the random initial positions of a Bohmian particle follow Born's statistical rule; see Dürr *et al.* 1992; Norsen 2018; Callender 2007; Colin 2012. In particular, the electron positions within the box are distributed according to the $\Psi^\dagger\Psi$ distribution, where Ψ is the ground-state wave function (16.10).[21]

Leveraging the QEH, Fig. 16.5 (left panel) presents a ToF histogram generated from $\approx 1.25 \times 10^5$ Bohmian trajectories, the flight times of which were acquired by numerically integrating the guiding equations. The main lobe of the histogram is preceded by many smaller, secondary lobes (seen magnified in the inset), which are caused by a small fraction of electrons that are rapidly transported by the diffraction-in-time wavelets discussed in Section 16.9 (see also Mousavi 2009; Das *et al.* 2019). ToF experiments involving cold neutrons and atoms have reported the presence of such secondary lobes Szriftgiser *et al.* 1996; Hils *et al.* 1998.

As in the experiments, we convert the computed Bohmian flight times into momenta using Equation (16.55), and the result is the momentum histogram displayed in Fig. 16.5 (right panel). The quantal prediction (16.56), given by

[20] Although, the momentum of a Bohmian particle defined along Newtonian lines as $m\dot{\mathbf{R}}$ is generally not conserved even in the absence of external potentials.

[21] In order for $\Psi^\dagger\Psi$ to be a legitimate probability distribution, $\chi^\dagger\chi = 1$.

$$\Lambda_{\mathrm{QM}}(p) \overset{(16.53)}{=} \frac{4\pi a}{\hbar} \frac{\sin^2(pa/\hbar)}{\left[\pi^2 - (pa/\hbar)^2\right]^2} \tag{16.60}$$

is fully in agreement with it. Since the dBB theory does not even utilize momentum-space wave functions to state its dynamical equations, this agreement is quite remarkable indeed.

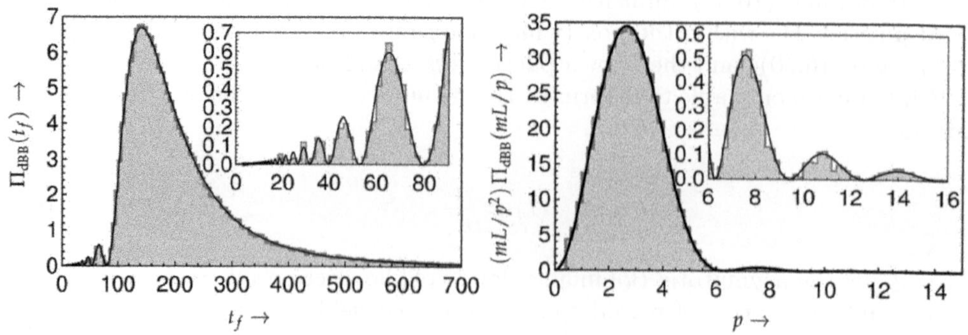

Fig. 16.5: Masses, lengths, and times are expressed in units of the electron mass m, the box radius a, and $ma^2/\hbar$, respectively. Left panel: ToF histogram made from $\approx 1.25 \times 10^5$ randomly sampled Bohmian trajectories emanating from the spherical box and arriving on a spherical detector of radius $L = 500$. Right panel: histogram of p reconstructed from these arrival times using formula (16.55). The solid line depicts $\Lambda_{\mathrm{QM}}(p)$ defined in (16.60).

We offer here some theoretical calculations that corroborate the numerical results, turning first to a consideration of the electron's arrival time t_1. The distribution of this quantity, $\Pi_{\mathrm{dBB}}(t_1)$, is usually very difficult to evaluate, but in the current scenario, where the Bohmian trajectories intercept the detection surface *at most* once, it is directly expressible in terms of the wave function, as explained, e.g. in Das and Nöth 2021, Section 6. We find for this distribution the exact result

$$\Pi_{\mathrm{dBB}}(t_1) = \frac{\hbar}{m}\,(4\pi L^2)\,\mathrm{Im}\left[\psi_>^* \frac{d\psi_>}{dr}\right](L, t_1). \tag{16.61}$$

Letting $t_1 \rightarrow t_f + t_0$ (cf. (16.54)), (16.61) becomes the ToF distribution depicted by the black curve in Fig. 16.5 (left panel). Finally, recalling our approximation (16.46), one can immediately see that the obtained ToF distribution would reproduce (16.60) exactly when incorporated into the left-hand-side of Equation (16.57) in place of $\Pi_{\exp}(t_f)$.

Our explanation of the ToF momentum spectroscopy technique for the particle-in-a-box scenario is now complete. Theoretical treatment proceeds along similar lines in most cases. That being said, it is important to be aware of exceptional situations in which the technique would not work. For instance, in situations where the initial wave functions are not well localized (Das and Struyve 2021), or when there is backflow (Das and Dürr 2021), or where more than one entangled particle is involved, e.g.

(e.g. Ayatollah Rafsanjani *et al.* 2024), the Bohmian trajectories are not necessarily asymptotically Newtonian, and the applicability of (16.55) is questionable.

16.7 Conclusion

The dBB theory views spin as an aspect of the movement of quantum particles dictated by guiding equations that incorporate spinor-valued wave functions. To illustrate this spin-aware motion of fermions, the first part of this chapter provided a comprehensive, step-by-step examination of the motion of a single electron confined in a spherical box using both relativistic and non-relativistic versions of the theory. Next, we investigated the reaction of the trapped electron to a sudden removal of the confining potential.

The chapter's second part focused on how the dBB theory is typically applied to actual experiments. For concreteness, we investigated the widely used ToF momentum spectroscopy method, which is grounded in trajectory-based assumptions. It was noted that standard quantum mechanics, which prohibits discussing particle trajectories and is beset with the long-standing 'arrival-time problem', makes such assumptions extremely difficult to defend. On the other hand, the dBB treatment of this experimental method is transparent and produces the anticipated results in a principled way, offering a solid theoretical basis for the heuristics that experimenters employ to interpret the measurements. A few potential situations in which the heuristics might be insufficient were also touched upon.

16.8 Appendix A: relativistic ground-state wave function

The electron's ground-state wave function assumes the form $\Psi(\mathbf{r},t) = \Psi(\mathbf{r})\, e^{-\,itE_{\mathrm{R}}/\hbar}$, where $\Psi(\mathbf{r})$ is a four-spinor satisfying Dirac's time-independent equation

$$\big(-\,i\hbar c\,\boldsymbol{\alpha}\cdot\boldsymbol{\nabla} + mc^2\beta\big)\,\Psi = E_{\mathrm{R}}\Psi \tag{16.62}$$

in the region $r < a$. Here, $\boldsymbol{\alpha} = (16.31)$, and $\beta = \mathrm{diag}(\mathbb{1}, -\mathbb{1})$. Denoting the upper and lower two-spinor components of Ψ by $\Psi_\pm$, Equation (16.62) yields the component equations

$$-i\hbar c\,(\boldsymbol{\sigma}\cdot\boldsymbol{\nabla})\,\Psi_\mp = \big(E_{\mathrm{R}} \mp mc^2\big)\,\Psi_\pm. \tag{16.63}$$

Eliminating Ψ_-, we obtain

$$-(\boldsymbol{\sigma}\cdot\boldsymbol{\nabla})^2\,\Psi_+ = \frac{E_{\mathrm{R}}^2 - m^2c^4}{\hbar^2c^2}\,\Psi_+. \tag{16.64}$$

The two-spinor Ψ_+ becomes the wave function of the electron in the non-relativistic limit. Noting that $(\boldsymbol{\sigma}\cdot\boldsymbol{\nabla})^2 = \nabla^2\mathbb{1}$, cf. Equation (16.9), we find that Equation (16.64) reduces to the usual time-independent Schrödinger–Pauli equation

$$-\frac{\hbar^2}{2m}\nabla^2\,\Psi_+ = E\,\Psi_+, \tag{16.65}$$

upon setting

$$\frac{E_{\mathrm{R}}^{2}-m^{2}c^{4}}{\hbar^{2}c^{2}}=\frac{2m}{\hbar^{2}}E. \tag{16.66}$$

The ground-state wave function solution of (16.65), which corresponds to $E=$ (16.12), has already been identified, *viz.*, $\Psi_{+}=\psi(\mathbf{r})\chi$ with $\psi(\mathbf{r})=$ (16.11), and χ being a constant unit spinor. Equation (16.66) then fixes E_{R} in terms of the non-relativistic ground-state energy E. Note that only $E_{\mathrm{R}}>0$ fulfilling (16.66) (Equation (16.34)) pertains to electron solutions. Finally, incorporating Ψ_{+} into Equation (16.63), we obtain Ψ_{-}, and hence the complete ground-state wave function (16.33).

16.9 Appendix B: time-evolution post switching

The free-particle Schrödinger equation (16.42) with initial condition $\psi_{>}(\mathbf{r},t_0)=$ (16.11) can be solved as follows: exploiting the spherical symmetry of the problem, express the solution as

$$\psi_{>}(\mathbf{r},t)=r^{-1}\varphi\Big(r,\frac{\hbar}{m}(t-t_0)\Big). \tag{16.67}$$

Consistency at $t=t_0$ requires that

$$\varphi(r,0)=r\,\psi_{>}(\mathbf{r},t_0)=\frac{H(a-r)}{\sqrt{2\pi a}}\sin\Big(\frac{\pi r}{a}\Big)e^{-it_0E/\hbar}. \tag{16.68}$$

($H(\cdot)$ denotes Heaviside's step function.) In order for Equation (16.67) to provide a solution $\psi_{>}$ of Equation (16.42), the function φ must satisfy the one-dimensional Schrödinger equation

$$i\frac{\partial\varphi}{\partial t}=-\frac{1}{2}\frac{\partial^{2}\varphi}{\partial r^{2}}. \tag{16.69}$$

Furthermore, given our *ansatz* (16.67), φ must vanish at $r=0$ in order for $\psi_{>}$ to be regular at the origin.

We then essentially seek the wave function of a free particle moving on the half-line ($r\geq 0$) subject to an impenetrable potential barrier at $r=0$, and specified initial condition at time zero. The solution can be expressed in the form

$$\varphi(r,t)=\int_0^{\infty}dr'\;\varphi(r',0)\,K(r,r',t), \tag{16.70}$$

where

$$K(r,r',t)=-\sqrt{\frac{2i}{\pi t}}\sin\Big(\frac{rr'}{t}\Big)\exp\Big[\frac{i}{2t}\left(r^{2}+r'^{2}\right)\Big] \tag{16.71}$$

is the the propagator of (16.69) with Dirichlet boundary condition at $r=0$; see Goodman 1981; Nevels *et al.* 1993. For $u=\pi/a$ and $v=r/t$, the trigonometric identity $2\sin(ur')\sin(vr')=\cos[(u-v)r']-\cos[(u+v)r']$, allows rewriting (16.70) as

$$\varphi(r,t) = N_0\left[\eta(r,t) - \eta(-r,t)\right], \tag{16.72}$$

where $N_0 = ie^{-iEt_0/\hbar}/(2\sqrt{2\pi a})$ and

$$\eta(r,t) = i\sqrt{\frac{2i}{\pi t}}\int_0^a dr' \exp\left[\frac{i}{2t}\left(r^2 + r'^2\right)\right]\cos\left(\frac{\pi r'}{a} - \frac{rr'}{t}\right).$$

The above Gaussian integral can be expressed in terms of the familiar error (or complementary error) function. However, it is convenient to express the final result in terms of the Moshinsky function $M(r,k,t)$ (Nussenzveig 1992) which is a judicious repackaging of the complementary error function:

$$M(r,k,t) = \frac{1}{2}\operatorname{erfc}\left(\frac{r-kt}{\sqrt{2it}}\right)e^{ikr - itk^2/2}. \tag{16.73}$$

(The function $M(r,k,t)$ itself satisfies Equation (16.69), reducing to the 'truncated plane wave' $H(-r)e^{ikr}$ in the limit $t \to 0^+$.) In any case, we have

$$\eta(r,t) = M\left(r - a, \frac{\pi}{a}, t\right) - M\left(r + a, \frac{\pi}{a}, t\right), \tag{16.74}$$

which via Equations (16.67) and (16.72) completely defines the time-evolved wave function $\psi_>(\mathbf{r},t)$, Equation (16.44).

For $|r - kt| \gg 1$, the Moshinsky function has the following approximation:

$$M(r,k,t) \approx \sqrt{\frac{it}{2\pi}}\,\frac{e^{ir^2/(2t)}}{r - kt} + H(kt - r)\,e^{ikr - itk^2/2}, \tag{16.75}$$

from which it follows, for $r \gg a$ and $r \gg \pi t/a$, the approximations

$$\eta(\pm r,t) \approx \sqrt{\frac{2t}{i\pi}}\,\frac{\sin(ra/t)}{r \mp \pi t/a}\,e^{i(r^2+a^2)/(2t)}, \tag{16.76}$$

in turn, using Equations (16.72) and (16.67), the result (16.46).

References

Aharonov, Y., Oppenheim, J., Popescu, S., Reznik, B., and Unruh, W.G. (1998). Measurement of time of arrival in quantum mechanics. *Physical Review A*, **57**: 4130–4139.

Allcock, G.R. (1969). The time of arrival in quantum mechanics III. The measurement ensemble. *Annals of Physics*, **53**(2): 311–348.

Aristarhov, S. (2022). Heisenberg's uncertainty principle and particle trajectories. *Foundations of Physics*, **53**(1): 7.

Ayatollah Rafsanjani, A., Kazemi, M., Hosseinzadeh, V. and Golshani, M. (2024). Non-local temporal interference. *Scientific Reports*, **14**(1): 3615.

Bell, J.S. (1971). Introduction to the hidden variable question. In d'Espagnat, B. (ed.) *Proceedings of the International School of Physics Enrico Fermi.* Course IL. Foundations of quantum mechanics. Course held at Varenna, Italy, June 29–July 11, 1970.

Berndl, K., Dürr, D., Goldstein, S., Peruzzi, G. and Zanghì, N. (1995). On the global existence of Bohmian mechanics. *Communications in Mathematical Physics*, **173**(3): 647–673.

Bohm, D. (1952a). A suggested interpretation of the quantum theory in terms of "hidden" variables. I. *Physical Review*, **85**(2): 166–179.

Bohm, D. (1952b). A suggested interpretation of the quantum theory in terms of "hidden" variables. II. *Physical Review*, **85**(2): 180–193.

Bohm, D. (1953). Comments on an article of Takabayasi concerning the formulation of quantum mechanics with classical pictures. *Progress of Theoretical Physics*, **9**: 273–287.

Bohm, D. and Hiley, B. J. (1993). *The Undivided Universe: An Ontological Interpretation of Quantum Theory.* London: Routledge.

Bricmont, J. (2016). *Making Sense of Quantum Mechanics.* Berlin: Springer.

Callender, C. (2007). The emergence and interpretation of probability in Bohmian mechanics. *Studies in History and Philosophy of Science Part B: Studies in History and Philosophy of Modern Physics*, **38**(2): 351–370.

Colijn, C. and Vrscay, E.R. (2002). Spin-dependent Bohm trajectories for hydrogen eigenstates. *Physics Letters A*, **300**(4): 334–340.

Colijn, C. and Vrscay, E.R. (2003a). Spin-dependent Bohm trajectories associated with an electronic transition in hydrogen. *Journal of Physics A: Mathematical and General*, **36**(16): 4689.

Colijn, C. and Vrscay, E.R. (2003b). Spin-dependent Bohm trajectories for Pauli and Dirac eigenstates of hydrogen. *Foundations of Physics Letters*, **16**(4): 303–323.

Colin, S. (2012). Relaxation to quantum equilibrium for Dirac fermions in the de Broglie–Bohm pilot-wave theory. *Proceedings of the Royal Society A: Mathematical, Physical and Engineering Sciences*, **468**(2140): 1116–1135.

Das, S. and Dürr, D. (2019). Arrival time distributions of spin-1/2 particles. *Scientific Reports*, **9**: 2242.

Das, S. and Nöth, M. (2021). Times of arrival and gauge invariance. *Proceedings of the Royal Society A*, **477**(2250): 20210101.

Das, S., Nöth, M. and Dürr, D. (2019). Exotic Bohmian arrival times of spin-1/2 particles: An analytical treatment. *Physical Review A*, **99**: 052124.

Das, S. and Struyve, W. (2021). Questioning the adequacy of certain quantum arrival-time distributions. *Physical Review A*, **104**: 042214.

Daumer, M., Dürr, D., Goldstein, S. and Zanghì, N. (1996). On the flux-across-surfaces theorem. *Letters in Mathematical Physics*, **38**: 103–116.

Daumer, M., Dürr, D., Goldstein, S. and Zanghì, N. (1997). On the quantum probability flux through surfaces. *Journal of Statistical Physics*, **88**(3–4): 967–977.

de Broglie, L. (1960). *Non-Linear Wave Mechanics, A Causal Interpretation.* Amsterdam: Elsevier Publishing Company.

del Campo, A., García-Calderón, G. and Muga, J.G. (2009). Quantum transients. *Physics Reports*, **476**(1): 1–50.

Dewdney, C. and Hiley, B.J. (1982). A quantum potential description of one-dimensional time-dependent scattering from square barriers and square wells. *Foundations of Physics*, **12**(1): 27–48.

Dewdney, C., Holland, P.R., Kyprianidis, A. and Vigier, J.P. (1988). Spin and non-locality in quantum mechanics. *Nature*, **336**(6199): 536–544.

Dürr, D., Goldstein, S. and Zanghì, N. (1992). Quantum equilibrium and the origin of absolute uncertainty. *Journal of Statistical Physics*, **67**(5–6): 843–907.

Dürr, D., Goldstein, S. and Zanghì, N. (2004). Quantum equilibrium and the role of operators as observables in quantum theory. *Journal of Statistical Physics*, **116**(1–4): 959–1055.

Dürr, D. and Teufel, S. (2009). *Bohmian Mechanics: The Physics and Mathematics of Quantum Theory.* Berlin: Springer.

Finley, J.P. (2021). Developments of Bohmian mechanics. Available at https://arxiv.org/abs/2110.01143

Freire Junior, O. (2019). *David Bohm: A Life Dedicated to Understanding the Quantum World.* Cham: Springer.

Godoy, S. (2003). Diffraction in time of particles released from spherical traps. *Physical Review A*, **67**: 012102.

Godoy, S. (2007). Transient stationary waves and angular momentum. *Physica B: Condensed Matter*, **390**(1): 112–117.

Godoy, S. and Villa, K. (2016). A basis for causal scattering waves, relativistic diffraction in time functions. *Journal of Modern Physics*, **7**(10): 1181–1191.

Goldstein, S. (2021). Bohmian Mechanics. In Zalta, E.N. (ed.) *The Stanford Encyclopedia of Philosophy.*

Goodman, M. (1981). Path integral solution to the infinite square well. *American Journal of Physics*, **49**(9): 843–847.

Greiner, W. (1992). *Relativistic Quantum Mechanics–Wave Equations*, 3rd edn. Texts and Monographs in Physics. Berlin Heidelberg: Springer.

Griffiths, D.J. and Schroeter, D.F. (2018). *Introduction to quantum mechanics*, 3rd edn. Cambridge: Cambridge University Press.

Heisenberg, W. (1987). *The Physical Principles of the Quantum Theory.* Cambridge: Cambridge University Press.

Hils, T., Felber, J., Gähler, R., Gläser, W., Golub, R., Habicht, K. and Wille, P. (1998). Matter-wave optics in the time domain: Results of a cold-neutron experiment. *Physical Review A*, **58**: 4784–4790.

Holland, P.R. (2003). Uniqueness of conserved currents in quantum mechanics. *Annalen der Physik*, **515**(7–8): 446–462.

Holland, P.R. (1992). The Dirac equation in the de Broglie-Bohm theory of motion. *Foundations of Physics*, **22**: 1287–1301.

Holland, P.R. (1993). *The Quantum Theory of Motion: An Account of the de Broglie-Bohm Causal Interpretation of Quantum Mechanics.* Cambridge: Cambridge University Press.

Holland, P.R. (1999). Uniqueness of paths in quantum mechanics. *Physical Review A*, **60**: 4326–4330.

Holland, P.R. and Philippidis, C. (2003). Implications of Lorentz covariance for the guidance equation in two-slit quantum interference. *Physical Review A*, **67**: 062105.

Horn, R.A. and Johnson, C.R. (2013). *Matrix Analysis*, 2nd edn. New York: Cambridge University Press.

Kofink, W. (1937). Über das magnetische und elektrische Moment des Elektrons nach der Diracschen Theorie. *Annalen der Physik*, **422**(1): 91–98.

Kurtsiefer, C., Pfau, T., Ekstrom, C.R. and Mlynek, J. (1995). Time-resolved detection of atoms diffracted from a standing light wave. *Applied Physics B. Lasers and Optics*, **60**: 229–232.

Kurtsiefer, C., Pfau, T. and Mlynek, J. (1997). Measurement of the Wigner function of an ensemble of helium atoms. *Nature*, **386**: 150–153.

Landau, L.D. and Lifshitz, E.M. (1977). *Quantum Mechanics: non-relativistic theory*, volume 3 of *Course of Theoretical Physics*, 3rd edn. Moscow: Pergamon Press.

Mermin, N.D. (1993). Hidden variables and the two theorems of John Bell. *Reviews of Modern Physics*, **65**: 803–815.

Moshinsky, M. (1952a). Diffraction in time. *Physical Review*, **88**: 625–631.

Moshinsky, M. (1952b). Efectos transitorios en la dispersión causada por una esfera rigída. *Revista Mexicana de Física*, **1**(1): 28–37.

Mousavi, S.V. (2009). Diffraction in the time of a confined particle and its Bohmian paths. *Journal of Physics A: Mathematical and Theoretical*, **43**(3): 035304.

Muga, J.G. and Leavens, C.R. (2000). Arrival time in quantum mechanics. *Physics Reports*, **338**(4): 353–438.

Muga, J.G., Sala, R. and Palao, J. (1998). The time of arrival concept in quantum mechanics. *Superlattices and Microstructures*, **23**(3): 833–842.

Myrvold, W.C. (2003). On some early objections to Bohm's theory. *International Studies in the Philosophy of Science*, **17**(1): 7–24.

Nevels, R.D., Wu, Z. and Huang, C. (1993). Feynman path integral for an infinite potential barrier. *Physical Review A*, **48**: 3445–3451.

Norsen, T. (2013). The pilot-wave perspective on quantum scattering and tunneling. *American Journal of Physics*, **81**(4): 258–266.

Norsen, T. (2014). The pilot-wave perspective on spin. *American Journal of Physics*, **82**(4): 337–348.

Norsen, T. (2018). On the explanation of Born-rule statistics in the de Broglie–Bohm pilot-wave theory. *Entropy*, **20**(6).

Nussenzveig, H.M. (1992). Moshinsky functions, resonances and tunneling. In Frank, A. and Wolf, K.B. (eds.) *Symmetries in Physics*, pp. 293–310. Berlin, Heidelberg: Springer.

Oriols, X. and Mompart, J. (2012). *Applied Bohmian Mechanics: From Nanoscale Systems to Cosmology*, 2nd edn. New York: Jenny Stanford Publishing.

Philippidis, C., Dewdney, C. and Hiley, B. (1979). Quantum interference and the quantum potential. *Il Nuovo Cimento B*, **52**(1): 15–28.

Reed, B.C. (2019). *Keplerian Ellipses.* 2053-2571. Morgan & Claypool Publishers.

Römer, S., Dürr, D. and Moser, T. (2005). Asymptotic behaviour of Bohmian trajectories in scattering situations. *Journal of Physics A: Mathematical and General*, **38**(39): 8421.

Struyve, W. (2005). The de Broglie-Bohm pilot-wave interpretation of quantum theory. Available at https://arxiv.org/abs/quant-ph/0506243

Symon, K. R. (1971). *Mechanics*, 3rd edn. Reading MA: Addison-Wesley.

Szriftgiser, P., Guéry-Odelin, D., Arndt, M. and Dalibard, J. (1996). Atomic wave diffraction and interference using temporal slits. *Physical Review Letters*, **77**: 4–7.

Tastevin, G. and Laloë, F. (2018). Surrealistic Bohmian trajectories do not occur with macroscopic pointers. *The European Physical Journal D*, **72**(183): 033018.

Teufel, S. and Tumulka, R. (2005). Simple proof for global existence of Bohmian trajectories. *Communications in Mathematical Physics*, **258**(2): 349–365.

Ullrich, J., Moshammer, R., Dorn, A., Dörner, R., Schmidt, L.P.H. and Schmidt-Böcking, H. (2003). Recoil-ion and electron momentum spectroscopy: reaction-microscopes. *Reports on Progress in Physics*, **66**(9): 1463–1545.

Valentini, A. (2010). De Broglie–Bohm Pilot-Wave Theory: Many Worlds in Denial? In Saunders, S., Barrett, J., Kent, A. and Wallace, D. (eds.) *Many Worlds?: Everett, Quantum Theory, and Reality.* Oxford: Oxford University Press.

Van Oosterhout, G. (1966). The identical relations between the bilinear covariants of Dirac's theory of the electron. *Physica*, **32**(6): 1090–1096.

Wilkes, J.M. (2020). The Pauli and Lévy-Leblond equations, and the spin current density. *European Journal of Physics*, **41**(3): 035402.

Wiseman, H.M. (2007). Grounding Bohmian mechanics in weak values and Bayesianism. *New Journal of Physics*, **9**(6): 165.

Wolf, S. and Helm, H. (2000). Ion-recoil momentum spectroscopy in a laser-cooled atomic sample. *Physical Review A*, **62**: 043408.

Index